*The Sago Palm*

*The Food and Environmental Challenges
of the 21st Century*

Frontispiece 1 Adult plants of the genus *Metroxylon*
Top left: *M. sagu*, East Sepik, Papua New Guinea; top right: *M. amicarum*, inflorescences at the mid-level axils are bearing fruits. Chuuk, Micronesia.; center left: *M. salomonense* at the fruiting stage, Singapore Botanic Gardens; center right: *M. vitiense*, Viti Levu Island, Fiji.; bottom left: *M. waburgii*, Upolu, Samoa.: bottom right: *M. Paulcoxii* (*M. upoluense*), Upolu, Samoa.

Frontispiece 2 Various sago palm leaves

Top left: no banding on rachis (kanduan); top center: dark grey banding on rachis (kunangu); top right: brown banding on rachis (kosogu) (East Sepik, PNG).; center left: spines on petiole from left to right – spineless (roe), sparse short spines (*rui*), dense long spines (runggamanu).; center right: vestigial spines on petiole from a relatively early stage (roe me eto) (Southeastern Sulawesi, Indonesia).; bottom left: smooth spineless petioles.; bottom right: traces of spines at petiole base and leaf sheath and spineless rachis after trunk formation (kanduan) (East Sepik, PNG).

Frontispiece 3 Various leaves of the section *Coelococcus* palms
Top left: *M. amicarum*, Chuuk, Micronesia; center left: adaxial surface of rachis of *M. warburgii*; bottom left: abaxial surface of rachis of *M. warburgii* (Upolu, Samoa).; top right: *M. salomonense*, Singapore Botanic Gardens.; bottom right: *M. vitiense*, Vanua Levu, Fiji.

Frontispiece 4 Inflorescences of the genus *Metroxylon*
Top left: *M. sagu*, East Sepik, Papua New Guinea; top right: *M. amicarum*, Chuuk, Micronesia; center left: *M. salomonense* in the fruiting stage, Singapore Botanic Gardens; center right: M. vitiense at the fruiting stage, Viti Levu, Fiji; bottom left: *M. warburgii* at the flowering stage, Upolu, Samoa; bottom right: *M. paulocoxii* (*M. upoluense*) at the flowering stage, Upolu, Samoa.

Frontispiece 5 Rachises of inflorescence of the genus *Metroxylon*
Top left: *M. sagu*, East Sepik, Papua New Guinea; top right: *M. amicarum*, Chuuk, Micronesia; center left: *M. salomonense*, Gaua, Vanuatu; center right: *M. vitiense*, Viti Levu, Fiji; bottom left: *M. warburgii* at the flowering stage, Esprit Santo, Vanuatu; bottom right: *M. paulocoxii* (*M. upoluense*), Upolu, Samoa.

*M. sagu*

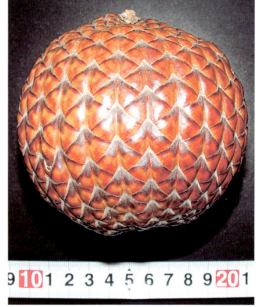
*M. amicarum*

*M. salomonense*

*M. vitiense*

*M. warburgii*

Frontispiece 6 Fruits of the genus *Metroxylon*

Frontispiece 7 Tropical peat soil (Selangor, Malaysia)

Frontispiece 8 Acid sulfate soil (Selangor, Malaysia)

Frontispiece 9 Jarosite in acid sulfate soil

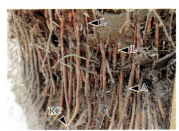

Frontispiece 10 Stem surface near ground level
A: adventitious roots, L: lateral roots, RC: root cap
Source: Nitta and Matsuda (2005).

Frontispiece 11 Sago palms distributed along the coast

Frontispiece 12 Lateral roots emerging from ground surface and growing upward
L: lateral roots
Source: Nitta et al. (2002)

Frontispiece 13 Sago palm leaf

Frontispiece 14 Leaf scars after leaves are removed

Frontispiece 15 Trunk cross section at approx. 1 m above ground of a palm about 8 years after trunk formation

Frontispiece 16 Sago palm inflorescence

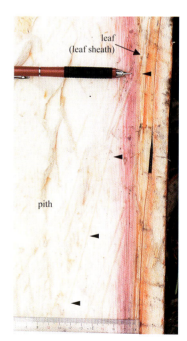

Frontispiece 17 Tracing direction of leaf vascular bundle running into pith

Frontispiece 18 Fruiting configuration

Frontispiece 19 Cross section of fruit with seed having horseshoe-shaped endosperm

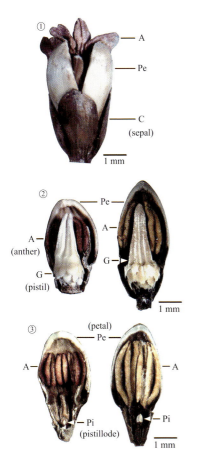

Frontispiece 20 Exterior view of flower ① and internal views of hermaphrodite and staminate flowers ②, ③

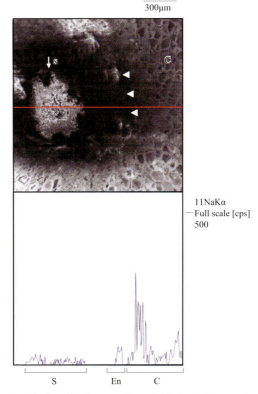

Frontispiece 21 Cross section of salt-treated sago palm (*M. sagu*) root and sodium distribution by energy dispersive X-ray spectroscopic analysis

The bottom charts show the distribution of sodium along the red line in the top images. C: cortex, S: stele, En: endodermis, [white triangle symbol]: outer edge of endodermis.

Source: Ehara et al. (2006b)

Frontispiece 22 Trimmed and prepared suckers

Frontispiece 23 Nursing collected suckers in a creek

Frontispiece 24 Established sucker after planting

Frontispiece 25 Well-managed sago palm farm (Mukah, Sarawak, Malaysia)

Frontispiece 26 Sago pith crushing work using a chipping axe

Frontispiece 27 Crushing work using a chipping axe

Frontispiece 28 Grater-type pith crushing work

Frontispiece 29 Extraction work: Sago pith washing by hands

Frontispiece 30 Extraction work: Washing by hands and an extractor

Frontispiece 31 Extraction work: Pith washing by feet

Frontispiece 32 Typical sago dish 'Papeda'

Frontispiece 33 Sago lempeng

Frontispiece 34 Sago pearls

Frontispiece 35 Sago noodles

Frontispiece 36 Sago cakes (unbaked)

Frontispiece 37 Sago cookies (baked)

Frontispiece 38 Atap (roofing material)

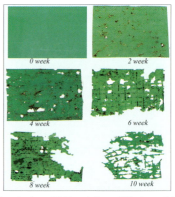
Frontispiece 39 Starch-based biodegradable plastic: stages of degradation in the ground (Courtesy of Dr. Pranamuda Hardaning)

*The Sago Palm*

*The Food and Environmental Challenges
of the 21st Century*

THE SOCIETY OF SAGO PALM STUDIES

First published in 2015 jointly by:
Kyoto University Press
69 Yoshida Konoe-cho
Sakyo-ku, Kyoto 606-8315, Japan
Telephone: +81-75-761-6182
Fax: +81-75-761-6190
Email: sales@kyoto-up.or.jp
Web: http://www.kyoto-up.or.jp

Trans Pacific Press
PO Box 164, Balwyn North, Melbourne
Victoria 3104, Australia
Telephone: +61-3-9859-1112
Fax: +61-3-8911-7989
Email: tpp.mail@gmail.com
Web: http://www.transpacificpress.com

Copyright © Kyoto University Press and Trans Pacific Press 2015.

Edited by Dr Karl Smith

Set by Jodie McLean, Melbourne. Email: jodie@bookprinter.com.au

Printed in Melbourne by BPA Print Group.

**Distributors**

**James Bennett Pty Ltd**
Locked Bag 537
Frenchs Forest NSW 2086
Australia
Telephone: +61 2 8988 5000
Fax: +61 2 8988 5031
Email: info@bennett.com.au
Web: www.bennett.com.au

**USA and Canada**
International Specialized Book Services (ISBS)
920 NE 58th Avenue, Suite 300
Portland, Oregon 97213-3786
USA
Telephone: (800) 944-6190
Fax: (503) 280-8832
Email: orders@isbs.com
Web: http://www.isbs.com

**Asia and the Pacific except Japan**
Kinokuniya Company Ltd.

*Head office:*
38-1 Sakuragaoka 5-chome
Setagaya-ku, Tokyo 156-8691
Japan
Telephone: +81-3-3439-0161
Fax: +81-3-3439-0839
Email: bkimp@kinokuniya.co.jp
Web: www.kinokuniya.co.jp

*Asia-Pacific office:*
Kinokuniya Book Stores of Singapore Pte., Ltd.
391B Orchard Road #13-06/07/08
Ngee Ann City Tower B
Singapore 238874
Telephone: +65-6276-5558
Fax: +65-6276-5570
Email: SSO@kinokuniya.co.jp

All rights reserved. No reproduction of any part of this book may take place without the written permission of Kyoto University Press or Trans Pacific Press.
ISBN 978–1–920901–13–4

# Contents

| | |
|---|---|
| Frontispiece | iii |
| Tables | xix |
| Figures | xxi |
| Preface | xxix |
| Preface to the English Edition | xxxiii |
| Dr Isao Nagato's contributions to sago palm research | xxxv |
| Contributors | xxxvii |
| 1 Origin, Dispersal and Distribution | 1 |
| 2 Ecology of the Sago Palm | 41 |
| 3 Morphology | 61 |
| 4 Growth Characteristics | 93 |
| 5 Physiology | 127 |
| 6 Cultivation and Management | 157 |
| 7 Starch Productivity | 199 |
| 8 Starch Extraction and Production | 235 |
| 9 Starch Properties and Uses | 253 |
| 10 Diversity of Uses | 297 |
| 11 Cultural Anthropological Aspect | 321 |
| 12 Future of Sago Palm in the 21st Century | 341 |
| Appendix: List of the International Sago Symposia and Proceedings | 353 |
| Bibliography | 355 |
| Index | 403 |

# List of Tables

Table 1-1 Arecaceae plants that store starch in the trunk
Table 1-2 Beccari's classifcation of section *Metroxylon* (*Eumetroxylon*) (1918)
Table 1-3 Beccari's classifcation of section *Coelococcus* (1918)
Table 1-4 Classification criteria of the *M. Sagu* species
Table 1-5 Morhological characteristics of the genus *Metroxylon*
Table 1-6 Comparison of yield and yield components of the *Mextroxylon* species
Table 1-7 Vernacular names and morhologial characteristics of specimens

Table 3-1 Diameter of adventitious root primordia (mm)
Table 3-2 Adventitious root primordia density on stem surface (per 100 $cm^2$)

Table 4-1 Sago palm growth stages in Sarawak, Malaysia
Table 4-2 Total number of ax1, ax2 & ax3 and estimated number of flower buds

Table 5-1 Photosynthetic properties of sago palm (seedling-stage)
Table 5-2 Difference among folk varieties in photosynthetic rate, stomatal conductance, mesophyll conductance, SPAD value and leaf thickness
Table 5-3 Changes in dry matter ratio (%) in above-ground parts by years after trunk formation
Table 5-4 Changes in dry weight percentage (%) in above-ground parts by years after trunk formation
Table 5-5 Varietal differences in dry matter production properties related to sago palm starch production
Table 5-6 Acetylene reducing activity in different parts of sago palm collected from the Philippines (nitrogen fixing activity)
Table 5-7 Nitrogen fixing activity enhancement by co-culture of nitrogen fixing bacteria and starch degrading bacteria
Table 5-8 Estimated biologically fixed nitrogen in mature sago palm stand

Table 6-1 The survival rate of sago palm suckers stored for different periods under shaded and unshaded conditions
Table 6-2 The survival rate of sago palm suckers planted at different depths
Table 6-3 Effect of shading on survival of suckers during wet and dry seasons
Table 6-4 Effect of groundwater level on root growth in sago palm suckers 1 year after planting
Table 6-5 Sago palm growth 2 years after planting
Table 6-6 Methane and $CO_2$ fluxes from sago palm cultivation soil with different fertilizer treatments

Table 7-1 Starch distribtion in sago palm pith at five sampling positions
Table 7-2 Starch density of sago palm pith at five sampling positions
Table 7-3 Variation in mean starch content and mean starch density in the pith and trunk of sago palms at different growth stages
Table 7-4 Comparison of starch content (%) in mature palms (stage 11) with normal and abnormal fruit development
Table 7-5 Starch yield of sago palms at diffrent development stages
Table 7-6 Sugar and starch contents in sago palm pith
Table 7-7 Feature of plastid-amyloplast system at growing point and in proximal tissue

Table 7-8 Starch yield per palm in Malaysia, Indonesia and Papua New Guinea
Table 7-9 Effect of spacing treatments on trunk production and pith volume of sago palms at 9 years after planting
Table 7-10 Number of clumps and palm population by development stage (per hectare) in natural (wild), semi-cultivated and cultivated sago palm stands
Table 7-11 Number of clumps and ratio of constituent varieties in natural and semi-cultivated sago palm stands
Table 7-12 Annual sago palm starch production (per ha)
Table 7-13 Production capacity of sago palm
Table 7-14 Comparison of productivity of the main moisture-rich starchy staples
Table 7-15 Comparison of starch productivity between sago palm and major tuber and root crops
Table 7-16 Yield level and yield potential of sago palm

Table 8-1 Patterns of sago pith crushing and starch extraction work
Table 8-2 Estimates of sago palm growing areas
Table 8-3 Sago palm cultivation areas and starch production in Indonesia (2006)
Table 8-4 Sago palm growing areas in central and eastern Indonesia
Table 8-5 Estimated sago palm growing area in Indonesia
Table 8-6 Sago palm production by province in PNG

Table 9-1 Dynamic viscoelasticity of various starch gels
Table 9-2 Food uses of sago starch
Table 9-3 Starch characteristics of folk varieties by group
Table 9-4 Total demand and supply for starches for starch year 2000 (thousand tons)
Table 9-5 Classification of modified starches by modification method
Table 9-6 Functions and uses of modified starch
Table 9-7 Classification of biodegradable plastics
Table 9-8 Physical properties of polylactic acid and general-purpose plastics
Table 9-9 Sago starch characteristics
Table 9-10 Physical properties of extruded pellet (EP) for fish culture

Table 10-1 Sago starch exports from Sarawak State, Malaysia, by destination country
Table 10-2 Chemical composition of sago palm pith and sago residue

Table 11-1 Folk classification of crops in Wanjeaka Village, Papua New Guinea
Table 11-2 Folk classification of *M. sagu Rottb.* in Wanjeaka Village, Papua New Guinea

Table 12-1 Biomass use in Japan
Table 12-2 Fields in which the practical use of biodegradable plastics is expected to expand

# List of Figures

Figure 1-1 Mature *Metroxylon* palms
Figure 1-2a Variation in the characteristics of sago palm leaf
Figure 1-2b Variation in the characteristics of the section *Coelococcus* palms
Figure 1-3 Galela classification of sago palms, Halmahera Island, Indonesia
Figure 1-4 Inflorescences of the genus *Metroxylon*
Figure 1-5 Branches on the inflorescence of the genus *Metroxylon*
Figure 1-6 Fruits of the genus *Metroxylon*
Figure 1-7 Specimen collection locations
Figure 1-8 UPGMA dendrogram based on RAPD data
Figure 1-9 Estimated dispersal routes of sago palms
Figure 1-10 Sago palm vegetation in North Sulawesi
Figure 1-11 Sago palm stand near Kendari, Southeast Sulawesi
Figure 1-12 Sago palm stand near Angoram, East Sepik Province, Papua New Guinea
Figure 1-13 Distribution of poor drainage, swamp and alluvial soil areas in Papua New Guinea
Figure 1-14 Sago palm vegetation zones in the floodplains of the Sepik River Basin
Figure 1-15 Dense natural sago palm stands with no trunked palms distributed along a river
Figure 1-16 Sago palm stand in East Sepik Province, Papua New Guinea
Figure 1-17 Semi-cultivated sago palm stand with many trunked palms
Figure 1-18 Thin trunks of wild varieties left to wither
Figure 1-19 Managed homestead stand of high-yielding sago palms and a trunk being felled with an axe for harvesting

Figure 2-1 Sago palm distribution
Figure 2-2 Coastal low land (Sarawak, Malaysia)
Figure 2-3 Köppen's climate classification
Figure 2-4 Annual mean temperature and minimum temperature (Semongok Agriculture Research Centre, Malaysia)
Figure 2-5 Solar radiation and chloroplast
Figure 2-6 Precipitation (top: Kuching, Borneo Is.) and relative humidity (bottom: modified version from Andriesse 1972) in the zone within 10 degrees north and south of the equator
Figure 2-7 Areas of typhoon formation
Figure 2-8 Tropical peat soil (Selangor, Malaysia)
Figure 2-9 Peat soil formation
Figure 2-10 Interrelations between landform units in tropical peat swamps wedged between 2 rivers
Figure 2-11 Pyrite ($FeS_2$) in acid sulfate soil
Figure 2-12 Acid sulfate soil (Selangor, Malaysia)
Figure 2-13 Thiobacillus, a type of sulfur oxidizing bacteria
Figure 2-14 Jarosite in acid sulfate soil
Figure 2-15 Pattern diagram of acid sulfate soil profile
Figure 2-16 Groundwater level in sago palm growing area
Figure 2-17 Ground subsidence by peat contraction
Figure 2-18 Sago palm roots
Figure 2-19 Sago palms distributed along the coast
Figure 2-20 Salt tolerance of sago palm

Figure 3-1 Stem surface near ground level
Figure 3-2 Adventitious root primordia on stem surface
Figure 3-3 Stem surface several meters above the ground surface
Figure 3-4 Lateral roots emerging from ground surface and growing upward
Figure 3-5 Scanning electron microscopic photographs of cross section of adventitious root
Figure 3-6 Scanning electron microscopic photographs of cross section of lateral root
Figure 3-7 Pattern diagram of sago palm leaf and part names
Figure 3-8 Leaflet attachment positions
Figure 3-9 Edges of leaflets on rachis
Figure 3-10 Leaf immediately after opening
Figure 3-11 Leaf immediately after opening
Figure 3-12 Leaf pattern diagram (a) and leaf image after all leaflets are converted into rectangles with equivalent areas (b)
Figure 3-13 Schematic representation of leaf as a trapezium in the proximal half and a half ellipse in the apical half
Figure 3-14 Cross section of petioles showing positional relations between leaves
Figure 3-15 Deviating phyllotaxis
Figure 3-16 Stomatal arrangement on adaxial surface (A) and abaxial surface (B) of leaflet
Figure 3-17 Internal structure of sago palm leaflet (cross section)
Figure 3-18 Cross sections of leaflets attached to apical (a), middle (b) and proximal (c) parts of mid-level leaf
Figure 3-19 Cross section of leaflet attached to middle part of high-level leaf
Figure 3-20 Cross section of leaflet attached to middle part of low-level leaf
Figure 3-21 Leaf base scars after removal of green leaves
Figure 3-22 Schematic representation of sago palm trunk (8 years after estimated trunk formation)
Figure 3-23 Emerged suckers
Figure 3-24 Schematic representation of sucker (grown about 3 m in nearly 5 years, about 1 year after trunk formation)
Figure 3-25 Tip of a sucker continuing to creep sideways
Figure 3-26 Creeping sucker is being cut (left). Cross section of sucker stem (right).
Figure 3-27 Trunk cross section at about 1 m above ground of sago plant about 8 years after trunk formation
Figure 3-28 Vascular bundle distribution in sago palm trunk cross section
Figure 3-29 Number of vascular bundles in trunk cross section at about 1 m above ground (trunk radius being 1)
Figure 3-30 Types of vascular bundles
Figure 3-31 Fibrous tissues of vascular bundle: cross section (a) and longitudinal section (b) of peripheral vascular bundles
Figure 3-32 Fibrous tissue on the periphery of a trunk of approximately 12 years of age
Figure 3-33 Leaf vascular bundle insertion into trunk (arrow)
Figure 3-34 Leaf vascular bundle insertions into trunk
Figure 3-35 The youngest part of vascular bundles constituting fibrous tissue on pith periphery
Figure 3-36 Tracing direction of leaf vascular bundle running into pith
Figure 3-37 Scanning electron microscopic photograph of pith cross section of a sago palm estimated 3 years after trunk formation growing in a dry area
Figure 3-38 Scanning electron microscopic photograph of pith cross section of sago palm variety Para Hongleu
Figure 3-39 Inflorescence at the fruiting stage
Figure 3-40 Exterior view of flower ① and internal views of hermaphrodite and staminate flowers ②, ③
Figure 3-41 Electron micrographs of pollens. Pollens of hermaphrodite flower of spineless type (folk variety Rumbio) from Siberut Island, West Sumatra

# List of Figures

Figure 3-42 Fruit attachment
Figure 3-43 Longitudinal section of fruit and schematic representation

Figure 4-1 (a) Rosette stage; (b) Sago palm initiating trunk formation
Figure 4-2 Sago palm life cycle
Figure 4-3 Differentiated bud inside the leaf margin
Figure 4-4 Position of bud differentiation
Figure 4-5 (a) Sago palm 5 years after trunk formation; (b) Pattern diagram showing the relationship between the trunk and leaf attachment positions
Figure 4-6 A large inflorescence emerged from this sago palm
Figure 4-7 Changes in germination rate (soaked in water, 25 °C)
Figure 4-8 State of germination (pattern diagram)
Figure 4-9 Germination process
Figure 4-10 Changes in water uptake by germinated seed (soaked in water, 25 °C)
Figure 4-11 Sago palm leaf appearance at the rosette stage
Figure 4-12 Leaf blade length and petiole length at the trunk formation stage
Figure 4-13 Sago palm leaf appearance at the trunk formation stage
Figure 4-14 (a) A prepared sucker with leaves trimmed off; (b) Young leaves inside
Figure 4-15 Leaf blade length and petiole length in a tree 8 years after trunk formation
Figure 4-16 Leaf length in trees approx. 2, 4 and 8 years after trunk formation
Figure 4-17 Pattern diagram of trunk
Figure 4-18 Length of each internode (exponential scale on vertical axis)
Figure 4-19 Diameter of each internode (exponential scale on vertical axis)
Figure 4-20 Shape of trunk apex
Figure 4-21 Node and internode positions on sucker longitudinal section
Figure 4-22 Sucker stem diameter at each node
Figure 4-23 Changes in root dry weight by tree age (Kendari, Indonesia)
Figure 4-24 Changes in root weight ratio between horizontal (a) and vertical (b) directions by age (Kendari, Indonesia)
Figure 4-25 Root ratio by diameter of sago palm in peat soil and alluvial mineral soil
Figure 4-26 Inflorescence at sago palm trunk apex
Figure 4-27 Second-order floral axis (ax2) and third-order floral axis (ax3)
Figure 4-28 Cup-shaped bracteoles covering flower buds on third-order floral axis (ax3)
Figure 4-29 Staminate flowers in full bloom
Figure 4-30 Nectar production by blooming flowers
Figure 4-31 Stamens and pistils of hermaphrodite flowers in full bloom
Figure 4-32 Fresh mature fruit of sago palm
Figure 4-33 Cross section of seedless fruit
Figure 4-34 Cross section of fruit with seed having horseshoe-shaped endosperm

Figure 5-1 Changes in transpiration rate and cuticular transpiration rate in sago palm seedlings treated with sodium chloride (black dot) and untreated (white dot)
Figure 5-2 Light-photosynthesis curve of sago palms grown under different light conditions (seedling stage)
Figure 5-3 Changes in leaf thickness and SPAD value by palm age
Figure 5-4 Changes in stomatal density and length by palm age
Figure 5-5 Relationship between years after trunk formation and total above-ground fresh weight
Figure 5-6 Helophytes
Figure 5-7 Cross section of sago palm root
Figure 5-8 Salt gland in *Avicennia marina*
Figure 5-9 Betaines
Figure 5-10 Membrane protein action
Figure 5-11 Model of aluminum-induced organic acid ion secretion from roots
Figure 5-12 Plant's aluminum stress tolerance mechanism

Figure 5-13 Influences of pH and aluminum on *stop1* variant
Figure 5-14 Sodium concentration in sago palm (*M. sagu*) parts treated with 342 mM NaCl
Figure 5-15 Sodium concentration in *M. warburgii* parts treated with 342 mmol NaCl
Figure 5-16 Cross section of salt-treated sago palm (*M. sagu*) root and sodium distribution by energy dispersive X-ray spectroscopic analysis
Figure 5-17 Phylogenetic relations among nitrogen fixing bacteria isolated from sago palms based on 16S rDNA gene sequences

Figure 6-1 Sucker placed in a transplant hole
Figure 6-2 NTFP sago plantation
Figure 6-3 Multi-purpose canal at NTFP plantation
Figure 6-4 Single-track railway laid at the plantation
Figure 6-5 Mechanically weeded transplanting row and suckers immediately after transplantation
Figure 6-6 Germinated seedlings grown in coconut palm fiber medium
Figure 6-7 Seedlings nursed in pots
Figure 6-8 Palm transplanted in the field 20 months earlier after a nursing period of 10 months in a pot
Figure 6-9 Nursing collected suckers in a creek (top) and a swamp (bottom)
Figure 6-10 Sucker nursery rafts made of sago palm rachises and petioles
Figure 6-11 Trimmed and prepared suckers
Figure 6-12 Forms of collected suckers
Figure 6-13 Effect of sucker weight on the number of new roots during the nursing period
Figure 6-14 Changes in the number of new roots, leaves and dry matter, total sugar and starch percentage during nursing period on rafts (Riau, Indonesia, 1999)
Figure 6-15 A planted sucker
Figure 6-16 Sago palm sucker management method
Figure 6-17 Potassium deficiency
Figure 6-18 Groundwater level and trunk height increase rate
Figure 6-19 Groundwater level and trunk volume increase rate
Figure 6-20 Termites: a pest affecting sago palms
Figure 6-21 Sago palm stump infested by sago beetles and fungi
Figure 6-22 Petiole damaged by monkeys
Figure 6-23 Sago palm logging using the chainsaw
Figure 6-24 The sago palm trunk cut to length for transportation
Figure 6-25 Cut log
Figure 6-26 Sago logs are rolled away on a track of leaf sheaths laid on the ground
Figure 6-27 Hooks are attached to cut ends for ease of transport
Figure 6-28 Sago logs assembled into a raft in the canal
Figure 6-29 Logs gathered in the sea
Figure 6-30 Logs tugged by boats in the sea
Figure 6-31 The chamber section of the automatic measuring device for $CO_2$ fluxes from soil set up in a sago palm field, Tebing Tinggi (bottom) and diurnal changes in $CO_2$ fluxes and soil temperature (5 cm deep) (29 September–2 October 2006: top)
Figure 6-32 Relationships between methane and $CO_2$ fluxes and groundwater table
Figure 6-33 Relationships between DOC and Ca concentrations in canal water from sago palm cultivation soil and months after transplantation
Figure 6-34 Ash contents and pH of sago palm cultivated soils and their adjacent secondary forest soils

Figure 7-1 Difference in coarse starch content by pith cross section position in sago palms at different palm ages (estimates)
Figure 7-2 Correlation between percentage of dry matter and starch content in pith
Figure 7-3 Palm age, trunk weight, coarse starch content and yield
Figure 7-4 Difference in pith total sugar content by log position in sago palms at different palm

ages (estimates)
Figure 7-5 Correlation between total sugar content and starch content in pith
Figure 7-6 Example of sugar measurements by high-performance liquid chromatography
Figure 7-7 Scanning electron micrograph of ground parenchyma cells at stem center in Rotan variety at early growth stage
Figure 7-8 Scanning electron micrographs of ground parenchyma cells at stem center in Rotan variety at middle growth stage
Figure 7-9 Schematic diagram of amyloplast separation/division process
Figure 7-10 Scanning electron micrograph of ground parenchyma cells at stem center in Rotan variety at middle growth stage
Figure 7-11 Scanning electron micrograph of ground parenchyma cells at stem center in Rotan variety at late growth stage
Figure 7-12 Major axis (A) and minor axis (B) of amyloplast by stem portion and variation coefficient
Figure 7-13 Number of amyloplasts per unit cross section area of ground parenchyma cells at stem center by stem portion and variation coefficient
Figure 7-14 Relationship between starch content (yield) and pith dry weight (A) or pith starch content (B)
Figure 7-15 Relationship between palm age and trunk length, trunk diameter or trunk volume
Figure 7-16 Frequency distribution of trunk lengths by study plot (palms/ha)
Figure 7-17 Frequency distribution of trunk lengths (50 cm intervals)
Figure 7-18 Frequency distribution of trunk lengths (1 m intervals)

Figure 8-1 Sago pith crushing work using a chipping axe
Figure 8-2 Extraction work: Sago pith washing by hands
Figure 8-3 Extraction work: Washing by hands and an extractor
Figure 8-4 Grater-type pith crushing work
Figure 8-5 Crushing work using a rasper machine
Figure 8-6 Extraction work: Washing by feet using pumping water
Figure 8-7 Crushing work using a chipping axe
Figure 8-8 Extraction work: Pith washing by feet
Figure 8-9 Pith crushing, Mindanao Island
Figure 8-10 Starch extraction, Mindanao Island
Figure 8-11 Distribution of different sago starch extraction methods and boundaries
Figure 8-12 Distribution of genetic groups in *M. sagu*
Figure 8-13 Type of sago starch extraction methods
Figure 8-14 Sago logs
Figure 8-15 Rotary cutter
Figure 8-16 Rasper
Figure 8-17 Hammer mill
Figure 8-18 Sieve bend
Figure 8-19 Rotary screen
Figure 8-20 Screw press
Figure 8-21 Multistage hydrocyclone
Figure 8-22 Super decanter
Figure 8-23 Manufacturing process in a sago starch factory

Figure 9-1 Starch granules in cells of the sago palm pith (SEM) x200
Figure 9-2 Photopastegram of sago, mung bean, potato and corn starches
Figure 9-3 Rapid Visco Analyzer (RVA) curves of various starches
Figure 9-4 Gel hardness and adhesiveness of various starches measured by Tensipresser
Figure 9-5 Changes in Hunter whiteness of various starch gels during storage at room temperature
Figure 9-6 Dendrogram by cluster analysis of physiochemical properties of starch (basal)

Figure 9-7 Manufacturing method of sago starch noodle *Sohun*
Figure 9-8 *Sohun*
Figure 9-9 *Mie sagu*
Figure 9-10 *Kue(h) bangkit*
Figure 9-11 *Kue(h) pisang*
Figure 9-12 *Kerupuk sagu*
Figure 9-13 Sensory evaluation of warabimochi made of sago starch
Figure 9-14 Sensory evaluation of kuzuzakura made of sago starch
Figure 9-15 Sensory evaluation of biscuits, in which wheat flour was partially substituted by sago starch
Figure 9-16 Starch dextrinization mechanism
Figure 9-17 Functional group formation by pH of oxidation reaction
Figure 9-18 Acetylated reaction by acetic anhydride
Figure 9-19 Acetylated reaction by vinyl acetate
Figure 9-20 Carboxymethyl reaction of starch by monochloroacetic acid
Figure 9-21 Hydroxyethyl starch formation reaction
Figure 9-22 Total plastics production in Japan (2005)
Figure 9-23 Manufacturing of polylactic acid from starch
Figure 9-24 Starch filtration residue
Figure 9-25 Disc dryer
Figure 9-26 Various characteristics of sago starch on starch diagram
Figure 9-27 Atomic force microscope (AFM) image of the structure of the outermost surface of sago starch
Figure 9-28 Scanning electron microscopic image of an enzymatically degraded sago starch granule during starch production

Figure 10-1 Articles of everyday use made of sago palm leaves
Figure 10-2 A variety of *upak* made of the skin of the sago leaf sheath base
Figure 10-3 The vascular bundle (circled) is removed from the base of leaf sheath to make a blow dart (indicated by the arrow)
Figure 10-4 A fishing tool called *biga* made of leaflet midribs
Figure 10-5 Roofing mats called *sapau* (*atap* in Malay)
Figure 10-6 Sago palm bark after pith removal
Figure 10-7 Sago palm bark used for flooring (and a sago crushing hammer on it)
Figure 10-8 *Songa* shoot of *M. vitiense*
Figure 10-9 Germinating fruits of *M. warburgii*
Figure 10-10 *M. amicarum* fruits (left) and longitudinal sections (right)
Figure 10-11 Folk crafts made of the fruit and endosperm of the section *Coelococcus* plants
Figure 10-12 Sago palm fruits used in flower arrangement
Figure 10-13 Sago starch manufacturing flow chart
Figure 10-14 Cross section of sago palm pith
Figure 10-15 Manual debarking
Figure 10-16 Dry sago residues after extracting starch from sago pith
Figure 10-17 Grindability of sugi wood and sago residue
Figure 10-18 Thermal softening curves of untreated, acetylated and lauroylated sago residue
Figure 10-19 Relationship between compressive stress and static cushion factor of foams with various sago residue contents
Figure 10-20 *Rhychophorus ferrugineus* (red palm weevil) adult
Figure 10-21 Mature larva of *R. ferrugineus*
Figure 10-22 Hacking a sago trunk away with a bush knife
Figure 10-23 A boy swallowing a *R. ferrugineus* larva
Figure 10-24 Satay made with *R. ferrugineus* larvae

Figure 11-1 Sago extraction work in Papua New Guinea

Figure 11-2 Bagged sago palm starch in Papua New Guinea
Figure 11-3 Sago palm starch served at a feast in Papua New Guinea
Figure 11-4 Cooked sago jellies (sago dumplings) in Papua New Guinea
Figure 11-5 *Lakatoi* boat at Hiri Moale Festival in Papua New Guinea

Figure 12-1 Change in world population
Figure 12-2 Sago starch
Figure 12-3 Sago palm plantation (Mukah, Sarawak, Malaysia)
Figure 12-4 Change in ethanol production in the United States
Figure 12-5 Rise in agricultural commodity prices
Figure 12-6 Biomass use
Figure 12-7 Bioethanol production in the world
Figure 12-8 Plastics production and waste in Japan
Figure 12-9 Biodegradable plastic production in Japan

# *Preface*

The twenty-first century has been characterized by four keywords: population, food, environment and resources. The subject of this publication, *Metroxylon sagu* or sago palm, is a plant which literally holds the key to resolving the major issues implied in these four keywords.

There are as many as 200 genera and 2,600 species of plants in the world that belong to the Arecaceae (palmae) family. A majority of them are distributed in the tropics and the subtropics. The most symbolic of them is perhaps the coconut palm, which conjures an image of tropical landscapes. The habitats of the coconut palm mostly coincide with the climatologically defined tropical regions. The coconut palm plays an important role as an agricultural and industrial crop whose fruit endosperm produces oil.

Another oil crop that is as important as the coconut palm is the oil palm, which has sometimes been criticized for the excessive deforestation associated with its plantation development. Other palms are largely unknown to the public except for the small percentage of people who actually utilize them. However, there are many palms in the tropics that have great potential as food crops or resource/material crops. The sago palm is one of these 'other' palms.

The sago palm is believed to originate in New Guinea Island and accumulates a large amount of starch in the trunk. People have been harvesting the sago palm for food for millennia. Starch is extracted from the sago palm by crushing the trunk pith into fine fragments and then kneading them in water to extract starch. Along with banana, breadfruit and taro, sago starch is one of the oldest staple foods for human beings. This is evident from the fact that 'sago' means bread in the Papuan language and 'sagu' means edible flour in the Malayan language. Sago palms are currently found in areas centering around its place of origin within latitudes 10 degrees north and south of the equator in Melanesia and Southeast Asia, up to an altitude of 700 m above sea level. The main countries within the distribution areas include Indonesia, Papua New Guinea, Malaysia, Thailand, the Philippines and the Solomon Islands. Sago starch is still used as a food staple in some of these areas and as a raw material for confectionery in all of them. The sago palm's high starch productivity has recently been attracting interest for possible industrial applications.

In academia, the sago palm has been studied and reported on mainly in the field of ethnological or anthropological area studies. Very few studies have been carried out in the fields of basic plant genetics, physiology, ecology, agronomy and starch properties and utilization.

Against this background, the Department of Agriculture of Sarawak State, Malaysia, designated the sago palm as the state's starch resource crop and began to conduct research on the sago palm growing area, growing environment and starch productivity from the 1970s. Sg. Talau Deep Peat Research Station (Sago Research Center) was set up in Dalat, Sarawak, in the early 1980s and systematic sago palm research was launched. The center was built on a site with thick peat soil layers partly because the utilization and development of deep peat soil sites throughout the Mukah-Dalat district was one of its goals and partly because the sago palm was the only crop that could be cultivated economically without significant soil improvement under the marshy, oligotrophic and low pH conditions which characterize peat soils.

A few years earlier (1977), a research group for the utilization and development of sago palm and sago starch was established within the Tropical Bio-Resources Research Committee of the Japanese Society for Tropical Agriculture chaired by the late Dr. Isao Nagato. At that time, Dr. Nagato was the president of the Research Center for Tropical Plant Resources. The research group hosted a symposium on the development of sago palm and the utilization of its products at the 45th conference of the Japanese Society for Tropical Agriculture in 1979. The proceedings were published in the society's journal, *Nettai nōgyō* (Japanese Journal of Tropical Agriculture), Vol. 23, No. 3 (September 1979), providing valuable information about sago palm for the first time in Japan. The research group subsequently undertook activities such as gathering information about the sago palm overseas and organizing study meetings.

Dr. Nagato strongly believed that the sago palm was the only crop that could solve the anticipated issues of the twenty-first century such as food crises arising from population growth in developing countries and the economic development of the tropical low land areas. Out of his strong desires to alert Japanese university students to the existence of this plant and to promote studies of the sago palm, he used his own funds to set up the Japan Fund for Sago Palm Research Promotion in 1979 and called on university and college students throughout Japan to form sago palm research clubs at their campuses. As a result, the sago palm research clubs were established at many universities and members were able to learn about the sago palm, many gaining firsthand knowledge through field studies. I was one of those students who were given a chance to study the sago palm through the sago palm research clubs. The management of the fund was delegated to the Japan Society for the Promotion of Science (JSPS) in 1986 and continued to provide support under the banner of the Tropical Bio-Resources Research Fund until 2008.

Thanks to Dr. Nagato's determination, Sago Yashi Sago Bunka Kenkyū Kai (Society for the Studies of Sago Palm and Sago Culture; the predecessor of the Japan Society of Sago Palm Studies) was founded in 1992. The academic society dedicated solely to the study of the sago palm, a tropical crop, is very unique in Japan. About 200 members joined in support of Dr. Nagato's aspirations. The society hosts annual conferences and seminars, issues its journal *SAGO PALM*, and maintains close exchange with other sago palm-related societies and researchers in Japan and overseas. Many of the studies that were conducted with the support of the aforementioned JSPS Tropical Bio-Resources Research Fund have been published in *SAGO PALM*.

These activities, which are represented by the founding of the Japan Society of Sago Palm Studies and the publication of *SAGO PALM*, stimulated sago palm studies and pushed Japan to the forefront of global sago palm research. Synchronous with these advances, sago palm research in the sago palm growing tropical countries began to gain momentum, especially in Malaysia, Indonesia and the Philippines. International symposiums have been organized biennially or triennially in recent years. Hosting these international symposiums is arousing the interest of many researchers in the host and other tropical countries as well as many government officials.

Obviously, the four keywords for the present century mentioned at the start – population, food, environment and resources – are closely interlinked with one another. The sago palm can be regarded as an 'environmentally sound starch resource crop for the twenty-first century' in that it is the only crop that can be economically cultivated in deep peat soils, it requires very little resource input for cultivation, and it boasts high starch productivity. Meanwhile, soaring crude oil prices have turned the world's attention toward biofuel production and major food crops such as corn, wheat and sugar cane have been diverted to bioethanol production. There have been corresponding rises in the prices of these crops and accompanying food shortages in developing countries. Even though the sago palm continues to be used for food by many people, the vast unused wild sago palm forests can supply starch resources without impacting its availability for use as food. One of the advantages of the sago palm is that it is less susceptible to meteorological disaster damage than seed crops as it accumulates starch inside its trunk. The sago palm can also make a contribution to food supply if it is introduced as a food crop to developing countries in the tropics where population growth is anticipated to cause food shortages in this century. In fact, the sago palm research group at Kyoto University has succeeded in introducing the sago palm to Tanzania in Africa where planted palms have already grown to the trunk formation stage.

This publication positions the sago palm as an 'environmentally sound starch resource crop for the twenty-first century' and attempts to bring together past studies on the sago palm, many of which have been undertaken by the members of the Japan Society of Sago Palm Studies, into one comprehensive volume. It covers a broad range of topics, including sago palm genetics, morphology, physiology, ecology, growing environment, cultivation, starch productivity, starch properties and applications, ethnology and anthropology. Nothing can give this group of contributing authors more pleasure than seeing this book play a role in stirring the interest of the readership in the sago palm.

I would like to express my gratitude to the members of the steering committee of the Tropical Bio-Resources Research Fund, Japan Society for the Promotion of Science, for supporting the publication of this book. On behalf of all contributing authors, I would also like to thank the editorial office of Kyoto University Press for their hard work. Finally, my heartfelt gratitude goes to the late Dr. Isao Nagato for his passion for the advancement of sago palm studies and the generous financial support he provided for that purpose.

Yoshinori Yamamoto
Chair, The Sago Palm Editorial Committee

# *Preface to the English Edition*

The sago palm is a kind of palm grown in the wet lands of Southeast Asia. It cannot grow in other places of the world even where the same environmental conditions are found. The sago palm breeds mainly by means of vegetative propagation. There is very limited sexual propagation by seeds because of the low germination rate of around 5% or so. Accordingly, the sago growing areas are quite restricted. In Southeast Asia, the sago palm is important for much more than food. The entire biomass of the sago palm is usable. Furthermore, the sago palm can grow very large, fixing correspondingly large quantities of carbon dioxide in the process. It can thus make an important contribution to abating the harmful effects of greenhouse gases and play a significant role in curbing global warming.

Few people know that Japan imports around 20 thousands ton of raw sago starch from Malaysia and Indonesia. Sago starch is typically used as dusting starch. The granules of sago starch are relatively large and uniform, contributing to its high price.

This book is a translation of the Japanese book *Sago Yashi*, originally published in 2010 by Kyoto University Press (Chair of Editorial Board, Yoshinori Yamamoto). It is the first multidisciplinary study of systematically collected data on sago and sago-related topics.

In the "Energetic Discussion" page of the *Mainichi Shimbun* (newspaper) on December 7, 2010, Mr. Masaharu Sakakibara, its editorial writer, introduced this publication to the general public, stating "This book points to the possibility of joint work of some 30 sago palm researchers in the fields of agriculture, anthropology and so forth." He wrote further that

> sago palm can grow around the equator of Southeast Asia. In 'La Description du Monde' or 'Il Milione' written by Marco Polo, sago palm is described as "a tree of excellent starch". However, sago palm is not widely known around the world, compared to other palms such as the coconut and the oil palm. It yields 200 to 300 kg of starch per palm, a much higher starch yield than corn or potato. Much attention has been paid to sago palm, with attempts to grow it in wide areas of wet land and/or acidic peat in Southeast Asia. Endeavoring to devise strategies to solve food crisis in developing countries with burgeoning populations, Japanese sago research groups have been steadily conducting research for over 30 years. In contrast to those who are simply interested in 'how to get food or money', these researchers have seriously engaged in studies about food in the 21st century.

This short article captures the essence of our sago palm research project and provides encouraging music to our ears.

Sago palm research is a scientific field that is inspiring young generations. The specific research fields are wide-ranging: from genetic information on large amount of carbon dioxide fixation to starch accumulation in the trunk. Younger scholars are now being invited to engage in studies about how to utilize the lowland areas in Southeast Asia to develop the technology for utilizing sago biomass totally.

The late Dr. Isao Nagato donated his properties to young scholars who aim to conduct sago research, to help them to realize his life-long dreams. Many Japanese sago scientists owe much to his generous funds. In addition to the original Japanese book mentioned above, his funds have been instrumental in expanding the activities of The Society of Sago Palm Studies (internationally known as the Japanese Society of Sago Palm Studies) and, in particular, the International Sago Symposium, held in 1985 and 2001 in Tokyo. Japanese sago scientists continue to be beneficiaries of his philanthropic endowment, making them the envy of sago scientists around the world. Having held many meetings about publishing books on the sago palm in English, The Japanese Society of Sago Palm Studies is very pleased to publish this volume, a task that sago scientists in other countries have longed to accomplish.

We acknowledge generous financial support proffered by the Japan Society for the Promotion of Science (JSPS) (Grant number 256004: Masanori Okazaki as representative) which provided us a Grant-in-Aid for Publication of Scientific Research Results. We are thankful to two professionals at Trans Pacific Press: Ms Minako Sato for translation from the original Japanese text into clear English and Dr Karl Smith for excellent editorial work. Thanks are also due to Kyoto University Press for their support throughout this project. We hope this book contributes to sago palm studies and serves as a stepping-stone for young sago scientists for further research.

Masanori Okazaki
Chair, The Sago Palm Editorial Committee

# Dr. Isao Nagato's contributions to sago palm research

The Late Dr. Isao Nagato

Japan has come to play a leading role in international research on sago palms. Much water has flowed under the bridge since the first International Sago Symposium was held in 1976.

The support of the late Dr. Isao Nagato played a very important part from 1976 to the present time. The author was contacted by Dr. Nagato after attending the symposium held in Kuching on Borneo Island in 1976 and discussed the status of sago palm research in the world and the cultivation and processing of sago palms in Sarawak. Dr. Nagato subsequently developed the belief that 'the sago palm ... will contribute to resolving the possible food crisis in the twenty first century' and invested large sums of his own funds in sago palm research. He built the foundations of sago palm research in Japan and served as the founding chairperson of The Society for the Studies of Sago Palm and Sago Culture, the predecessor to The Society of Sago Palm Studies.

Dr. Nagato's contributions are immeasurable. He took notice of the potential of sago palm as a starch producing crop in the tropics very early and dedicated himself to the promotion of sago palm research. Now that the whole world is joining the trend toward the production of bioethanol as a petroleum substitute and beginning to use starch in large-scale alcohol production, the exploitation of the highly productive sago palm is considered to be of great help in solving the world's food and energy problems. As one of the few crops that are cultivated in the vast tropical low lands and brackish-water zone, the sago palm offers tremendous potential and possibilities.

Even though no naturally growing sago palms exist in Japan, The Society of Sago Palm Studies has been able to send researchers to international sago symposiums, provide financial support for sago palm studies, host the International Symposium on Sago in Tokyo in 1985 and in Tsukuba in 2001, and publish the proceedings of the symposiums thanks to the support of the Nagato Fund for the Sago Palm Research Promotion.

Keiji Kainuma
Former President of The Society of Sago Palm Studies
April, 2001

# Contributors

**Editorial Committee**

**Chair**
Okazaki, Masanori
Faculty of Bioresources and Environmental Science
Ishikawa Prefectural University

Yamamoto, Yoshinori
Faculty of Agriculture, Kochi University

**Chief editor**
Ehara, Hiroshi
Graduate School of Bioresources, Mie University

**Editors** (alphabetical order)

Ando, Ho
Faculty of Agriculture, Yamagata University: Chapter 6

Ehara, Hiroshi
Graduate School of Bioresources, Mie University: Chapters 1 & 8

Goto, Yusuke
Graduate School of Agricultural Science, Tohoku University: Chapter 4

Hirao, Kazuko
Aikoku Gakuen Junior College: Chapter 9

Nitta, Youji
College of Agriculture, Ibaraki University: Chapter 3

Okazaki, Masanori
Faculty of Bioresources and Environmental Science, Ishikawa Prefectural University: Chapter 2 and 12

Ohmi, Masaharu
Graduate School of Agriculture, Tokyo University of Agriculture and Technology: Chapter 9

Toyoda, Yukio
College of Tourism, Rikkyo University: Chapters 10 & 11

Toyota, Koki
Graduate School of Bio-Applications and Systems Engineering, Tokyo University of Agriculture and Technology: Chapter 5

Yamamoto, Yoshinori
Faculty of Agriculture, Kochi University: Chapter 7

**Contributors** (order of writing)

Yamamoto, Yoshinori
Kochi University

Okazaki, Masanori
Tokyo University of Agriculture and Technology

Kainuma, Keiji
Nihon University

Ehara, Hiroshi
Mie University

Takamura, Tomoki
formerly Kyoto University

Shimoda, Hiroyuki
formerly Tokyo University of Agriculture and Technology

Kimura, Sonoko Dorothea
Tokyo University of Agriculture and Technology

Nitta, Youji
Ibaraki University

Goto, Yusuke
Tohoku University

Nakamura, Satoshi
Miyagi University

Watanabe, Manabu
Iwate University

Jong, Foh Shoon
formerly PT. National Timber and For4est Produce, Indonesia

Miyazaki. Akira
Kochi University

Naito, Hitoshi
Kurashiki University of Science and the Arts

Toyota, Koki
Tokyo University of Agriculture and Technology

Ando, Ho
Yamagata University

Kakuda, Kenichi
Yamagata University

Sasaki, Yuka
Yamagata University

Watanabe, Akira
Nagoya University

Yoshida, Tetsushi
Kochi University

Nishimura, Yoshihiko
University of the Ryukus

Mishima, Takashi
Mie University

Takahashi, Setsuko
formerly Kyoritsu Women's University

Kondo (Hamanishi), Tomoko
Kyoritsu Women's University

Hirao, Kazuko
Aikoku Gakuen Junior College

Ohmi, Masaharu
Tokyo University of Agriculture and Technology

Konoo, Shigeki
formerly Nippon Starch Chemical Co., Ltd.

Toyoda, Yukio
Rikkyo University

Mitsuhashi, Jun
formerly Tokyo University of Agriculture and Technology

Kamimura, Toru
Kobe City College of Nursing

# 1
# Origin, Dispersal and Distribution

## 1.1 Taxonomy

### 1.1.1 Starch accumulating palms

The Arecaceae family consists of six subfamilies, containing approximately 200 genera and 2,600 species. The subfamilies Calamoideae genera *Metroxylon*, *Eugeissona*, *Raphia* and *Maurutia*, Corypha genera *Corypha*, *Phoenix* and *Borassus*, and Arecoideae genera *Arenga*, *Caryota*, *Wallichia*, *Roystonea*, *Butia*, *Syagrus* and *Bactris* are known to produce starch in the trunk (Table 1-1).[1] Based on starch yield, the genus *Metroxylon* is the most productive among them and *M. sagu* Rottb. (true sago palm) of the section *Metroxylon* (*Eumetroxylon*) is considered to be the most promising. In Southeast Asia and Melanesia, the starch produced from sago palms has long been used for food (Takamura 1990) and remains an important starch resource in the region to this day. Starch obtained from the trunk of *A. pinnata* and its close relative (*A. microcarpa*) is also called sago (while sago originally means starch obtained from sago palms, the term is often used loosely to refer to starch obtained from the trunk of other palm species and non-palm plants) and starch from *A. pinnata* is sold under the name of sago in parts of Indonesia (specifically: Java and Sulawesi), while starch from *A. microcarpa* and *C. utan* is sold as sago in local markets in the Sangihe and Talaud Islands, Indonesia, and Mindanao Island, the Philippines. Yatay starch from *Butia* yatay (yatay palm) is also available in the region but production is much smaller than that of sago starch.

In Japan, the true sago palm is called 'masago yashi' or 'seigo yashi'. It is an evergreen tree that belongs to the subfamily Calamoideae genus *Metroxylon* section *Metroxylon* (*Eumetroxylon*), and the only species of this section (see details below). The chromosome number is n=16. It was first mentioned in Japanese literature as 'sago bei' in *Yamato sōhon* (Japanese botany) by Ekken Kaibara (1709). Sago is still an important staple food for the residents of New Guinea, the Maluku Islands (Moluccas), Sulawesi, Borneo (Kalimantan) and Siberut Island of the Muntawai Islands west of Sumatra.

The word 'sago' appears to derive from a Javanese word that means 'starch obtained from palm pith' but it has become a common name for starch in general in many Southeast Asian languages. As mentioned above, starch sourced from the trunk of other palm species or cycads, or cassava and arrowroot (*Maranta arundinacea*) is often called sago.

Table 1-1 Arecaceae plants that store starch in the trunk

| Subfamily | Tribe | Subtribe | Genus | Species |
|---|---|---|---|---|
| Coryphoideae | Corypheae | Coryphinae | *Corypha* | *C. utan* Lamarck gebang. *C. umbraculifea* L. talipoto. |
| | Phoeniceae | | *Phoenix* | *P. paludosa* Roxb., mangrove date palm. |
| | Borasseae | Lataniinae | *Borassus* | |
| Calamoideae | Calameae | Eugeissoninae | *Eugeissona* | *E. utilis* Becc., wild Borneo sago palm chirimen uroko yashi. *E. insignis* Becc. |
| | | Metroxylinae | *Metroxylon* | *M. sagu* Rottb. |
| | | Raphiinae | *Raphia* | |
| | Lepidocaryeae | Lepidocaryum | *Mauritia* | *M. flexuosa* L. f., miriti palm, moriche palm. |
| Arecoideae | Caryoteae | | *Arenga* | *A. pinnate* (Wurmb) Merr. *A. microcarpa* Becc. |
| | | | *Caryota* Fishtail palms | *C. urens* L., fishtail palm, toddy or kittool palm, bastard sago, wine palm, solitary fishtail palm, jaggery palm. |
| | | | *Wallichia* Wallich palms | *W. disticha* T. Anderson |
| | Areceae | Roystoneinae | *Rostonea* Royal palms | |
| | Cocoeae | Butiinae | *Butia* | *B. yatay* Becc., yatay palm jelly palm, butia palm. |
| | | | *Syagrus* | Syagrus palm. |
| | | Bactridinae | *Bactris* | Peach palm. |

## *1.1.2 Taxonomy*

The genus *Metroxylon* is divided into two subgroups: section *Metroxylon* (*Eumetroxylon*) and section *Coelococcus*. Classification by Beccari (1918) is shown in Tables 1-2 and 1-3. Images of mature palms belonging to these sections are shown in Figure 1-1. Plants belonging to this genus produce fruits that are covered with scales. Under this classification, *Eumetroxylon* bears fruits with 18 rows of longitudinally-arranged scales and *Coelococcus* 24 to 28 rows. Beccari divided the section *Eumetroxylon* into three types: spineless *M. sagus* Rottb., spiny *M. rumphii* Mart., and spineless *M. squarrosum* Becc. and placed 2 varieties under *M. sagus*, and 7 varieties and 6 subvarieties under *M. rumphii*, based on morphological characteristics such as the shape and size of fruit, leaf sheath, petiole, and the presence, length and density of spines growing on the rachis of the inflorescence, and the area of distribution. For the section *Coelococcus*, he designated 6 species and 2 varieties based on the area of distribution and the shape and size of fruit and flower. However, it was later pointed out that Beccari's classification was based on a small number of fruit specimens and inadequate classification criteria. It was later confirmed by Whitemore (1973) that *M. salomonense* and *M. bougainvillense* are of the same species. Rauwerdink (1986) argued that there was no substantive difference between *M. sagus* and *M. rumphii* because they could be cross-pollinated and segregation for spininess and fruit color was observed at a certain rate in the resultant seeding. He proposed classification of the genus *Metroxylon* into 5 species, namely, *M. sagu*, *M. amicarum*, *M. vitiense*, *M. salomonense* and *M. warubugii*. Rauwerdink then divides *M. sagu* into 4 forms – *M. sagu* forma *sagu*, f. *tuberosum*, f. *micracantum* and f. *longispinum* according to criteria such as spininess, the timing of appearance/disappearance of spines, and spine length (Table 1-4). Considering that spines are generally longer at the juvenile stage and become shorter as the palm matures (Sastrapradja 1986), Rauwerdink's 4 formae are defined according to the characteristics of their spines: spineless, spine traces at the base of petiole (leaf sheath), short spines not longer than 4 cm on the leaf sheath and petiole, and spines 4 to 20 centimeter long. Figure 1-2 shows some examples of leaf characteristics. As mentioned later, variation in the color of petiole of spineless plants is observed as well as the level of spininess.

It has been reported that the spine traits of seedlings from the same mother palm are heterogeneous and that spineless plants can occur from the seeds obtained from a spiny mother palm (Jong 1995a). Ehara et al. (1998) report that 28% of the progeny of spineless mother palms were found to be spiny. In view of these findings, it must be said that Rauwerdink's classification of *M. sagu* formae is also problematic. According to recent studies, no obvious relationships are found between the morphological characteristics and genetic distance of *M. sagu* in Malaysia, Indonesia, the Philippines and Papua New Guinea as discussed in the next section.

In communities that are highly dependent on the sago palm, however, more diverse types are recognized based on morphological characteristics, pith color, yield and other traits. For example, the Galela folk classification on the island of Halmahera, Maluku Province, Indonesia, studied by Yoshida (1980) has 8 types (Figure 1-3). All of them have spines at the juvenile stage and lose

Table 1-2 Beccari's classification of section *Metroxylon* (*Eumetroxylon*) (1918)

| | Species/variety | Distribution/vernacular name etc.[1] | Characteristics |
|---|---|---|---|
| 1. | *M. sagus* Rottb. (forma typica). | [Malay Archipelago*] | Leaf sheath, petiole, bract, and first- and second-order branches are all spineless. |
| 1a. | var. *molat* Becc. | [Seram Island] sagu molat, sagu malat: West Seram, Halmahera. | Globose fruit, dented at base, smaller than forma typica fruit (diameter 2.8 cm). |
| 1b. | var. *peekelianum* Becc. | [German New Guinea] bia tun: Namatani near Salai. | Small globose fruit, length 2–2.3 cm, diameter 2–2.2 cm, scale border (1/2 mm) coloration. |
| 2. | *M. rumphii* Mart. (forma typica) | [Malay Archipelago*] | Spines on leaf sheath, petiole, first-order branch of inflorescence; large globose fruit, diameter just over 2.5 cm; small spines on leaflet midrib. |
| 2a. | var. *rotang* Becc. | [West Seram] sagu rotang, Rumph's sagu duri rotang: equivalent of var. *micrachanthum*. | Slightly smaller fruit than 2, with short spines on petiole. |
| 2b. | var. *longispinum* Becc. | [Amboyna (Ambon Island)] lapia macanaru, leytiomor, macanalo, macanalum: Hitoe, equivalent of var. *micracanthum* subvar. *makanaro*. | Larger fruit than 2, growing very long spines on petiole sparsely. |
| 2c. | var. *sylvestre* Becc. | [West Seram] lapia ihur, ihul sagu ihor: Seram Island (uncommon on Ambon Island). | Slightly squat globose fruit, length 3–3.5 cm, diameter 3.5–3.8 cm; spines on leaflet midrib. |
| 2d. | var. *ceramense* Becc. | [Seram Island] sagu ceram, sagu merah, sagu putih, sagu hitam (4 types). | Medium-sized globose or elliptical fruit, smaller than *M. sagus* and larger than var. *micracanthum*; large and wide leaflet. |
| 2d'. | var. *ceramense* subvar. *platyphyllum* Becc. | [Amahai (Seram Island)] sagu ceram: Amahai on the central southern coast of Seram Island. | Ovoid and elliptical fruit, length 3.7 cm, diameter 2.6 cm; very large and wide leaflet 12 cm wide. |
| 2d". | var. *ceramense* subvar. *rubrum* Becc. | [Amahai (Seram Island)] sagu merah. | Elliptical fruit, length 3 cm, diameter 2.2 cm. |
| 2d"'. | var. *ceramense* subvar. *album* Becc. | [Amahai (Seram Island)] sagu hitam. | Globous fruit, rounded at top and dented at base, diameter 3.2 cm. |
| 2d"". | var. *ceramense* subvar. *nigrum* Becc. | [Amahai (Seram island)] | Globous fruit, rounded at top, smaller than 2d"', diameter 3 cm. |
| 2e. | var. *micracanthum* Becc. | [Seram Island] Rumphius sago duri rottang, lapia luliuwe: Ambon name (common in Humohela, Seram). | Very small obovoid fruit, narrow and not depressed at top, thin pericarp at top and thick, porous and hard at base. |
| 2e'. | var. *micracanthum* subvar. *tuni* Becc. | [West Seram] | Fruit diameter 2.3 cm. |
| 2e". | var. *micracanthum* subvar. *makanaro* Becc. | [West Seram] Related to var. *ceramense*. Different from Rumphius lapia macanaru. Equivalent of Martius *M. longispinum* (var. *longispinum* Becc.). | Fruit 2.7–2.8 cm. |
| 2f. | var. *buruense* Becc. | [Buru Island] | Very small fruit, globose, 1.8–2 cm. |
| 2g. | var. *flyriverense* Becc. | [New Guinea, Fly River] | |
| 3. | *M. squarrosum* Becc. | [East Seram] | Inflorescence and bract are spineless; rachilla is flat; leaf margin is spineless. |

1) Distribution area is shown in square brackets [ ]. Italicized terms that follow are either vernacular names or folk variety names in the area. Specific places of distribution follow the colon (:).

*: Malay Islands in the original. Partially supplemented by author.

Table 1-3 Beccari's classification of section *Coelococcus* (1918)

| Species/variety | Distribution | Characteristics |
|---|---|---|
| 1. *M. warburgii* Heim. | New Hebrides (now Vanuatu) | Fruit length 10–12 cm, diameter 7–9 cm, 24 rows of scales; back of leaflet is pale blue-green color (wax bloom)*. |
| 2. *M. upoluense* Becc. | Upolu Island, Samoa | Small fruit narrowing toward base, length 3.3 cm, diameter 2.5 cm, 24 rows of scales. |
| 3. *M. vitiense* Benth. Et Hook. | Fiji Islands | Fruit is globose, conical with round base and wide top, length 5.5–6.5 cm, diameter 4.5–7 cm. |
| 4. *M. amicarum* Becc. | Caroline Islands | Large globose fruit, diameter 8 cm, slightly flat at top, no dent at base. |
| 4a. var. *commune* Becc. | | Flower length 8–8.5 mm, diameter 3.5–4 mm; fruit diameter of 8–9 cm is slightly larger than length. |
| 4b. var. *maius* Becc. | | Flower length 12 mm, diameter 5–6 mm; fruit diameter 11–13 cm. |
| 5. *M. salomonense* Becc. | Solomon Islands, German New Guinea, New Britain Island | Fruit diameter 7 cm, slightly squat globose, no dent at base; seed diameter 4 cm; pericarp thickness 5–6 mm; pointed scales. |
| 6. *M. bouganvillense* Becc. | Bougainville Island. | Fruit diameter 5.5 cm, slightly squat globose, dented at base; pericarp thickness 10–12 mm; seed diameter 2.5 cm; pointed scales. |

*: The abaxial leaflet surface (back) is slightly white-tinged and less glossy than the adaxial surface probably due to different levels of cuticle accumulation (Ehara et al. 2003b). This characteristic is distinct from other species.

Table 1-4 Classification criteria of *M. sagu*

| | Morphological characteristics | Variety (vernacular name) by Rauwerdink |
|---|---|---|
| 1 | Leaf sheaths and petioles are completely spineless at all stages. | forma *sagu* (PNG: ambutrum, kaparang, awirkoma) |
| 1 | Leaf sheaths and petioles are covered with spines or not completely smooth. | 2 |
| 2 | The base of the petiole (leaf sheath) has knob-like structures that are vestigial spines at all stages. | forma *tuberatum* (PNG: koma, oliatagoe) |
| 2 | Leaf sheaths and petioles are covered with spines. | 3 |
| 3 | Leaf sheaths and petioles are covered with spines that are not longer than 4 cm. | forma *micracanthum* (PNG: makapun, waipi, kangrum, mandam) |
| 3 | Leaf sheaths and petioles are covered with spines that are 4–20 cm long. | forma *longispinum* (PNG: wakar, ketro, anum, ninginamé, nago, tring, passin, kangrum, wombarang, moiap) |

Prepared and partially supplemented by author based on Rauwerdink (1986)

Figure 1-1 Mature *Metroxylon* palms

Top left: *M. sagu*, East Sepik, Papua New Guinea; top right: *M. amicarum*, inflorescences at the mid-level axils are bearing fruits. Chuuk, Micronesia; center left: *M. salomonense* at the fruiting stage, Singapore Botanic Gardens; center right: *M. vitiense*, Viti Levu Island, Fiji; bottom left: *M. waburgii*, Upolu, Samoa: bottom right: *M. Paulcoxii* (*M. upoluense*), Upolu, Samoa.

them as they mature but the spine trait is one of the key criteria for this folk classification:
1. Presence of spines on petiole and rachis
2. Length of spine and leaflet
3. Color of back of petiole and rachis (green without banding, dark grey band, brown band)
4. Red base of leaflet
5. Spines on leaflet (thick or thin)
6. Width of leaflet (wide or narrow)

Since juvenile leaflet characteristics such as spines and petiole/rachis band colors disappear as the sucker develops through growth stages and cannot serve as classification criteria for mature plants, leaf color and length, petiole color and starch color are used as criteria at the mature stage:
1. Traces of spines on petiole and rachis
2. Leaf length
3. Leaf color (deep or pale green)
4. White tinge at base of petiole (leaf sheath)
(from Yoshida's report with author's addition)

While Yoshida reports that the coloring of the banding on the back of petiole and rachis disappears as the tree develops, this feature can be observed in some plants even at the harvesting stage (Figure 1-2).

In Indonesia's eastern islands, 5 types are recognized under the vernacular names of sagu molat, sagu tuni, sagu ihur, sagu makanaro and sagu ikau (Flach 1980) while 8 types are reported in Irian Jaya Province (now Papua Province) (JICA 1981b) and 15 types in East Sepik Province, Papua New Guinea (Shimoda and Power 1992a). It has been reported, however, that a simple sequence repeat analysis of chloroplast DNA of plants collected from various locations in Papua Province has found that they are divided into 3 groups with about 77% belonging to one group (Abbas et al. 2006). Thus, the extent of genetic variation among various local sago palm types under folk classification is not yet clear. In view of this situation, it was agreed at the 8th International Sago Symposium held in Jayapura, Papua, Indonesia, in 2005 that various sago palm types classified by local peoples based on morphological characteristics and growth habits were to be treated as folk varieties for the time being (Ehara 2005).

Figure 1-2a Variation in the characteristics of sago palm leaf

Top left: no banding on rachis (*kanduan*); top center: dark gray banding on rachis (*kunangu*); top right: brown banding on rachis (*kosogu*) (East Sepik, PNG); center left: spines on petiole from left to right – spineless (*roe*), sparse short spines (*rui*), dense long spines (*runggamanu*); center right: vestigial spines on petiole from a relatively early stage (*roe me eto*) (Southeastern Sulawesi, Indonesia); bottom left: smooth spineless petioles; bottom right: traces of spines at petiole base and leaf sheath and spineless rachis after trunk formation (*kanduan*) (East Sepik, PNG).

Figure 1-2b Variation in the characteristics of the section *Coelococcus* palms
Top left: *M. amicarum*, Chuuk, Micronesia; center left: adaxial surface of rachis of *M. warburgii*; bottom left: abaxial surface of rachis of *M. warburgii* (Upolu, Samoa); top right: *M. salomonense*, Singapore Botanic Gardens; bottom right: *M. vitiense*, Vanua Levu, Fiji.

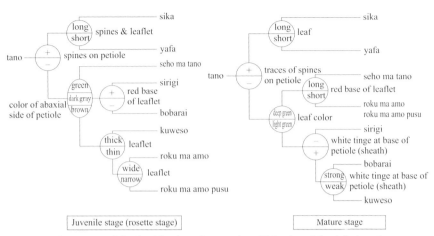

Figure 1-3 Galela classification of sago palms, Halmahera Island, Indonesia
Source: Yoshida (1980)

Rauwerdink proposes that section *Coelococcus* should include 4 species as mentioned above but the current classification includes 5 species, namely, *M. salomonense* (Warb.) Becc. found in the Solomon Islands and northern and central Vanuatu, *M. warburgii* (Heim) Becc. in Vanuatu, Fiji and Samoa, *M. vitiense* (H. Wendl.) H. Wendl. ex Hook. in Fiji, *M. paulocoxii* McClatchey in Samoa, and *M. amicarum* (H. Wendl.) Becc. in Micronesia (Barrau 1959; Beccari 1918; Dowe 1989; Ehara et al. 2003d; McClatchey 1999; Rauwerdink 1986). Beccari (1918) classified specimens from Upolu Island, West Samoa (now Samoa), as *M. upoluense* whose inflorescence and leaf samples are held at Royal Botanic Gardens, Kew, England. However, McClatchey (1998) argues that the description of Samoan sago palms by Beccari is not detailed enough to be used for species identification and reports a new species, *M. paulcoxii*, in addition to *M. warburgii* within section *Coelococcus* found in Samoa. The presence of *M. warburgii* in West Samoa has also been reported by Rauwerdink (1986) and confirmed by the authors.

Morphological characteristics of each *Metroxylon* species are shown in Table 1-5. The *Coelococcus* palms are markedly different from *Metroxylon sagu* in that they are non-suckering. Furthermore, *M. amicarum* grows lateral inflorescences from leaf axils while *M. salomonense*, *M. warburgii*, *M. vitiense* and *M. paulocoxii* produce a terminal racemose inflorescence just as *M. sagu* does (Figure 1-4). The lateral inflorescence-producing *M. amicarum* is plenanthic (polycarpic) but the other terminal inflorescence-producing *Coelococcus* palms are hapaxanthic (monocarpic) like *M. sagu*. And the racemose inflorescence-producing species also present variation in the branching system and pattern, fruit shape, or the color and rigidity of spines on the petiole/rachis. In *M. vitiense*, second-order branches on the inflorescence are pendulous. Rauwerdink (1986) reports that second-

Table 1-5 Morphological characteristics of the genus *Metroxylon*

| | Morphological characteristics, distribution | Species (vernacular name) |
|---|---|---|
| 1 | Fruit is covered with 18 rows of vertically arranged scales, 5.2–5.6 cm in diameter (approx. 3 cm for non-starch-producing type). Clump forming. Spineless or spiny, with varying spine length and density (section *Metroxylon*). New Guinea, Indonesia, Mindanao, Malaysia. | *M. sagu* |
| 1 | Fruit is covered with 24–28 rows of scales. Solitary, non-suckering, non-clumping (section Coelococcus). | 2 |
| 2 | Lateral inflorescence. First-order branches emerge at leaf axils.* Cultivated in Guam, Caroline Islands and the Philippines.<br>Fruit is globose, 26–29 rows of scales, 7–10.3 cm in diameter (Micronesia); 24–25 rows (Singapore Botanic Gardens**); 26 rows (Friendly Islands/ Tonga Archipelago**); 28 rows (the Philippines**).[1] | *M. amicarum* (Moen & Uman Islands: *rwung, foun rupwung*, Pohnpei Island: *oahs, ohs*)[1] |
| 2 | Terminal inflorescences, first-order branches, arising from axils of bracteal leaves at crown. | 3 |
| 3 | Second-order branches on inflorescence droop and very short (20 cm), Fiji Islands, spines on petiole and rachis are black, fruit is globose, 27–32 rows of scales, 5.5 cm in diameter.[1] | *M. vitiense* (*songa*, sago, *seko*)[1] |
| 3 | Second-order branches on inflorescence do not droop. | 4 |
| 4 | Third-order branches on inflorescence droop and are very long (20 cm), pubescence on the adaxial side of all pedicel bracts, Solomon Islands, Bougainville Island, Santa Cruz Islands.<br>Inflorescence show 2 or 3 orders of branching, fruit is globose, Vanuatu.[2] Spines on petiole and rachis are soft, rachillae arising as both second-order branching on first-order branch of inflorescence and third-order branching after second-order branching (Vanuatu),[3] 27–28 rows of scales on fruit (Vanuatu), 27–28 rows, diameter 7 cm (Singapore Botanic Gardens), 24–27 rows (Shortland Islands, PNG**).[1] | *M. salomonense* |
| 4 | All branches on inflorescence arising upright. No pubescence on the adaxial side of all pedicel bracts, New Hebrides (Vanuatu), West Samoa (Samoa). Fruit is pyriform.[2] Vanuatu, Rotuma and Vanua Levu Islands (Fiji). Marked cuticle formation on the abaxial side of leaflet. 26 rows of scales on fruit, diameter 6–7.3 cm (Vanuatu), 24–25 rows, 7.3–8.1 cm (Rotuma), 24 rows, 7.9 cm (Samoa), 24–25 rows (Malekula Island, New Hebrides**).[1] | *M. warburgii* (Samoa: *niu o lotuma*, meaning 'palm from Rotuma') |
| 4 | Inflorescence show 2 to 3 orders of branching, rachillae arising non-uniformly (drooping or upright).[4]<br>Second-order branches arising between the center and tip of a first-order branch form rachillae, which droop, and third-order branches arising between the center and base of a second-order branch form rachillae, which grow upright (2 types of branching pattern coexist at the central segment).[1] | *M. paulcoxii* |

Source: Rauwerdink (1986)
1) Based on survey/observation by author.
2) Dowe (1989).
3) Ehara et al. (2003d).
4) McClatchey (1998).
*: In those species that form a terminal racemose inflorescence, the leaves set below the inflorescence become progressively shorter whereas in *M. amicarum*, leaves arising from axils on first-order branches of the inflorescence at the crown level are not any shorter than those at the lower levels.
**: Specimens held by Kew Gardens.

order branches on *M. salomonense*, *M. warburgii* and *M. paulcoxii* are not pendulous. However, Dowe (1989) has found variation in the second- and third-order branching system of the inflorescence of *M. salomonense*. In fact, where rachillae grow as third-order branches on a second-order branch, the second-order branch grows erect on the first-order branch. However, rachillae do form on first-order branches as second-order branches in some cases and these second-order branches (rachillae) are pendulous in such cases (Figure 1-5). These different second-order branching patterns can be found on a single first-order branch of *M. salomonense* (Ehara et al. 2003d). Regarding distinct characteristics of spines on the petiole and rachis, *M. vitiense* form black spines while *M. salomonense* has soft and flexible fibrous spines unlike any other species (Figure 1-2).

The rates of germination of *M. sagu* seeds are generally low (see Chapters 3 and 4 for descriptions of fruit/seed and germination) but the seeds of *M. warburgii*, *M. vitiense* and *M. amicarum* have high germinability. It is notable that *M. warburgii* produces viviparous seeds (Ehara et al. 2003d). All *Metroxylon* species produce fruits that are covered with scales on the outermost layer but the number of scale rows varies from one species to another ranging from 21 to 32 rows with intraspecific variation in some. All of the *Coelococcus* species produce fruits that are larger than *M. sagu* fruits; *M. amicarum* fruits in particular are sometimes over 10 cm in polar diameter (Figure 1-6).

*M. warburgii* has shorter trunk length and smaller trunk diameter than *M. sagu* but the trunk length of *M. salomonense*, *M. vitiense* and *M. amicarum* is comparable to or longer than that of *M. sagu* (Ehara et al. 2003c; Ehara 2006a). Their leaves are important as building and livingware material and the hard endosperm of *M. amicarum* and *M. warburgii* seeds is utilized as craftwork material. Pre-emergent young leaves around the growing point of *M. vitiense* are utilized as a vegetable. Regarding starch yield, *M. salomonense*, *M. warburgii*, *M. vitiense* and *M. amicarum* are all low in the dry matter and starch contents of pith compared with *M. sagu* (Table 1-6). For this reason, *M. salomonense* and *M. amicarum* have low yield despite the large size of their trunk. In contrast, *M. salomonense*, *M. warburgii* and *M. amicarum* have relatively higher total sugar content in pith palms even after the emergence of the inflorescence. In Micronesia, Melanesia and Polynesia, *Coelococcus* palms are mostly regarded as emergency crops and had been utilized when staple crops suffered climate damage up to the 1950s or 1960s (Ehara et al. 2003d). It has been reported that salt was collected from the ashes of burned leaves in Vanuatu during the 1940s; similar examples were observed in PNG as well (Cabalion 1989). Today, roof thatching is the most common use of the leaves and the domestication of *M. warburgii* is currently under way in Vanuatu and Samoa.

Regarding phylogenetic relationships between *M. sagu* and the *Coecoloccus* species, it has been reported that based on analysis of 5S nrDNA spacer sequence data, *M. salomonense* is genetically closer to *M. sagu* compared with *M. warburgii* of Vanuatu and *M. amicarum* of Micronesia (Ehara 2006a).

# Origin, Dispersal and Distribution

Figure 1-4 Inflorescences of the genus *Metroxylon* palms
Top left: *M. sagu*, East Sepik, Papua New Guinea; top right: *M. amicarum*, Chuuk, Micronesia; center left: *M. salomonense* in the fruiting stage, Singapore Botanic Gardens; center right: *M. vitiense* at the fruiting stage, Viti Levu, Fiji; bottom left: *M. warburgii* at the flowering stage, Upolu, Samoa; bottom right: *M. paulocoxii* (*M. upoluense*) at the flowering stage, Upolu, Samoa.

Figure 1-5 Branches on the inflorescence of the genus *Metroxylon* palms
Top left: *M. sagu*, East Sepik, Papua New Guinea; top right: *M. amicarum*, Chuuk, Micronesia; center left: *M. salomonense*, Gaua, Vanuatu; center right: *M. vitiense*, Viti Levu, Fiji; bottom left: *M. warburgii* at the flowering stage, Esprit Santo, Vanuatu; bottom right: *M. paulocoxii* (*M. upoluense*), Upolu, Samoa.

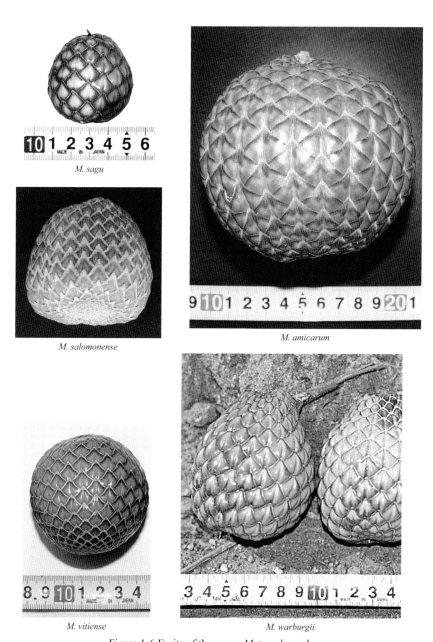

Figure 1-6 Fruits of the genus *Metroxylon* palms
*M. sagu*: Batu Pahat, Malaysia; *M. amicarum*: Chuuk, Micronesia; *M. salomonense*: Singapore Botanic Gardens; *M. vitiense*: Viti Levu, Fiji; *M. waburgii*: Malekula, Vanuatu.

Table 1-6 Comparison of yield and yield components of the *Mextroxylon* species

| Species (survey location) | Trunk length (m) | Trunk diameter (cm) | Pith density (g/cm³) | Pith dry matter (%) | Pith dry matter weight (kg) | Pith starch content (%) | Pith total sugar content (%) | Starch yield (kg) |
|---|---|---|---|---|---|---|---|---|
| *M. sagu* (Indonesia) | 8.6 | 45.2 | 0.770 | 41.1 | 413.9 | 77.1 | 4.9* | 309.8 |
| *M. salomonense* (Vanuatu) | 8.5 | 58.0 | 0.850 | 18.5 | 326.0 | 48.9 | 15.3 | 159.4 |
| *M. warburgii* (Vanuatu) | 5.3 | 32.6 | 0.937 | 33.2 | 131.4 | 36.4 | 13.1 | 35.9 |
| *M. amicarum* (Micronesia) | 10.7 | 43.8 | 0.794 | 16.0 | 179.4 | 38.8 | 10.0 | 71.8 |
| *M. vitiense*** (Fiji) | 8.3 | 39.0 | 0.903 | 25.0 | 190.5 | 27.2 | 7.9 | 60.6 |

Prepared by author based on Ehara (2006a)
*: The average of 2 folk varieties used for food in PNG.   **: see photos below.

*M. vitiense* stand near Navua, Viti Levu Island, Fiji

*M. vitiense* stand (marshland with some submerged areas even during the dry season)

## 1.2 Geographical origin, dispersal and distribution of the true sago palm

*M. sagu* Rottb. (true sago palm) is found in the Malay Peninsula from southern Thailand to western and eastern Malaysia, Brunei, Indonesia's Sumatra and the surrounds, Java, Kalimantan, Sulawesi, the Maluku Islands, West Papua (Irian Jaya), the central and southern Philippines, Papua New Guinea (PNG) and the Solomon Islands situated in a zone 10 degrees north and south of the equator. It has high environment adaptability and can grow in low land swamps, acidic soils and brackish-water regions in which other crops cannot survive. The true sago palm grows wild near lakes and rivers and is found up to an altitude of about 700 m above sea level around Lake Tondano in northern Sulawesi and about 1,000 m in PNG. Because wild stands of sago palms are found these days primarily in swamps and peat moors unsuitable for rice paddy development, such sites are generally understood to be its habitats. However, the true sago palm can also grow in well-drained soils where it in fact grows better than it does under flooded or submerged conditions. In other words, today sago palm stands are found growing in places that have not been subject to agricultural development, which does not mean that low land swamps are the 'natural' habitat for the true sago palm.

Starch yield from the true sago palm varies greatly depending on the habitat and the folk variety. Trunk diameter, which has a large impact on yield variation among yield components, is closely-connected to habitat environment, especially the natural soil fertility parameters (Ehara et al. 2000, 2005; Yamamoto et al. 2005c). Pith starch content shows a positive correlation with stomatal density on the abaxial leaflet surface within areas of relative geographical proximity, and variation in leaflet morphology reflects the environmental conditions of the habitat (Ehara et al. 2005). This suggests that differences in the growing environment exert a significant influence over starch production of the true sago palm through factors such as trunk size and leaflet morphogenesis. However, in order to identify determining factors of starch production, it is necessary to examine the effect of genetic background on growth and yield determination after ascertaining the genetic diversity of the true sago palm and the genetic correspondence between folk varieties growing in different localities. Securing stable production is also important for the development and utilization of sago (palm starch) resources. To this end it is important to increase our knowledge about speciation of *Metroxylon* species so that high-performance lines can be selected and developed. Against this background, the relationship between geographical distribution of true sago palms and their genetic distances have been studied (Ehara et al. 2002, 2003a, 2005). Below is the result of a random amplified polymorphic DNA (RAPD) analysis on a total of 38 population samples collected from 22 sites in the Malay Archipelago and 1 site in Papua New Guinea (PNG) (see Figure 1-7 and

Table 1-7 for collection sites, place names and morphological characteristics). Of the 38 samples, 16 are spineless and 22 are spiny. They presented diverse morphological characteristics, including black or brown banding on the abaxial side of petiole/rachis, no banding, white pith, reddish pith and so on.[2] The RAPD-PCR analysis produced 77 amplification products of which 5 products are found in all the populations while 72 products were polymorphic in all 38 populations. Figure 1-8 is the resultant UPGMA dendrogram, which divides the populations into two main groups. Group A includes subgroup A1 consisting of 9 populations from western Malaysia, 8 from Sumatra and the surrounds, 1 from western Java and 2 from Southeast Sulawesi and subgroup A2 consisting of 3 populations from Southeast Sulawesi and 2 from Mindanao. The populations belonging to subgroup A1 are mainly distributed over the western part of the Malay Archipelago. Group B includes 12 populations collected in the eastern part of the Malay Archipelago. It is divided into subgroup B1 consisting of 6 populations from Seram and subgroup B2 consisting of 2 from Seram and 4 from Ambon. The analysis has found a relationship between the genetic distance and geographical distribution of sago palms and a smaller genetic variation in the western part than in the eastern part of the Malay Archipelago and indicated that the more genetically varied populations distributed in the eastern area are possibly divided into 4 broad groups.

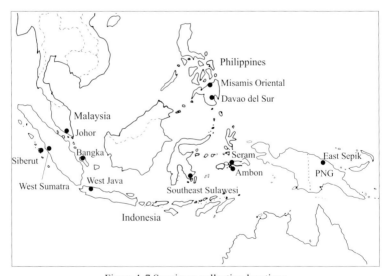

Figure 1-7 Specimen collection locations

Table 1-7 Vernacular names and morphological characteristics of specimens

| No. | Vernacular name | Collection site* | Morphological characteristic | | |
|---|---|---|---|---|---|
| | | | Spine | Banding† | Pith‡ |
| 1 | Ambtrung 1 | Batu Pahat, Johor (Jh1) | -- | DG | W |
| 2 | Ambtrung 2 | Batu Pahat, Johor (Jh1) | -- | DG | W |
| 3 | Ambtrung 3 | Batu Pahat, Johor (Jh1) | -- | DG | W |
| 4 | Ambtrung 4 | Batu Pahat, Johor (Jh1) | + | - | W |
| 5 | Ambtrung 5 | Batu Pahat, Johor (Jh1) | + | - | W |
| 6 | Ambtrung 6 | Batu Pahat, Johor (Jh1) | + | - | W |
| 7 | Ambtrung 7 | Batu Pahat, Johor (Jh1) | + | - | W |
| 8 | Ambtrung 8 | Batu Pahat, Johor (Jh2) | -- | DG | W |
| 9 | Ambtrung 9 | Batu Pahat, Johor (Jh2) | + | - | W |
| 10 | Rumbio 1 | Padang, W. Sumatra (WS1) | -- | DG | W |
| 11 | Rumbio 2 | Padang, W. Sumatra (WS2) | -- | DG | W |
| 12 | Rumbio 3 | Padang, W. Sumatra (WS2) | -- | DG | W |
| 13 | Sagu 1 | Siberut, W. Sumatra (Sb1) | -- | Br | W |
| 14 | Sagu 2 | Siberut, W. Sumatra (Sb3) | -- | Br | W |
| 15 | Gobia | Siberut, W. Sumatra (Sb2) | + | - | W |
| 16 | Marui | Siberut, W. Sumatra (Sb4) | + | - | W |
| 17 | Sagu 3 | Bangka, S. Sumatra (Bn) | -- | DG | W |
| 18 | Kiray | Bogor, W. Java (WJ) | -- | DG | W |
| 19 | Roe 1 | Konda, S. E. Sulawesi (SeS1) | -- | DG | W |
| 20 | Roe 2 | Totombe, S. E. Sulawesi (SeS2) | -- | DG | W |
| 21 | Runggumanu 1 | Totombe, S. E. Sulawesi (SeS3) | + | - | W |
| 22 | Runggumanu 2 | Lakomea, S. E. Sulawesi (SeS2) | + | - | W |
| 23 | Rui | Lakomea, S. E. Sulawesi (SeS3) | + | - | R |
| 24 | Molat 1 | Seram, Maluku (Sr1) | -- | DG | W |
| 25 | Tuni 1 | Seram, Maluku (Sr2) | + | - | W |
| 26 | Ihur | Seram, Maluku (Sr3) | + | - | R[1] |
| 27 | Tuni 2 | Seram, Maluku (Sr3) | + | - | W |
| 28 | Tuni 3 | Seram, Maluku (Sr3) | + | - | W |
| 29 | Molat 2 | Seram, Maluku (Sr3) | -- | DG | W |
| 30 | Makanaru 1 | Seram, Maluku (Sr4) | + | - | W |
| 31 | Makanaru 2 | Seram, Maluku (Sr4) | + | - | W |
| 32 | Tuni 4 | Ambon, Maluku (Am1) | + | - | W |
| 33 | Tuni 5 | Ambon, Maluku (Am2) | + | - | W |
| 34 | Makanaru 3 | Ambon, Maluku (Am3) | + | - | W |
| 35 | Makanaru 4 | Ambon, Maluku (Am3) | + | - | W |
| 36 | Saksak | Misamis Oriental, Mindanao (MO) | -- | DG | W |
| 37 | Lumbio | Davao del Sur, Mindanao (DdS) | + | - | W |
| 38 | Wakar | East Sepik (ESp) | + | - | R[2] |

* Jh: Johor, WS: West Sumatra, Sb: Siberut, Bn: Bangka, WJ: West Java, SeS: Southeast Sulawesi, Sr: Seram, Am: Ambon, MO: Misamis Oriental, DdS: Davao del Sur, ESp: East Sepik.

† Banding pattern on petiole and rachis (DG: dark gray, Br: brown, -: no banding).

‡ Pith color (W: whitish light brown, R: reddish brown).

1) Soerjono (1980). 2) Flach (1997).

Vavilov's concept of centers of origin of cultivated plants states that the origin of a plant taxon is the place where the highest diversity of that taxon is found. The origin of a plant species, accordingly, can be considered to be the place where the greatest number of varieties and other variants are found within the taxon. On this basis, the great genetic variation found in the eastern Malay Archipelago, including the Maluku Islands, supports the traditional hypothesis that the area from the Maluku Islands to New Guinea Island is the center of origin of the true sago palm.

Assuming that the origin lies somewhere in the area from the Maluku Islands to New Guinea Island, it is conceivable based on the genetic distance data of the populations found in this area that the species spread to the present-day Malaysia, Sumatra and the Philippines via the routes shown in Figure 1-9. The above analysis indicates that two of the populations belonging to subgroup A1 are from Southeast Sulawesi. The reason for this is unclear but there are cases of human-caused migration in Southeast Sulawesi when sago palm suckers were used as gifts for weddings and childbirths. Hence it is necessary to take into account large human influences on the distribution of sago palm trees.

Among the aforementioned clusters, A1, A2 and B1 include both spineless and spiny populations. This means that the genetic distance between the spineless and spiny populations is not necessarily farther than that between spineless populations or spiny populations. Consequently, the presence of spines on the petiole/rachis appears to be unrelated to genetic distance. This finding supports Rauwerdink's proposition (1986) that the spineless sago palms and the spiny sago palms which historically have been considered heterospecific should be lumped together into the same taxon as *M. sagu*. Also, in view of the cases in which seedlings germinated from spineless sago palm seeds sometimes form spines (Ehara et al. 1998) and vice versa (Jong 1995a), it is more appropriate to treat all sago palms as belonging to one species. Subgroup A1 includes 2 populations with brown banding on the back of petiole/rachis. Three populations with reddish pith were found in subgroups A2 and B1 and outside of the two main groups. However, no definite relationships were found between these traits, i.e., the banding pattern on petiole/rachis and the color of pith, and the genetic distances of populations. These findings indicate that there is no correspondence between morphological characteristics and genetic distance. Kjær et al. (2004) has examined the relationships between various characteristics representing morphological features and genetic distances using AFLP analysis of sago palm populations growing in Papua New Guinea and reported that no correspondence relationships were found between different morphological characteristics and genetic distances. As mentioned in the previous section, local communities that are heavily dependent on sago palm have what is called folk classification based on morphological features and yield properties under which many local types are identified. In field research,

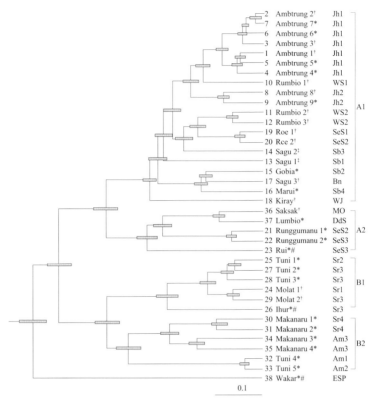

Figure 1-8 UPGMA dendrogram based on RAPD data
\*: spiny population, †: dark gray banding population, ‡: brown banding population,
#: reddish pith population, ▭: standard error. Ehara et al. (2003a)

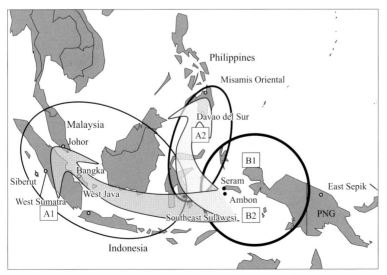

Figure 1-9 Estimated dispersal routes of sago palms

various types of sago palms are categorized by their vernacular names and each locality's folk classification criteria and it has been agreed that these types are treated as folk varieties for the time being. The past genetic diversity analyses conducted by Ehara et al. (2003a), Kjær et al. (2004) and Abbas et al. (2006, 2008) suggest that genetic diversity of the true sago palm is not very high. The true sago palm as a resource plant category is an underdeveloped economic plant (Ehara 2006b) and there is no cultivar as yet. No conclusion has been reached so far as to whether intraspecific variation can be treated as varieties in the taxonomic hierarchy or they must be regarded as formae. According to an analysis of the diversity of sago palms collected from all parts of Indonesia using the *Wx* gene marker designed from sequences involved in encoding starch synthesis genes, a total of 100 individual plants collected from 9 locations, including Sumatra, Java, Kalimantan, Sulawesi, Ambon and 4 locations in Papua (former Irian Jaya: western half of New Guinea Island), are divided into 4 groups (Abbas et al. 2008). Moreover, approximately 90% of the individual plants fall into two of the groups. This means that there may be local variation in the diversity of folk varieties but it may be logical to think that a majority of sago palms growing in various locations are genetically quite close on the population level and only a small number of types are specifically different. In any case, based on the genetic background, the true sago palms can be divided into 4 broad types.

So far, a certain relationship has been observed between leaflet morphology and growing environment in limited areas, including Sulawesi, northern Maluku, and the areas surrounding Sumatra (Ehara et al. 2005). Combined with the RAPD analysis finding of the relatively close genetic distances between populations within areas of geographic proximity, it is easy to understand the existence of relationships between the growing environment and the morphological characteristics involved in yield determination within limited areas. In order to ascertain the effect of genetic background on productivity, it will be necessary to conduct further analyses such as a comparative analysis of the growth and yield characteristics of genetically distant folk varieties under constant environmental conditions.

## 1.3 Sago palm stands in major sago-producing countries

### *1.3.1 Indonesia and Malaysia*
In Indonesia and Malaysia, it is difficult in many cases to determine whether a particular sago stand is a native, wild stand or a secondary forest grown from suckers planted by the locals and propagated under semi-cultivation management. As pointed out by JICA (1981a), when sago palm suckers are transplanted and grown under extensive crop management, they form a sago

palm stand that is no different from a native (wild) stand after a few decades because of their ecological characteristics. While Richards (1952), author of *The Tropical Rain Forest*, suggested that marshland vegetation, including sago palm stands, is found in New Guinea but very little data are available. Since then, only some fragmentary and small-area materials about sago palm vegetation distribution have become available.

The sago palm is said to be one of the oldest food plants used by the people of the Southwest Pacific islands (Bellwood 1985; Nakao 1983) but where is the location of its first use? The prevailing view originally was that the sago palm's center of origin would have been on New Guinea Island. Recent studies by genetic analysis have supported that (see 1.2). It is likely that the distribution mode of wild sago palm stands reflects the past human-caused selection and migration as well as the natural dispersion of the above variants.

This section shall give an overview of the distribution of various sago palm stands in Indonesia and Malaysia, including wild stands past and present, referring to recent surveys and findings on them with the aid of older materials where possible.

Looking back in history, according to *Zhufan zhi*, a late-thirteenth-century geography book, and *Dao yi zhi lue*, a mid-fourteenth-century history book from China, sago palms were a predominant food source in the region from the southern part of Mindanao Island, the Philippines, the northern part of Borneo down to northern Sulawesi and the Maluku Islands, while rice was dominant in Indochina, Malay Peninsula and eastern Java (Takaya 1983). It is difficult to know whether all sago palms were of *Metroxylon sagu* species but it is highly likely that it was the main species used for starch collection. Although these documents refer to the Malay Peninsula as a rice-producing area, sago palms were widely planted and utilized on the southwest coast during the seventeenth century as reported below.

Next is an overview of sago palm stands in Indonesia and Malaysia.

## 1.3.1.a Sago palm stands in Indonesia

### Northeast coast of Sumatra Island

On and around the Island of Sumatra, sago palm stands are found along the northeastern coast and on the Mentawai Islands off the western coast. The ecology of sago stands and their changes are described in this section based on studies of sago palm use, fishery and agriculture in swamps along the northern shore conducted in the 1980s.

In Mandah, Indragiri Hilir Regency, Riau Province, eastern Sumatra, sago collection and fishery were the main means of livelihood as agriculture was difficult due to saline water intrusion into coastal wetland forests. According to a mid-1980s survey, the total sago palm growing area was 2,862 ha in 9 desas (villages) belonging to Mandah District with a total population of just below 30,000. The port of Khairiah Mandah about 15 km upstream from the mouth of the Mandah River is the seat of the district office. Interviews with local elders revealed that in the 1910s this area was full of sago palm plantations; there were settlements of stilted houses along the river banks at intervals of 30 to 50 m with each house having a sago palm plantation to a depth of 150 to 200 m in the backyard (Takaya and Poniman 1986).

A very similar situation was found in the basin of the Batang Hari River in Jambi Province, about 200 km southeast of Mandah along the coast. The population of Tanjung Jabung on the coast was 320,000 in 1982 prior to the survey but it was only 33,000 in 1930. Slash-and-burn marsh cultivation of rice and sago palm cultivation would have been the primary means of livelihood in those days (Furukawa 1986). According to those who were born at the beginning of the twentieth century, their ancestors were active in this area and marshy forests in the tidal river basin were bordered by sago palm stands. The parit sago, or small canals for transporting sago, were dug in some places for the harvesting and transportation of sago palms which grew densely to 300 meters or so from the river banks. Both areas saw the construction of sago processing factories along the canals by Chinese merchants in the mid-1930s and many of the local villagers harvested mature sago palm trees from their own plantations, transported and sold them to these factories. However, they soon ran out of mature sago palms, the factories were closed, the residents moved out and the sago-growing settlements gradually disappeared, with some people dispersing across marshy forests to set up coconut palm plantations.

The two survey locations were approximately 200 km apart by linear distance but the people utilized sago palm plantations in a very similar manner. However, according to Batang Hari River residents, sago palms were more abundant in Riau, from where dry sago was exported to Singapore before WWII. These accounts confirm the existence of vast stands of sago palms in this region in the early twentieth century but their types are not clear.

Today, almost a century after sago harvesting had virtually ceased in these areas, it is delightful to be able to report that on the island of Tebing tinggi in Riau Province, Indonesia's National Timber and Forest Product Co. LTD (NTFP) is currently developing a 20,000 ha sago palm plantation (Jong 2002a).

**Sago palm distribution in Kalimantan**
According to a report from an investigation on the distribution of sago palm stands in South Kalimantan Province, small dense clusters of several sago palms each were observed among the miscellaneous tree forests in an area of about 150 $km^2$ between Banjarmasin and Amuntai in the north. However, it appears that these palms are grown for harvesting "atap" or roof thatch and

very few are used for starch collection. Furthermore, while planted stands of sago palm are dotted all over the area, no primary forests are found (JICA 1981a). Based on the size of the wild sago palm growing area (about 73,000 ha), the arable area for planting sago palm was estimated by the Department of Agriculture of South Kalimantan Province, to be about 13 times larger than the present growing area. Similarly, the distribution range of sago palms in West Kalimantan is also described (Rasyad and Wasito 1986) without detailed information.

On the higher slopes of the hilly areas in the upper river basin of the Kayan River (which originates near the Sarawak border in northeastern Borneo) young shoots and fruit endosperms of Nanga (*Eugeissona utilis*) are eaten. Sometimes the pith of the mature stem is crushed and washed for wet sago collection. Jaka (*Arenga undulatifolia*) are found to be growing on the levees of the Sega River a short distance south of this area where young shoots are eaten as a vegetable. Nanga also grow in this same habitat, but there has been no report of sago palm growth at all in this area (Yamada and Akamine 2002).

**Sulawesi**

According to a report by Rasyad and Wasito (1986), sago palm stands in Sulawesi are distributed across various locations from northeastern Manado on the Minahasa Peninsula to Palopo, on the eastern shore of the Gulf of Boni.

In Manado, northern Sulawesi, many small stands of spineless sago palms were observed from the hilly area to the shores of Tondano Lake, 650 meters above sea level, away from the flat land paddy field zone. Short sago palms were planted around fishponds by the lake (Figure 1-10). According to residents, erosion control was the primary purpose for planting sago palms here. This is also the case in river levees, where sago palms are principally an emergency crop. There was also a densely planted forest near the lake for the production of atap (roof thatch). These sago palms grow to a height of 8 to 10 meters on flat land but only to around 5 meters as the altitude increases (Figure 1-10). They were all a spineless type and tall isolated trees were also found in some locations (Takamura and Yuda 1985).

In an earlier study, Yoshida (1977) pointed out that there were marked differences in the relative dominance of spiny and spineless sago palms between the Maluku Islands and Sulawesi. This tendency was also confirmed in the southern part of Sulawesi as discussed below.

The Luwu region, adjacent to the deep part of the Gulf of Boni in southern Sulawesi Province, is the leading sago production region in the province. The area between Masamba, 100 km north of Palopo, and Watampone, more than 150 km south of Palopo, and the Malangke area to the east are dotted with large colonies or plantations of sago palms (Yuda et al. 1985). According to the form

Figure 1-10 Sago palm vegetation in North Sulawesi
Top: Sago palms for atap near LakeTondano
Center: Fishponds and sago palms on the shore of LakeTondano
Bottom: Large sago palms near Manado

Figure 1-11 Sago palm stand near Kendari, Southeast Sulawesi (in a backswamp behind a fishpond) (Photo: Hiroshi Ehara)

classification approach that was prominent at that time, spineless types of sago palm were dominant in this region. Semi-wild stands of sago palms were found around Kendari on the opposite side of the gulf in Southeast Sulawesi (Figure 1-11). They were not actively managed, however, except for some weeding and sucker thinning that took place at the time of leaf collection for atap. In some areas, the Bugis people (who migrated into the region around the mid-twentieth century) and the Javanese immigrants (who came after them) turned many sago palm stands into rice paddies. Hence all that remain today are the sago palms lining paddy water channels (Yamamoto 1999).

Osozawa (1988) conducted a survey in Pengkajoang Village, a traditional sago producing settlement in Malangke District, southern Sulawesi Province, and found that the local Bugis society has an established form of ownership of utilizable sago palms with substantial areas of land around these trees. Their practice of husbandry and post-harvest management is notably extensive. If they can earn a reasonable level of income by occasional logging and natural regeneration, then they have no motives to practice intensive farming by investing extra labor or material. Consequently, these sago palm stands appear to be wild growths although their constituent trees are in fact treated as property owned by residents. Osozawa confirmed that while sago palm stands in Palopo settlements amounted to a total of 150 ha or so, their stand composition was not homogeneous and therefore the precise area of sago palm plantations was almost impossible to determine.

**The Maluku Islands**
Information on the sago palm growing area for each district of the Maluku Islands was reported at the third International Sago Palm Symposium along with a distribution map (Rasyad and Wasito 1986). The data were largely identical to those presented at the FAO/BPPT Consultation on the Development of the Sago Palm and Its Products convened in Jakarta in 1984, based on a study carried out by BPPT (Indonesia's Agency for the Assessment and Application of Technology) but it was noted that the potential productivity of sago palm stands would vary depending on factors such as planting density, the properties of constituent trees, and distribution. Since there are recognized problems with the basis of calculation for the sago palm stand areas, it would be wise to use this information merely to surmise the state of wild and planted forests. For reference, the total sago palm stand area was reported to be 30,000 ha in the Maluku region and 4,180,000 ha in West Irian. For the Maluku Islands, the areas were approximately 18,000 ha on Halmahera and Bacan Islands, 11,000 ha on Seram Island and less than 1,000 ha on Buru Island. On Ambon Island, 4 to 5 types of sago palms, distinguished on the basis of spininess, spine forms and starch traits, were reported together with their distribution. Wild-like stands are distributed along the coast of Seram Island but sago palms are now planted in villages situated at around 750 m above sea level as part of farming trials which include sweet potato, taro and other tubers (Sasaoka 2007).

**Irian Jaya (Papua Province and West Papua Province)**
Irian Jaya is near the supposed place of origin of sago palms. Vast wild sago palm forests spread from the coastal to inland areas, including Merauke, Agats, Inanwatan and Bintuni. In many of the sago palm growing areas, sago starch is still an important staple food which maintains an important position in the local diet. Sago palm stands range from pure forests to mixed forests containing other trees and herbaceous plants to varying degrees. The estimated sago palm growing area in Irian Jaya is somewhere between 800,000 to 4,180,000 ha according to past reports but no scientific study has been undertaken for the province as a whole. Flach (1997) estimates that Irian Jaya has 1,200,000 ha of wild sago palm stands and 14,000 ha of semi-cultivated stands. A recent study by the University of Papua, however, using LANDSAT and surface surveys, estimates the sago palm growing areas in the Regencies of Waropen, South Sorong and Jayapura to be 255,000 ha, 150,000 ha and 682,000 ha respectively. Hence, the total area for these 3 regencies alone exceeds 1 million ha. Thus, further studies are needed to arrive at an accurate assessment of the extent of sago palm growing areas in Irian Jaya.

It has been reported that the numbers of folk varieties (or local varieties) in Irian Jaya range from 4 to 5 in Waropen, 10 in Salawati, 14 in Wasior, 9 in Inanwatan, 3 in Ongari, 35 in Sentani, 16 in Kaure, 5 in Wendesi, 1 in Timika, 2 in Agats, 6 in Sarmi, and 17 in Biak and Supiori Island. As we can see, the largest number is 35, found on the shores of Sentani Lake near Jayapura and the smallest number of varieties is found in Timika and Agats in the southeastern region, having only 1 to 2 varieties each. This broad range in the number of folk varieties is very interesting in relation to the center of origin of the sago palm.

While very little has been studied about genetic variation of these folk varieties, the following are some examples of their classification criteria.

1. Central Sentani District (Matanubun 2004)
   Spininess, spine growing mode and shape, pith color, starch color, papeda color, taste and viscosity, above-ground plant and young leaf color, leaf sheath color, leaf shape, leaflet width, tree trunk shape, starch yield.
2. Inanwatan (Renwarin et al.1998)
   Spininess, leaf sheath color and coloration level, ensiform and young leaf color, leaf length/width ratio, trunk diameter/length ratio, years to harvest, starch yield, starch physical properties and color.
3. Entire Irian Jaya (Widjono et al. 2000)
   Spininess, above-ground plant (young leaf) color, leaf sheath color, trunk size, starch color, tree crown shape, starch yield.

Widjono et al. (2000) collected 61 accessions of sago palms across Irian Jaya of which some are likely to be synonyms of the same type named by different ethnic groups. These require further investigation.

Shimoda and Power (1992a) discovered some wild types of sago palms in Papua New Guinea and reported that their starch yield was much lower compared with cultivated types. On the shores of Lake Sentani, Papua Province, Irian Jaya on the western part of the Island of New Guinea, Yanagidate et al. (2007) found a variety named Manno (a folk variety not actively utilized) that is said to be wild (Sagu hutan) and reported that starch yield was lower than cultivated types. Many of the palms appear to reach the flowering and fruiting stage without being harvested. Manno has a very high total sugar content in the pith even nearing the flowering stage, which is considered to be the right time for harvesting. It might be that a lower rate of starch synthesis from sugar compared with cultivated types (actively utilized folk varieties) is the reason for low starch productivity (Yanagidate et al. 2007). Manno palms are felled for sago grub collection.

In comparison, a survey of local sago palm growers on the subject of major cultivated types on the Lake Sentani shores revealed high yielding folk varieties such as Para, Yepha, Folo and Osukulu which produce 750 to 1,000 kg of 'wet sago' per palm. Suckers from these high yielding types are collected and transplanted near the local people's homes. Rondo is a unique cultivated type; a fast maturing variety with 10 to 12 years from sucker transplantation or emergence to flowering. Although starch productivity is low at 150 to 200 kg (dry starch), the pith can be extracted, boiled and eaten as it is. When the boiled pith is cut into cubes and smoked, it can keep up to 5 months. Variation in ecological adaptation of these varieties is also recognized. Rondo is planted on property boundaries as it is known to be highly fire resistant. Mongging, Hobholo, Osukulu and Para are well adapted to lowland swamps near the lake while Yepha is not. The leaves of Yepha and Para make excellent atap for roof thatching which lasts longer.

Matanubun and Maturbongs (2006) reported that the number of harvestable palms per hectare of sago palm stands in Irian Jaya ranges from 15 to 68 (42.2 palms on average). The average starch yield (dry starch) per palm ranges from 76 to 401 kg. Based on these numbers, the annual starch production per hectare is estimated to range from 1.1 to 27.3 tons. In contrast, Westphal and Jansen (1989) estimate the starch productivity of wild sago stands in Irian Jaya as 2.5 tons per ha/yr.

### 1.3.1.b Sago palm stands in Malaysia

**Peninsular Malaysia**
In his *Rice in Malaysia*, Hill (1977) reports that sago palms were used in some areas prior to the spread of rice farming across the Malay Peninsula and cites a historical record showing that the majority of cargos on ships leaving Johor during the 1640s consisted of sago starch, dried-salted fish and locally produced nuts. According to Hill, paddy rice growing and other farming activities were possible on the inland side of the coastal mangrove zone of northern provinces such as Perak and Kedah but most villages in Selangor in the south were dependent on the production of sago palms, betel nut, coconuts and sugarcanes even at the end of the nineteenth century as rice growing was not profitable enough. In the twentieth century, increases were reported in the sago palm plantation areas from 1966 to 1984 in the Provinces of Johor, Malacca and Kelantan. The frequent occurrence of semi-wild sago palm stands, both spiny and spineless, in swamps along river levees and paddy waterways was also noted (Othman 1991). Regarding vegetation prior to the development of a drainage system in the river basin of Batu Pahat, which was a major sago palm growing district until recently, coconut palms, rubber trees and fruit trees were grown in areas not affected by rising water while sago palms were grown in the areas that were flooded by rising water (Tan 1986). These reports suggest that there were indigenous sago palm stands as well as sago palm plantations in the Peninsula.

Based on the fact that the southwestern part of the Malay Peninsula and the northern coast of Sumatra Island are separated only by a narrow strip of sea water and have a long history of exchange, it seems natural to assume that the varieties of planted sago palm and the systems of managing them would have resembled each other in these two regions.

**Sarawak**
According to a study of the sago palm stands of Sarawak's Oya Dalat and Mukah River, conducted from 1975 to 1978 by collating aerial survey photographs with topological maps and the data of soil surveys (Tie et al. 1991), the total sago palm plantation in this area was 19,720 ha, including 6,400 ha in Oya Dalat 5,520 ha in Mukah, Sibu District, 3,240 ha in Pusa Saratok, Sri Aman District, and others. As organic soils (both deep and shallow) and clay soils accounted for 60% and 33%, respectively, and there was very little acid sulfate soils or podsols, it was confirmed that sago palms were once grown on the favorable soil in this area.

According to Osozawa (1982), who surveyed the downstream basin of the Mukah River, sago palms had been grown as a commercial crop before the early nineteenth century and the river-dwelling Melanau people cultivated them and exported sago starch. However, utilization was limited to areas within 1.5 km or so of the river and the discontinuous sago palm stands with varying stand densities were already not pure forests. While sago palm stands have existed in these areas in a semi-cultivated state for a long time, in recent years the areas of sago palm plantation were increased in response to the strong need for generating income in areas where soil properties are favorable for growing sago palm.

## *1.3.2 Papua New Guinea*

### 1.3.2.a Soils and vegetation of sago palm distribution areas
Sago palm stands in Papua New Guinea (PNG) are mostly distributed across swamps in the downstream basins of the Fly River and the Sepik River (Figure 1-12). Some are also found in areas with good soils and water throughout the year in the lower basin of the Kikori River, within inland forests below an altitude of 700 m, and along creeks and narrow ravines in topographically complex locations. The total sago palm growing area is estimated to be between 1 and 1.2 million ha (Flach 1983; Power 2002). Wild stands account for a large part of it.

This section will describe the state of wild sago palm stands in PNG based on studies conducted in and around Imbuando Village in the downstream basin of the Sepik River (Shimoda and Power 1990, 1992a, b). The first study looked at their conditions over three years from 1982 to 1985 and the follow-up studies were undertaken in 1992 and 2007.

According to Bleeker (1983), soils in sago palm stands in the lower Sepik River Basin include Fluvaquents, immature soils made of hydromorphic sediments, and Alluvials of poor drainage such as Hydraqents, peat and Histosols. A majority of them are located on natural river levees, lake or lagoon shores, and estuarine swamps (Figure 1-13).

Pure mangrove forests have developed in the salt water zone up to 15 km or so from the mouth of the Sepik River. From the transitional brackish water zone 2 to 3 km upstream, nipa palms begin to make their way into the mangrove forests first, followed by occasional sago palms. On both banks of the fresh water zone of the river up to 100 km from that point, almost pure forests of wild sago palms have developed deep into the hinterland.

In the areas where seawater used to be intrusive in the Purari Delta, some sago palms grow among the dominant species of nipa palms and mangroves. The sago palm has a high salinity tolerance; 7.5% is reportedly the upper limit of soil pore water salt content for its new planting sites (Ulijaszek 1991).

The sago palm habitat in the Sepik River Basin can be grouped into the following three types based on vegetation as illustrated in Figure 1-14 (van Kraalinngen 1983; Shimoda and Power 1990).

Figure 1-12 Sago palm stand near Angoram, East Sepik Province, Papua New Guinea (the sago palm vegetation extends over several kilometers across swamps along the Sepik River) (Photo: Hiroshi Ehara)

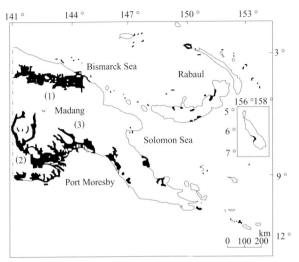

Figure 1-13 Distribution of poor drainage, swamp and alluvial soil areas in Papua New Guinea
Major rivers: (1) Sepik River (2) Fly River (3) Kikori River Center
Source: Bleeker (1983)

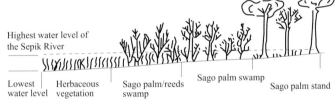

Figure 1-14 Sago palm vegetation zones in the floodplains of the Sepik River Basin

1. Behind the vegetation of sedges and reeds by the riverside or lakeside, sago palms grow among herbaceous plants such as wild sugarcane (*Sacchurum robstum*) and a type of reed (*Phragmites karka*). This is a permanent freshwater zone. The sago palms here are stunted, rarely forming trunks because they grow too thickly.
2. In areas where the ground water level is below the soil surface during short periods of low precipitation, sago palms form pure forests and their suckers grow so thickly that people cannot easily enter the forest. Some individual plants may form trunks and flower but such trunks yield little starch.
3. At the highest zone in the flood plain, the ground water level fluctuates throughout the year from about 70 to 80 cm above the ground (the rainy season around January and February) to about 80 cm below the ground

(the dry season around July and August). Sago palms proliferate in the understory of open forests of occasional dicotyledonous trees, including Pandanus, *Campnosperma spp.*, and Terminalia (Flach 1981: 250–700 plants/ha; Jong 2001: 599 plants/ha). Sago palms in this zone form tall and thick trunks which store large quantities of starch.

### 1.3.2.b Wild folk varieties (wild stands) and useful folk varieties (semi-cultivated stands) within natural forests

The local residents of the lower Sepik River Basin call the group of sago palms (folk variety group) with short and thin trunks that yield little starch '*wel saksak*' (wild sago) and the group of high-yielding folk varieties that are said to have been selected and propagated through sucker transplantation by their ancestors '*saksak tru*' (planted sago). Based on the degree of utilization and management, the former can be called 'wild folk varieties' (collectively 'wild stands') and the latter 'high-yielding planted folk varieties' ('semi-cultivated stands').

Ninety-eight percent of PNG's sago palm stands are reportedly natural (wild) forests (Jong 2002a). Except when leaves are harvested as roofing material or in extremely unusual cases of starch extraction by humans, natural forests maintain their vegetative balance under natural conditions without human intervention (Sim and Ahmed 1990). In sago palm stands on the flood plain, large amounts of organic matters flow in during seasonal river flooding which form organic peat soil deposits under prolonged submerged conditions. Figure 1-15 shows sago palm stands distributed along the river.

The sago palm stands that are called 'planted sago' (semi-cultivated stands) in Imbuando Village can be grouped into two types based on their location and utilization. One type consists of highly utilized sago palm stands of high-yielding folk varieties, which are said to have been collected and planted on natural levees and near the village by the ancestors of the villagers for utilization in daily life (Figure 1-16). They are small stands of sago palms ranging from 10 to 30 ares down to homestead woodlands of several palms. Very little management is carried out except the removal of dead leaves to clear work sites and the felling of wild folk varieties or other trees at the time of harvesting. A little more hands-on management activity may involve the removal of young suckers and dead leaves around trunked sago palms to create more space to allow more sunlight. Starch extraction is carried out throughout the year from such sago palm stands.

The other type of semi-cultivated stands are clusters of high-yielding folk varieties planted here and there inside vast unmanaged natural forests (Figure 1-17). On these spots, the underground water level drops up to 80 cm below ground for a considerable period during the dry season and people are able to enter. Sago palms grow in the understory below taller dicotyledonous trees that provide partial shade. These stands are found in locations where small

Figure 1-15 Dense natural sago palm stands with no trunked palms distributed along a river (Photo: Hiroshi Ehara)

Figure 1-16 Sago palm stand in East Sepik Province, Papua New Guinea (suckers of various ages are growing around the mother palm) (Photo: Hiroshi Ehara)

Figure 1-17 Semi-cultivated sago palm stand with many trunked palms

waterways or passages exist for transportation of harvested logs to the village or water is available for on- or near-site starch extraction from the harvested logs.

The latter type of semi-cultivated stands are considered to be outer areas for subsistence farming in contrast to small areas of slash-and-burn fields nearby. The sago palms for starch extraction in these stands probably supplement the more highly cultivated sago palm stands. There are no clear boundaries between the natural forest and the high-yielding folk variety clusters within it. A mixture of wild folk varieties and high-yielding folk varieties is found in the transition zone. These stands are largely unmanaged except for the removal of wild folk variety seedlings from the forest floor, the pruning of dead branches and leaves and the felling of other trees to clear the harvesting site.

The owners visit these sago palm clusters occasionally but they have no management program other than harvesting. This is no different from gathering wild food. Once suckers are planted, they do not require any special attention until harvesting. However, men sometimes inspect their sago palms when they come by for hunting or working in the field or while women from their own group are extracting sago starch. These occasional visits to sago palm clusters are useful in determining the proper time for harvesting (Ohtsuka 1983).

The thinning of excessive suckers within a clump or tree thinning to optimize the number of palms per area may be carried out for the purpose of facilitating harvesting work but rarely as a means to increase starch productivity per unit area.

## 1.3.2.c Semi-cultivated stands of diverse folk varieties

The charactaristics of wild stands and semi-cultivated stands are:
1. In wild stands sago palms grow densely and there are few trunked palms.
2. Wild folk varieties have thin and short trunks.
3. Wild folk varieties accumulate very little starch in the trunk (Figure 1-18).

Past sago palm studies in various parts of PNG have found 10 to 20 folk varieties that are known by different names in different localities (Shimoda and Power 1992a; Toyohara et al. 1994; Ulijaszek 1991). These folk varieties are distinguished on the basis of: (1) spininess, spine length; (2) color, pattern and size of leaf sheath, rachis, leaflet, trunk etc.; (3) angle of rachis and leaflet (drooping or erect); (4) starch yield; (5) starch color, taste, ease of extraction, etc.

The most remarkable characteristics among them are spininess and spine length. Four wild folk varieties are recognized but they are called by the same name (Wakar No. 1, 2, 3 and 4) and grow long spines. Only one of them (Wakar No. 1) is said to be used for starch extraction, and only on rare occasions.

High-yielding folk varieties present a continuous spectrum of spine length, ranging from 30 cm (vernacular names: Anun, Ketro, Kangrun), which is as long as some wild folk varieties, to a few millimeters (Mandam, Wayapee, Makapun), to spineless (Koma, Awir-koma, Ambtrum).

Figure 1-18 Thin trunks of wild varieties left to wither

A census of folk varieties within semi-cultivated stands in Imbuando Village has found that long-spine folk varieties accounted for 60 to 63% of the total population. The spiny folk varieties, including both long- and short-spine types, made up 76 to 86% whereas spineless folk varieties and wild folk varieties growing incidentally in semi-cultivated stands only accounted for 0 to 7% respectively (Shimoda and Power 1992a).

According to local residents, the long-spine folk varieties are dominant because of their short duration between the sucker emergence stage and the harvestable stage immediately before flowering (5 to 6 years after trunk formation), large trunk diameter and high starch yield. In fact, they have a high starch yield even when harvested early (3 to 4 years after trunk formation). For semi-cultivated stands under minimal management, spines are currently not regarded as a serious hindrance as hardly any work is carried out after sucker planting until harvesting. It is considered that spininess is a less important factor than starch yield for the residents of this village who only need to collect subsistence quantities of starch from the sago palms.

In addition, the color and taste of extracted starch vary between folk varieties and residents tend to enjoy eating different tastes of starch (Toyoda 1997). This would be one reason for the continuation of semi-cultivated stands comprised of diverse folk varieties of sago palm.

### 1.3.2.d Sago palm stands in homesteads

The most intensively managed sago palm stands in this locality are those planted in homesteads adjoining houses. One such example that the authors observed during our 2007 survey in this area is described here. When the residents migrated from another district and settled on the side of a road near Angoram, East Sepik Province, just over a decade earlier, they brought sago palm suckers with them, which they planted on their new properties. A small stream ran down the hills through the site. About 10 sago clumps were positioned randomly along this stream at intervals of 5 to 10 meters. They spoke of another cluster of a similar size in a slightly higher position up the hills.

Each clump consisted of 5 to 10 individual sago plants, including trunked individuals of various ages and suckers at the rosette stage. Dead leaves were pruned and the area around each clump was weeded (Figure 1-19). The sago palms appeared to grow quite vigorously. The residents of just over 10 years had begun harvesting 2 of these clumps two years earlier than our observations. There were 2 to 3 other clumps with trunked palms that would be harvestable within a year. Such sago palm clusters were found here and there adjoining people's homes on the natural levees of the Sepik River. This indicates that it is possible to extract stable quantities of starch from spacious sites that only require thinning of excessive suckers.

Figure 1-19 Managed homestead stand of high-yielding sago palms (left) and a trunk being felled with an axe for harvesting (right)

### 1.3.2.e Starch production from natural sago palm forests

Shimoda et al. (1994) found that sago palms in this area form trunks 4 to 6 years after suckers are planted and are harvestable 5 to 6 years after that. This means that the fastest growing plant will produce its first harvestable trunk around 10 years after planting and the slowest growing plant will take 16 years. The starch yield of the harvested trunk varies greatly between the different environmental conditions of the wild and high-yielding stands, depending on the folk variety and growing conditions. Even in high-yielding stands of this area, not to mention rarely managed wild stands, a clump does not necessarily produce harvestable trunks every 2 or 3 years. It is therefore extremely difficult to forecast the annual starch yield per area.

Earlier studies found vast differences in the dry starch yield per trunk, ranging from 0 to 41 kg for wild folk varieties and 100 to 300 kg for high-yielding folk varieties. The annual yield per hectare is estimated to be between 1 and 2 t/ha for wild stands and between 5 and 6 t/ha for high-yielding stands. It has been projected that increased husbandry for these natural sago palm forests through density optimization by thinning excessive clumps and suckers and ongoing sucker management could improve the yield to 7 to 10 t/ha/yr (Flach 1984; Shimoda and Power 1986; Uljaszek 1991; Power 2002) but there have been no reported cases in which this has been achieved.

During the author's survey of the Sepik River Basin in 2007, many of the sago palm stands appeared to be in more unfavorable conditions than previously observed, mostly because the local demand for, and the importance of, sago starch as food were decreasing due to the depopulation of villages in the area. There were strong calls among the local people for the amelioration and development of sago palm stands in relation to the expanding uses of sago starch, including ethanol production.

Authors:
1.1 and 1.2: Hiroshi Ehara
1.3.1: Tomoki Takamura, Yoshinori Yamamoto
1.3.2: Hiroyuki Shimoda

Notes

1. *Eugeissona*: The English name of *E. utilis* Becc. is a wild Borneo sago palm. Sago (starch extracted from the trunk) is a staple food of the Punan tribes. *E. insignis* Becc. is used in the same way.
   *Maurutia*: maurutia palms, moriche palms, buriti; The English names of *M. flexuosa* L. f. are miriti palm, moriche palm. In Amazon, sago is used to make bread (Ipurana).
   *Corypha*: *C. utan* Lamarck (geban), *C. umbraculifea* L. (talipoto).
   *Phoenix*: *P. paludosa* Roxb. (mangrove date palm).
   *Arenga*: *A. pinnata* (Wurmb) Merr., *A. microcarpa* Becc.
   *Caryota*: *C. urens* L. (fishtail palm, toddy palm, kittool palm, bastard sago, wine palm, solitary fishtail palm, jaggery palm).
   *Wallichia*: *W. disticha* T. Anderson.

2. Nine 10-base primers were used in the RAPD-PCR analysis (see Ehara et al. 1997, 2003a, for reaction liquid composition and amplification methods). PCR was repeated twice on all samples and reproducible amplification products were selected for analysis. Based on the presence or absence of amplification products, genetic similarity (S) was computed between all samples according to the Nei and Li (1979) model which was converted to diversity (D = - ln (S)). An UPGMA cluster analysis was undertaken and a rooted tree was constructed using the program PHYLIP (ver. 3.6: Felsentein 2001). The standard errors of branching points of the tree were calculated using a method developed by Nei et al. (1985).

# 2
# *Ecology of the Sago Palm*

This chapter presents an overview of the natural environment in sago palm (*Metroxylon sagu*) growing areas and its effect on sago palm growth. Understanding the relationship between *M. sagu*'s natural habitat and its rate of growth serves as a first step in the search for an optimum environment for sago palm production. The sago palm is primarily found growing in areas of high temperature, high humidity and strong sunlight up to an altitude of 700 m above sea level in the parts of Southeast Asia and Melanesia between the latitudes 10 degrees north and south of the equator (Flach 1977). However, sago palms are also found near latitude 12 degrees north in the Philippines (Okazaki et al. 2007) and up to 1000 m above sea level in New Guinea [see Chapter 1].

In the coastal wetland ecology, Rhizophoraceae forests (many types of *Rhizophora* trees are collectively called mangroves (Nakanishi 2005)) form on the foreshore, nipa palm (*Nypa fruticans*) stands are distributed behind them, and sago palm stands grow behind nipa palm stands on the backshore (Flach et al. 1986b; Okazaki 2006). Low land forests extend further behind the sago palm stands (Figure 2-2). Seawater contains 10.8 g/l of sodium ions and 19.39 g/l of chloride ions (Bowen 1979). Sago palms and other low land vegetation are growing in the brackish water zone of relatively low salt concentrations where seawater is frequently replaced by fresh water. The brackish water zone is comprised of soils developed in parent material deposited by streams and rivers (inorganic or mineral soils) and peat, Histosols, formed in organic parent material supplied by vigorous vegetation (organic soils). Soils consisting of river deposits contain some amount of pyrite ($FeS_2$), which is a product of bacterial reduction of seawater sulfate. Unless soils become dry and oxidized to allow the actions of sulfur-oxidizing bacteria or iron-oxidizing bacteria, even high sulfide-containing soils remain potential acid sulfate soils and do not change into actual acid sulfate soils (Kyuma 1986a). However, once sulfur- or iron-oxidizing bacteria become activated, soils were acidified rapidly and turned into actively acid sulfate soils.

In contrast, newly formed soils containing only low levels of sulfides such as pyrite are classified as part of Fluvisols or Gleysols according to the WRB or as part of Entisols or Inceptisols according to Soil Taxonomy. Histosols, which are widely distributed along the coastlines of islands to a thick layer of up to 20 m, have developed due to the small size of island streams and rivers

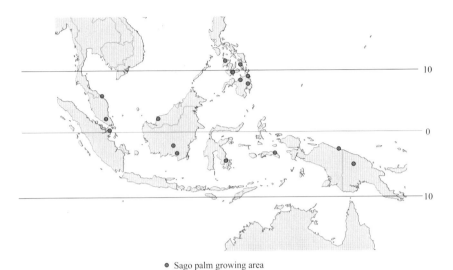

• Sago palm growing area
Figure 2-1 Sago palm distribution

Figure 2-2 Coastal low land (Sarawak, Malaysia)

that supply only small amounts of deposits. Mineral soils with a sufficient supply of nutrients are no doubt the optimal growing soils for the sago palm based on its nutrient requirements. However, the cultivation of sago palms has commenced in the coastal organic soil areas on Southeast Asian islands because the rapidly expanding global population demands the development of areas of historically low crop yields and poor accessibility. Other starch storing plants such as cassava, maize and potato cannot grow in overhydrated soils. Of course, even the sago palm appears to prefer intermittent water supply to permanent submersion, which reduces leaf emergence and delays or diminishes starch accumulation (Yamamoto 1998a).

## 2.1 Temperature and solar radiation

According to the Köppen climate classification, many of the Southeast Asian islands are located in the tropical rainforest climate (Af) or tropical monsoon climate (Am) zones (Figure 2-3). While the savanna climate (Aw) and the temperate oceanic climate (Cfb) are found in some parts of low lands and highlands respectively, most of the sago palm distribution areas up to an altitude of 700 m are associated with the tropical rainforest climate or the tropical monsoon climate. Based on long-term meteorological observations, the tropical rainforest climate is characterized by year-round high temperatures with a small annual temperature range due to high solar altitudes throughout the year. However, the closer the area is to the coast, the more oceanic the climate becomes, presenting higher temperatures throughout the day with a relatively small daily range. In general, the optimum temperature range for plants is from 10 to 35 °C approximately. The minimum temperature is considered to be more important than the mean temperature for the sago palm. The minimum temperatures above herbaceous vegetation and under tree canopies at Malaysia's Semongok Agriculture Research Centre ranged from

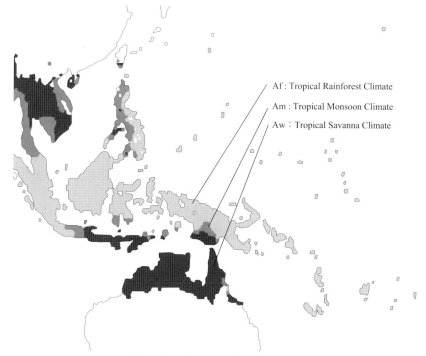

Figure 2-3 Köppen's climate classification

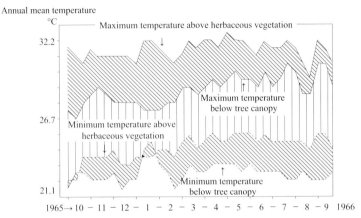

Figure 2-4 Annual mean temperature and minimum temperature (Semongok Agriculture Research Centre, Malaysia)

Source: Modified version from Andriesse (1972). Temperature scale has been changed from Fahrenheit to Celsius.

21 to 25 °C (Figure 2-4).

The habitat of the sago palm is dictated by temperature, and the sago palm is extremely sensitive to cold stress. Research on low temperature damage to tropical and subtropical plants has found that it occurs to the photosynthetic as well as non-photosynthetic organs and progresses from the initial reversible stage to irreversible injuries over time. The decreased photosynthetic function of the photosynthetic organs due to low temperature are considered to be initiated by the inhibition of electron acceptors within photosynthetic system I and the subsequent deactivation of P-700 followed by the partial degradation of large binding subunits PS1-A/B at the reaction center (Yoshida 2002). In non-photosynthetic organs, cold stress at 0 °C for 24 to 48 hours is said to cause the deactivation of tonoplast $H^+$-ATPase ahead of other biomembrane enzymes and promote simultaneous intravacuolar alkalization and cytoplasmic oxidization within a short period of time (Yoshida 1994). Cold-responsive genes (*cor*) and low-temperature-induced genes (*lti*) which are linked to cold stress have been identified in recent years (Shinozaki 1995). Like drying stress and freezing stress, cold stress is a kind of hyperosmotic stress (salt stress) for plants, which means that water is lost and concentrations of dissolved substances increase in them. The cDNA microarray analysis of *Arabidopsis* gene expression under cold stress has confirmed the expression and induction of these genes at low temperatures. There is a need for molecular biological research of cold damage to the sago palm. To date there have been no studies of cold damage to the sago palm for field surveys.

In the tropical rainforest climate, sunshine duration is not very long due to cloudiness and the shorter summer days (compared with high-latitude regions). Therefore the amount of solar radiation (the quantity of incoming solar energy per unit area on a horizontal plane near the Earth's surface measured by megajoules per square meter ($MJ/m^2$)) is around 20 $MJ/m^2$/day.

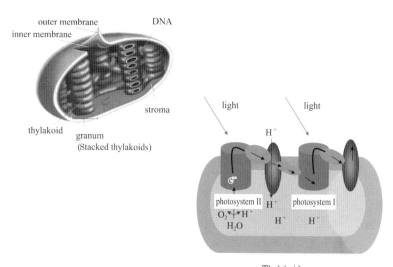

Figure 2-5 Solar radiation and chloroplast

The mean solar radiation for 1984 was 17.6 MJ/m$^2$/day in the sago palm growing area of Mukah, Sarawak, Malaysia, and 19.2 MJ/m$^2$/day in Roxas, Panay, the Philippines. Flach et al. (1986a) reported based on the rate of leaf emergence of sago palm seedlings in laboratory experiments that an air temperature of 25 °C and above, a relative humidity of 90%, and a solar radiation amount of 9 MJ/m$^2$/day provide the optimal growing conditions for the sago palm. This solar radiation amount is said to correspond to that on a cloudy day.

Radiation from the sun stimulates photosystems I and II, which consist of chlorophylls, other photosynthetic pigments and proteins inside a thylakoid (thylakoids stack on top of each other to form a granum) within a chloroplast of a leaf (Figure 2-5). This, in turn, activates chlorophyll at the reaction center to release an electron, and initiates a photochemical reaction. The sago palm increases its photosynthetic production by augmenting its leaves and increasing the number of chloroplasts per plant (Yamamoto et al. 2002a). Uchida et al. (1990) found that about 37 to 45 days after leaf development in the sago palm the maximum photosynthesis rate is 13–15 mgCO$_2$/dm$^2$/h and the light saturation point is 600–750 μmol/m$^2$/s and that the light saturation point is higher in leaves under unshaded conditions than those growing under 80% shaded conditions. In view of its maximum photosynthesis rate, light saturation point, CO$_2$ compensation point and optimum photosynthetic temperature range, the sago palm has physiological activity that is adapted to the tropics and the subtropics but is regarded as a C$_3$ plant (CO$_2$ taken into mesophyll cells reacts with a C$_5$ compound (ribulose bisphosphate) in the Calvin-Benson cycle and produces a C$_3$ compound (phosphoglycerate)) (Yamamoto 1998a).

## 2.2 Precipitation and humidity

The sago palm growing zone between latitudes 10 degrees north and south has high temperatures throughout the year. It also has high precipitation, amounting to more than 2,000 mm per year as it is subject to the effects of the intertropical convergence zone (equatorial low pressure belt) in which ascending air currents form a low pressure system (Figure 2-6). High temperatures cause high rates of evaporation, which leads to high humidity. Consequently, cumulonimbus clouds develop over the ocean, mostly during the day, often bearing a squall, which is a combination of gusty winds and torrential rain. After the squall, descending cold air and moderate winds cool the temperature.

Many typhoons are born in the middle latitude zone (Mori 2007) (Figure 2-7). Figure 2-7 shows the locations of formation of all typhoons from 1970 to 1997. A large majority of them formed in the area between latitudes 5 and 35 degrees north and longitudes 100 and 180 degrees east. None were formed between the equator and latitude 5 degrees north. Typhoon formation requires Coriolis force, which is generated by the rotation of the Earth. Coriolis force is 0 on the equator and it is considered to remain too small to form a typhoon up to around latitude 5 degrees north. The large body of the sago palm tree is vulnerable to high winds. Accordingly, the sago palm habitat is defined not only by temperature and precipitation; the climatic conditions that deter the formation of typhoons is also an important factor in sago palm growth.

Relative humidity is also considered to influence sago palm growth. Flach et al. (1986a) report that while the optimal relative humidity in terms of the leaf emergence rate was 90%, there was no change in the leaf emergence rate in seedlings when the relative humidity levels were between 70% and 50%.

## 2.3 Soil type

### 2.3.1 Peat soils

While continental Southeast Asia in the tropical monsoon climate has large complicated river systems extending into the interior, Southeast Asian islands in the tropical rainforest climate have much shorter rivers which transport smaller amounts of soil. The islands therefore have the right conditions for the formation of peat soils called Histosols.

In areas of high groundwater levels, water-loving plants thrive. Dead plants are usually broken down by decomposers such as microbes but the decomposition process is inhibited in areas of high groundwater. Regardless of air temperature, plant residues accumulate to form peat soils (Histosols: Fibrists or Hemists) where plant reproduction exceeds plant decomposition (Figure 2-8). Accordingly, peat soils form in the tropics with high temperatures and vigorous woody plant production (tropical peat soils) as well as they

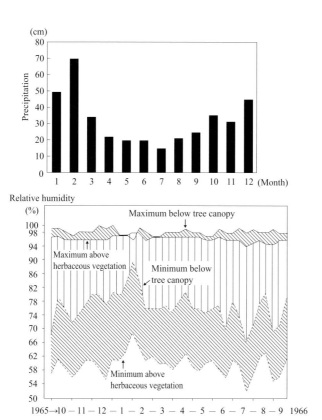

Figure 2-6 Precipitation (top: Kuching, Borneo Is.) and relative humidity (bottom: modified version from Andriesse 1972) in the zone within 10 degrees north and south of the equator

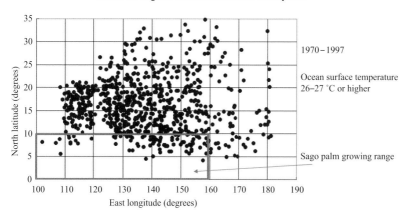

Figure 2-7 Areas of typhoon formation

Source: Mori 2007. 77 typhoons formed in the sago palm growing areas from 1970 to 1997 (http://www8.ocn.ne.jp/~yohsuke/taihuu_2.htm).

do in the cold or alpine region with low temperatures and inhibited plant decomposition. The tropical peat soils are primarily composed of woody plants (low land forests) and therefore called woody peats (Okazaki 1998). The peat soils are classified into Fibrists, Hemists or Saprists depending on the degree of decomposition.

In the cool temperate zone, low moor peat made of tall herbaceous vegetation develops gradually. This is then followed by transitional peat due to the proliferation of plants that require relatively small amounts of water to grow. Transitional peat continues to accumulate to form a dome-shaped surface. The supply of nutrients from below to the plants growing on the surface dwindles and the lower layer of peat begins to breakdown to some degree, returning

Figure 2-8 Tropical peat soil (Selangor, Malaysia)

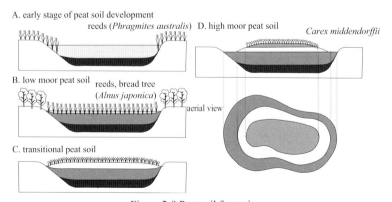

Figure 2-9 Peat soil formation

Source: Shoji 1976
Peat soil develops in stages from low moor peat to high moor peat.

moisture to the peat. Thus, plants that can survive on precipitation alone will proliferate and form high moor peat (Shoji 1976) (Figure 2-9). Peat soil is classified as a type of intrazonal soil which contains more than 20% of organic matter and identifiable plant residues, accumulates in the upper 50 cm of soil, and has a peat deposit of over 25 cm in thickness (United States Department of Agriculture 1975; Kyuma 1986b) but the definition differs slightly from one country to another. Tropical peat soil is also known to form a dome-shaped surface at the final stage of its development (Scott 1985) (Figure 2-10).

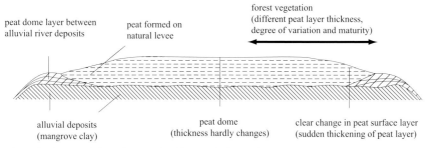

Figure 2-10 Interrelations between landform units in tropical peat swamps wedged between 2 rivers
Source: Anderson (1961)

As peat soil is mostly composed of plant residues, it has a small apparent specific gravity of 0.07–0.3 and a small bearing capacity. Drainage causes marked contraction and land subsidence (Kyuma 1986b). Peaty land developments involving drainage can experience land subsidence up to, or sometimes exceeding, 1 m per year. Murayama (1995) reports that peat soil in Johor, Malaysia, has subsided by 3 cm per year on average over a period of 35 years. Peat soil exhibits strong to weak acidity and, compared with common soil, lacks elements such as calcium, potassium, phosphorus, iron, copper and zinc (Tie et al. 1991; Yamaguchi et al. 1994). The estimated area of peat soil distribution in the world varies from around 240 million ha (Driessen 1980) to 325–375 million ha (FAO 2006) and in Asia the area is estimated to be between 23.5 million ha (Driessen 1980) or 20 million ha (FAO 2006). The peat soil in East Asia is primarily woody peat. The constituent parts of water flowing into peat soil ensure the normal growth of the sago palm (Yamaguchi et al. 1998).

Fukui (1984) states that the sago palm is mostly distributed across areas of peat soils rather than other soil types. Considering the nutrient requirements, however, fluvial deposit-based soil is the optimum soil type for the sago palm. It is likely that sago palms are planted in peat soil areas primarily because other crops grow so poorly in that soil. Sato et al. (1979) report that the thicker the peat deposit in soil, the slower the sago palm grows, and that it grows better in soil with a thinner peat layer and an underlying argillaceous deposit. The argillaceous subsoil contains many elements that are considered to be beneficial to the growth of the sago palm. Tie et al. (1991) have also reported on the distribution of sago palms and sago palm growing areas in Sarawak, Malaysia, which consist of 62.0% peat soil, 33.4% gley soil and 3.1% acid sulfate soil. They state that soils in the sago palm growing areas are 'problem soils' which require sufficient attention for development. Traditional sago palm cultivation is carried out under minimal management, including sucker control, without using fertilizers. A comparison of sago palms at the same growing stage has shown that the starch yield of the sago palms planted in peat soils is only 23% of those planted in mineral soils (Jong and Flach 1995). Malaysia embarked on the development of a large-scale sago palm monoculture plantation in the peat soil area of Sarawak State for the first time in the world, based on various experiments (Kueh et al. 1991). The initial plantation area of several thousand hectares expanded to 15,740 ha by 2001 (Hassan 2002) and reached 40,000 ha by 2007 (Sahamat 2007).

Kakuda et al. (2000) compared the nitrogen supply in peat soils and mineral soils. The amount of ammonium nitrogen released from peat soils through mineralization on the 50th day of incubation was approximately 5.8 mg/kg in Tebing Tinggi, Indonesia, and 4.7 mg/kg in Mukah, Malaysia. They concluded that it is a property of peat soils that mineralization and nitrogen supply occur more readily compared with mineral soils. Based on an analysis of peat soils and soil solutions from Mukah, Malaysia, Kawahigashi et al. (2003) argued that the peat soil horizon exhibits different characteristics depending on the peat accumulation process.

## *2.3.2 Acid sulfate soil*

Fluvial sediment-based soils are known as mineral soils and are classified into Entisols, Inceptisols and so on. The fluvial sediment-based mineral soils generally contain many elements that are supportive of sago palm growth. However, soils found in coastal swamps contain sulfur in the form of pyrite ($FeS_2$) derived from sulfate ions in seawater. Strongly acidic soils due to the occurrence of sulfuric acid through oxidization of pyrite in marine or brackish lacustrine sediments or volcanic ejecta (Figure 2-11) are called acid sulfate soils (Thionic Fluvisols; Sulfaquents) (Figure 2-12). Soils that are already strongly acidic are called actual acid sulfate soils and those that become acidic after oxidization are called potential acid sulfate soils.

Pyrite in marine sediments is produced by the actions of sulfate-reducing bacteria as follows (Kyuma 1986a).

$$SO_4^{2-} + 8H^+ + 8e^- \rightarrow S(-II) + 4H_2O \quad S(-II) : H_2S, HS^-, S^{2-}$$
$$Fe^{2+} + S(-II) \rightarrow FeS$$
$$2Fe^{3+} + S(-II) \rightarrow 2Fe^{2+} + S^0$$
$$FeS + S^0 \rightarrow FeS_2$$
$$Fe_3S_4 \rightarrow FeS_2 + 2FeS$$

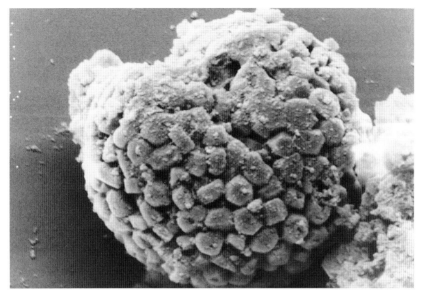

Figure 2-11 Pyrite (FeS$_2$) in acid sulfate soil

If pyrite formed thus stays as it is, soils do not become strongly acidic. When pyrite is oxidized by iron-oxidizing bacteria or sulfur-oxidizing bacteria (Figure 2-13), the following process takes place to produce acids and acidify soils.

goethite

Figure 2-12 Acid sulfate soil (Selangor, Malaysia)

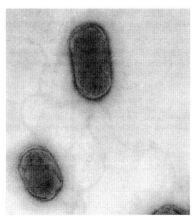

Figure 2-13 Thiobacillus, a type of sulfur oxidizing bacteria
(Photo: Courtesy of Yoko Katayama)

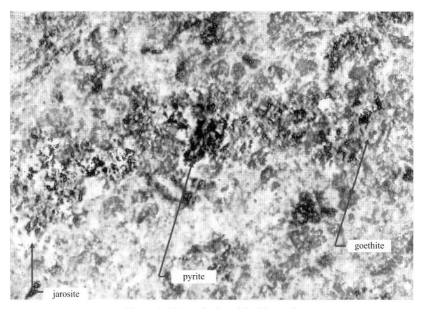

Figure 2-14 Jarosite in acid sulfate soil

$2FeS_2 + O_2 + 4H^+ \rightarrow 2Fe^{2+} + 4S^0 + 2H_2O$
$FeS_2 + 2Fe^{3+} \rightarrow 2S^0 + 3Fe^{2+}$
$4Fe^{2+} + O_2 + 4H^+ \rightarrow 4Fe^{3+} + 2H_2O$
$2S^0 + 12Fe^{3+} + 8H_2O \rightarrow 12\ Fe^{2+} + 2SO_4^{2-} + 16H^+$
$2S^0 + 3O_2 + 2H_2O \rightarrow 2SO_4^{2-} + 4H^+$

As reaction progresses, substances such as jarosite and goethite, which are characteristic of acid sulfate soils, are formed (Figure 2-14).

$$Fe^{3+} + 3H_2O \rightarrow Fe(OH)_3 + 3H^+$$
$$Fe^{3+} + 2H_2O \rightarrow Fe(OH)_2^+ + 2H^+$$
$$3Fe(OH)^{2+} + 2SO_4^{2-} + K^+ \rightarrow KFe_3(SO_4)_2(OH)_6 \text{ (jarosite)}$$

When pH reaches 4 or higher, jarosite becomes hydrolyzed and produces goethite and acid as follows.

$$KFe_3(SO_4)_2(OH)_6 \rightarrow 3FeOOH(\text{goethite}) + 2SO_4^{2-} + K^+ + 3H^+$$

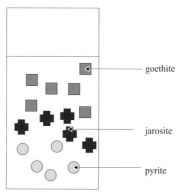

Figure 2-15 Pattern diagram of acid sulfate soil profile

The yellow to yellowish brown argillaceous sedimentation containing jarosite is called cat clay, which derives from the Dutch term 'Katteklei'. Its soil profile is shown in Figure 2-15. Due to the process of formation and breakdown of pyrite, a horizon of accumulated hydrated iron oxide mottles such as goethite beneath the topsoil, a horizon of accumulated jarosite, and a horizon of accumulated pyrite can be observed in a typical acid sulfate soil profile.

The area of acid sulfate soil distribution in the world is 12.6 million ha, 53%, or 6.7 million ha, of which is situated across Asia (van Breemen 1980). They are classified under Thionic Fluvisols according to the WRB soil classification system and under Sufaquents, Sulfimemists, Sulfaquepts and Sulfohemists according to Soil Taxonomy.

Jalil and Bahari (1991) compared the starch yield of sago palms growing in plantations 0.5 km, 0.7 km and 3.0 km from the seashore (soil pH: 3.3–3.8 in acid sulfate soils) and found that the starch yield was extremely high near the coast and lower in the inland plantation.

When the acid enters the soil and increases its acidity, it dissolves aluminum from the soil. This means that plants growing in acidic soils are simultaneously affected by acid ($H^+$) and aluminum (including monomeric forms of aluminum $Al^{3+}$, $Al(OH)^{2+}$, $Al(OH)^{2+}$ and multinuclear aluminum polymers). While plants' acid tolerance must be separated between hydrogen ion tolerance and aluminum ion tolerance, previous studies of sago palm's acid tolerance are primarily based on outdoor observations; there have been no known studies that strictly distinguished between the impacts of hydrogen ions and aluminum ions. In general, acid stress (high hydrogen ion concentrations) on plants causes growth inhibition. The increase in the hydrogen ion concentration in the cytoplasm is brought about by the plant's own metabolic reaction as well as environmental factors such as:

1. When plants absorb ammonium ions, they produce hydrogen ions in the nitrogen assimilation process.
2. The absorption of sugar, amino acids, nitrate ions, chloride ions, phosphate ions and so on increases hydrogen ions in the cytoplasm for hydrogen ion co-transport.
3. Hydrogen ions produced by the light reactions of photosynthesis migrate from the chloroplast to the cytoplasm.
4. Lactic fermentation under reduced oxygen conditions increases the lactic and hydrogen ion concentrations.
5. Increased gas concentrations of carbon dioxide, sulfur dioxide and nitrogen dioxide cause them to dissolve in the cytoplasm and increase hydrogen ions.
6. Increased hydrogen ion concentrations in the soil solution ultimately increase the amounts of hydrogen ions absorbed into the cytoplasm.

Aluminum tolerance varies among different species and varieties. *Melastoma*, *Camellia sinensis*, *Hydrangea*, *Melaleuca*, rice and buckwheat exhibit high degrees of aluminum tolerance; cucumber, peas and tomato exhibit medium level aluminum tolerance; and barley, burdock, carrot, beet and alfalfa have low aluminum tolerance. In the case of sago palm aluminum tolerance is provided by the suppression of aluminum transformation from root cell membrane to stele and the aluminum stock in root and petiole.

Since aluminum has a very small effective ionic radius of 0.0535 nm and a large atomic valence of +3, it readily binds to various other substances. The components formed through aluminum binding are hard to metabolize and therefore accumulate in situ. Increased hydrogen ion concentrations in the soil cause aluminum ions to dissolve from the soil. Aluminum ion stress on plant cells varies depending on the aluminum ion species, complex formation and coexistent ions. Monomeric aluminum ions ($Al^{3+}$, $Al(OH)^{2+}$, $Al(OH)_2^+$) have strong phytotoxicity but the $Al_{13}^{(7+)}$ polymer is said to be most toxic among them. However, not all ion species have been tested for toxicity. It would be more accurate to state that the phytotoxic property of the tridemeric polymer was confirmed because it was the resonant aluminum ion that could be analyzed by nuclear magnetic resonance.

Symptoms of aluminum ion stress manifest in plant roots in the form of dehydration in crown roots, or numerous dimples in epidermal cells of the proximal root several millimeters from the root tip, or transverse cracks extending to the cortex (Wagatsuma 2002). The aluminum binding sites in roots include pectin and proteins in the cell wall, phospholipids and proteins in the plasma membrane, nucleic acids in the nucleus, and phosphorus compounds, proteins, carboxylic acids, phenols and organelles in the cytoplasm. Their aluminum binding makes aluminum difficult to metabolize. When aluminum accumulates in the nucleus of the dividing cell in the root tip and binds to phosphoric acid in DNA, the template activity of DNA is inhibited and root elongation is disrupted. The low pH tolerance mechanism of the sago palm is described in Chapter 5.

## *2.3.3 Other Entisols and Inceptisols*

Soils washed down by rivers and streams form sedimentation at the seashore at various rates. When fluvial sediments deposit too quickly for sulfate reducers to work, they form a soil that is free of pyrite. The parent soils come from many different sources, ranging from sand to clay. A soil that is not supplied with water containing dissolved oxygen forms a blue-gray colored reduced layer (gley horizon) due to the actions of soil microbes. A majority of soil microbes obtain energy by using substances that can replace oxygen as the terminal electron acceptor in the respiratory chain. When iron (Fe III) is used as the terminal electron acceptor, it is reduced to iron (Fe II).

$$Fe(III) + e^- \rightarrow Fe(II)$$

Under laboratory conditions, adding an alkaline solution to an Fe(II) solution to produce $Fe(OH)_2$ deposition forms a white precipitate, not blue-gray. Up to the present date, the average composition of the blue-gray precipitate points to $Fe_3(OH)_8$ (ferrous / ferric iron), a composite of Fe(II) and Fe(III), as the product. When the reduced soil is oxidized by oxygen in the air or iron-oxidizing bacteria, Fe(II) changes into Fe(III) and produces compounds such as non-crystalline hydrated iron oxide and crystalline ferrihydrite $Fe_5HO_8 \cdot 4H_2O$, goethite α-FeOOH, lepidocrocite γ-FeOOH and maghemite γ-$Fe_2O_3$. These iron compounds congregate respectively and form mottling (brown to yellow) in the soil profile. At high temperature, hematite α-Fe$_2$O$_3$ (red) may form out of these iron compounds.

Although sago palm growing areas are mostly composed of low land soils, the sago palm can grow well in other soils (Okazaki 2000). Creating an environment in which water is readily available for the sago palm ensures its growth.

## 2.3.4 Groundwater level

Although water saturated land with a high water table is suitable for sago palm growth, growth is inhibited and starch yield declines markedly or even to zero in permanently submerged or waterlogged land (Takaya 1983; Shimoda and Power 1990). Flach et al. (1977) also found in their sago palm seedling culture experiment that the rate of leaf emergence slows under waterlogged conditions.

Peat soils form in areas of excessive water from frigid to temperate zones. Even in the tropics, peat soils form where plant production exceeds decomposition. Figure 2-16 shows changes in the groundwater level in the tropical peat area in Riau Province, Indonesia, where sago palm plantations are situated (Sasaki et al. 2007). Rapid drainage by the development of the dome-

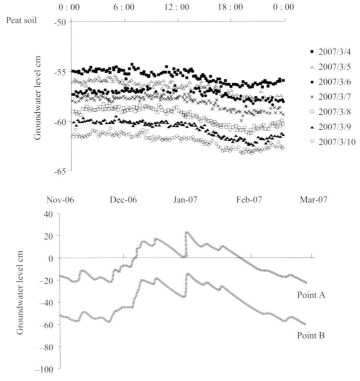

Figure 2-16 Groundwater level in sago palm growing area
Source: Sasaki et al. (2007)
Groundwater level indicates diurnal and seasonal variations.

shaped peat soil causes sinking of the ground itself (Figure 2-17). Hence, measurement of the water table from the ground does not always accurately reflect the groundwater level; a more relevant reference point therefore needs to be established. Sasaki et al. (2007) have demonstrated that the water table in peat soils is changed by seasonal flooding and that sago palm growing sites near the coast are subject to groundwater level variations on an even shorter tidal cycle. Hashimoto et al. (2006) found that, 8 years after trunk formation, both the number of leaves and the diameter at chest height of sago palms growing in tropical peat soils with an average water table of 57–68 cm are smaller than those in the areas of lower water table. Jalil and Bahari's (1991) finding that the sago starch yield varies depending on the distance from the coast was mentioned earlier. Yamamoto et al. (2004a) have studied the trunk length, trunk weight and leaf scar number of six individual sago palms growing at various distances from the seashore and found that the leaf scar interval (= trunk length/leaf scar number) was shorter in coastal sago palms than inland sago palms and that the trunk weight per leaf scar interval (= trunk weight /leaf scar number) was smaller in coastal sago palms than inland sago palms. These findings all indicate that both high groundwater levels and salt concentrations affect the growth of sago palms in coastal habitats with sustained high water tables.

Figure 2-17 Ground subsidence by peat contraction
The ground level dropped by about 1 m due to peat drying and shrinking.

fine root  thick root

Figure 2-18 Sago palm roots

As a response to these high water tables, the sago palm maintains roots with highly developed aerenchyma. Kasuya (1996) reports that the sago palm forms both definite roots and adventitious roots and has as much as 4.7 t/ha in peat soils of roots less than 5 mm in diameter and 11.4 t/ha of such roots in alluvial soils. Yamamoto (1997) has grouped the sago palm roots into fine roots (diameter 4–7 mm) and thick roots (diameter 7–11 mm) (Figure 2-18). The fine root arises from the thick root in waterlogged ground or during the wet season. It has no pith cavity or large vessels in the stele but has small vessels and sieve tubes around the stele. The thick root forms the lysigenous aerenchyma on the inside of the exodermis and has pith cavities and large vessels at the center of the stele as well as small vessels and sieve tubes outside the large vessels. The sago palm's ability to grow in saturated soils appears to stem from the structures of its fine and thick roots. Nitta et al. (2002a) reports that the thick roots 6 to 11 mm in diameter are suited for the transportation of air, nutrients and water while the medium roots 2 to 5.5 mm in diameter are suited for the transportation of air. The sago palm develops its root system primarily in a soil layer 0 to 30 cm below ground both in peat soils and mineral soils. However, as the palm grows older it increases its root mass in the soil, the ratio of vertically growing root mass to total root mass, and the ratio of root mass of roots 2 mm to 5.5 mm in diameter (Miyazaki et al. 2003).

## 2.3.5 Brackish water and seawater

Brackish water is generally defined as a mixture of seawater and freshwater having a salt concentration of 0.2 to 30 parts per thousand (salt content of seawater is generally 33 to 34‰) but some consider 17‰ to be the upper limit. Only plants equipped with a mechanism to eliminate salt or limit salt absorption in some way (halophytes) (Matoh 1999) can grow in the brackish water zone (Figure 2-19).

Figure 2-19 Sago palms distributed along the coast
(Photo: Courtesy of Hiroyuki Shimoda)

Sago palm growth was not inhibited when a solution of about 6–7 millisiemens (mS) per cm conductivity (Hoagland solution) was added, but the growth rate slowed when this range was exceeded (Flach et al. 1977). Ehara et al. (2006a) studied the sodium content of each part of sago palm seedlings collected after treating them with 0–2.0% sodium chloride solutions. They found that the sago palm absorbs sodium ions from the root and translocates them to the petiole and the leaflet, but it retains sodium ions in the root and gradually transfers them to the lower leaves. Yoneta et al. (2004, 2006) demonstrate that sago palm seedlings grow best at a salt concentration of around 10 mmol while growth is inhibited in water or in a salt concentration range from 10 to 20% and that the salt concentration inside the sago palm root increases but salt is not transported to the trunk or the leaf. They also point to the possibility that the plant uses osmoregulation by increasing the potassium concentration in the leaf in response to higher sodium chloride concentrations (Figure 2-20). Ehara

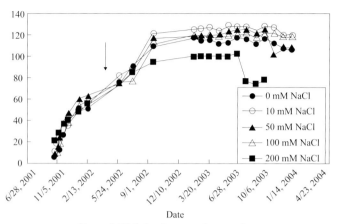

Figure 2-20 Salt tolerance of sago palm

et al. (2008a, b) cultured *Metroxylon* plants (*M. sagu* Rottb. and *M. vitiense*) hydroponically for one month, then cultured them for another month in a 2% sodium chloride solution, and found that the transpiration rate declined in all sago palms and that the potassium content increased as the sodium content increased in the petiole. This result showed a close involvement of potassium in the salt tolerance of the section *Metroxylon* plants and reproduced the result previously demonstrated by Ehara et al. (2006a). The salt tolerance mechanism of the sago palm is described in detail in Chapter 5.

Authors:
Masanori Okazaki,
Sonoko Dorothea Kimura

# 3
# *Morphology*

Sago palm has an enormous above-ground body that can grow to a height of 20 m or more. While secondary auxetic growth takes place to produce the vascular cambium inside the stem or root in the relatively large body of gymnosperms or dicotyledons, the sago palm is a monocotyledon which does not have such a tissue. Its above-ground body is comprised entirely of primary tissues that originate in the apical meristem of the stem apex. The 'stem' is an atactostele, which has many collateral vascular bundles distributed irregularly instead of in a circular pattern in the stele (where vascular bundles are located inside the stem), rather than an eustele (which has many collateral vascular bundles arranged in a circular pattern in the stele) commonly found in gymnosperms and dicotyledons. As the morphogenesis of the sago palm is clearly different from that of gymnosperms and dicotyledons, even though it is a very large plant it is not appropriate to call its 'stem' a 'trunk', which refers to the main shaft of arboreous plants. Nevertheless, the sago palm stem is commonly referred to as 'trunk' in everyday language and therefore the term 'trunk' is used in this article for practicality.

## 3.1 Roots

### *3.1.1 Types of roots*

Each individual sago plant has 2 types of roots that are readily distinguishable with the naked eye, based on their thickness (Yamamoto 1998a; Nitta et al. 2002a). One is the 'thick root' ranging from approximately 6 to 11 mm in diameter. The other is the 'thin root' which is typically between 4 and 6 mm in diameter (Nitta et al. 2002a; Nitta and Matsuda 2005).

The thick roots are adventitious roots, also called primary roots (Figure 3-1). Their primordia form just inside the epidermis of the stem. They are wart-like protuberances a few millimeters in diameter which are observable when the petiole is removed (Figure 3-2). The primordia of the adventitious roots grow and emerge from the stem surface, elongating downward to reach the ground (Figure 3-1). They continue to elongate in a downward or transverse direction in the soil.

Figure 3-1 Stem surface near ground level
Adventitious roots emerged and extended downward. Some of them produced lateral roots immediately after emergence. Some lateral roots grew upward.
A: adventitious roots, L: lateral roots, RC: root cap
Source: Nitta and Matsuda (2005)

Figure 3-2 Adventitious root primordia on stem surface
Removal of the petiole reveals wart-like adventitious root primordia a few millimeters in diameter. Some grew several millimeters and emerged from the surface.
Source: Nitta et al. (2002a)

Figure 3-3 Stem surface several meters above the ground surface
Protuberances of adventitious root primordia are observable.
Source: Nitta et al. (2002a)

Figure 3-4 Lateral roots emerging from ground surface and growing upward
Such lateral roots are often observed in areas prone to submersion. About 10 cm of lateral roots are exposed to the air.
L: lateral roots
Source: Nitta et al. (2002a)

The adventitious roots emerging from nodes on the stem of monocotyledons are sometimes called nodal roots (Nitta 1998) but the adventitious roots of the sago palm form in all parts of the stem, including both the node-like leaf joints and the internodes (Yamamoto 1998a; Nitta et al. 2002a). Hence, it is inappropriate to call the sago palm's adventitious roots nodal roots.

Adventitious root primordia form both at the basal part of the stem and at the apical side of the stem, several meters above the ground, where multiple petioles overlap (Figure 3-3). While adventitious root primordia sometimes elongate and emerge on the inside of the petiole, on the basal side of the stem, they often remain as buds on the stem surface on the apical side of the stem.

The thin roots are lateral roots. They are also called branch roots. Their primordia form both on the adventitious roots (thick roots) and on lateral roots (thin roots) which elongate horizontally in the soil. They elongate downward or obliquely downward in mineral soils and deep peat soils, but can also be found growing upward in especially deep peat soils. Figure 3-4 shows the lateral roots (thin roots) of sago palms elongating upward and growing out of the ground into the air by about 10 cm. This kind of upright elongation and exposure of lateral roots to the air is often observed in areas of submerged or wet soils. These roots act as aerial roots, taking in and feeding air to the plant body as in the case of mangroves. The lateral roots that branch from adventitious roots are primary lateral roots. Those that branch out from primary lateral roots are secondary lateral roots. Secondary lateral roots 1 to 2 mm in diameter emerge in some cases.

By comparison, in the case of tree roots or sweet potato tubers, enlarged roots are sometimes called thick roots and unenlarged roots are called fine roots (Shimizu 2001). These are different from the abovementioned thick roots and fine roots.

## 3.1.2 Thickness and density of adventitious root primordia on the stem

There have been very few studies on the thickness and density of adventitious root primordia. Nitta et al. (2002a) surveyed sago palms growing in a local farmer's sago palm plantation in Mukah, Sarawak, Malaysia, and reported that the diameters of adventitious root primordia were greater at the stem base than at the stem apex regardless of the number of years of trunk formation (Table 3-1). The number of adventitious root primordia per unit surface area of the stem was greater at the stem apex than at the stem base regardless of the number of years of trunk formation (Table 3-2). The number of adventitious root primordia at the stem base was greater in individual plants with a greater number of years of trunk formation. When the average total numbers of adventitious root primordia were calculated and compared between individual plants with different lengths of time of trunk formation, they, too, were greater in the older plants (Table 3-2).

Table 3-1 Diameter of adventitious root primordia (mm)

| Stem position[1] | Estimated years after trunk formation | | |
|---|---|---|---|
| | 7 | 5 | 2 |
| Apical | 1.8 ± 0.1 (9) | 2.9 ± 0.1 (5) | 2.9 ± 0.1 (4) |
| Middle | 2.6 ± 0.1 (5) | 2.9 ± 0.1 (3) | 3.2 ± 0.1 (3) |
| Basal | 3.5 ± 0.1 (1) | 3.2 ± 0.1 (1) | 3.6 ± 0.1 (1) |
| Overall mean | 2.7 ± 0.1 | 3.0 ± 0.0 | 3.3 ± 0.0 |

Numbers in the table represent mean ± standard error.
1) Stem was cut into 90 cm-long sections from the base. Numbers in brackets ( ) indicate the positions of stem sections from stem base.
Source: Nitta et al. (2002a)

Table 3-2 Adventitious root primordia density on stem surface (per 100 $cm^2$)

| Stem position[1] | Estimated years after trunk formation | | |
|---|---|---|---|
| | 7 | 5 | 2 |
| Apical | 202.9 (9) | 159.4 (5) | 124.5 (4) |
| Middle | 175.0 (5) | 139.0 (3) | 69.0 (3) |
| Basal | 108.2 (1) | 85.1 (1) | 56.9 (1) |
| Overall mean | 162.0 | 127.8 | 84.5 |

Numbers in the table represent mean values.
1) Stem was cut into 90 cm-long sections from the base. Numbers in brackets ( ) indicate the positions of stem sections from stem base.
Source: Nitta et al. (2002a)

### 3.1.3 Internal structure of root

Seen in cross-section, the adventitious root (thick root: Figure 3-5) and the lateral root (thin root: Figure 3-6) are both comprised, from the outermost layer to the center, of epidermis, exodermis, suberized and thickened sclerenchyma, cortex, and stele. Vascular bundles run inside the stele and consist of xylem containing vessels and phloem containing sieve tubes. These tissues are larger and more numerous in adventitious roots than lateral roots. In both types of root, the cortex is formed of schizogenous aerenchyma, the intercellular space created by the separation of adjoining cell walls as the plant grows, and lysigenous aerenchyma, the intercellular space created by the loss of adjoining tissue cells as the plant grows. These tissues develop an aeration function and adapt to growth in submerged conditions (Nitta et al. 2002a).

This internal structure of sago palm roots is very similar to the crown roots of paddy rice (Nitta 1998).

## MORPHOLOGY

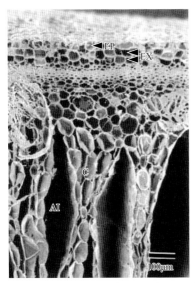
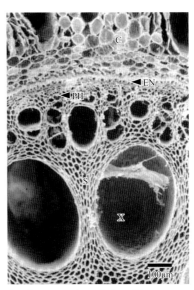

Figure 3-5 Scanning electron microscopic photographs of cross section of adventitious root
Left: Outer part. Right: Stele part inside. Cortical aerenchyma is very well developed. The tissue remains in the few layers near the exodermis. Xylem vessels inside the stele are quite large with a diameter of about 500 μm.
AI: aerenchyma, C: cortex, EN: endodermis, EP: epidermis, EX: exodermis, PH: phloem, X: xylem.
Source: Nitta and Matsuda (2005)

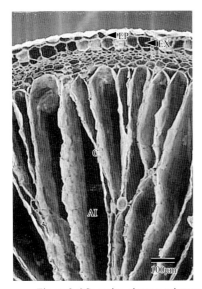
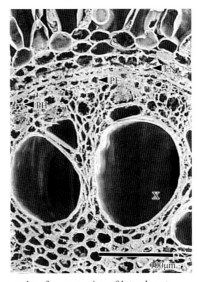

Figure 3-6 Scanning electron microscopic photographs of cross section of lateral root
Left: Outer part. Right: Stele part inside. Cortical aerenchyma is very well developed. Xylem vessels inside the stele have a diameter of about 100 μm.
AI: aerenchyma, C: cortex, EN: endodermis, EP: epidermis, EX: exodermis, PE: pericycle, PH: phloem, X: xylem.
Source: Nitta and Matsuda (2005)

## 3.2 Leaf

The Arecaceae family includes both fan palms with palmate compound leaves which are attached to a short rachis and feather palms with pinnate compound leaves which are attached to an elongated rachis. Some of the latter group have bipinnate compound leaves (*Caryota* spp.).

The sago palm has large pinnate compound leaves which have many leaflets attached to a rachis. Figure 3-7 is a diagram of a sago palm leaf. It depicts an 8.2 m long leaf from a plant shortly after trunk formation. Leaflets are represented by line segments, reflecting the exact measurements of leaflet length as well as their insertion positions and angles on the rachis.

The mid rib is sometimes treated as having two parts – petiole and rachis – but there is no distinct demarcation on the smooth surface; the attachment of leaflets is the only recognizable external morphological boundary between them (Figure 3-7). Accordingly, the insertion position of the most proximal leaflet is regarded as the boundary between petiole and rachis. The base of the petiole is sometimes treated as a distinct part called the leaf sheath (Jones 1995). The leaf sheath is the extended base of the petiole, which wraps around the trunk and fuses on both ends. This fused state is observable from the leaf primodium stage but no morphological boundaries between the petiole and leaf sheath are found on the topside or underside of the petiole (leaf mid rib). The leaf sheath length is included in the petiole length in Figure 3-7. Those which have no spines on the abaxial surface of the petiole are called 'hon sago' or 'mushi sago' (true sago or spineless sago) and those which have spines are called 'toge sago' (spiny sago).

The youngest visible leaf has a pointed stick form and emerges along a groove running longitudinally in the petiole of the previous leaf. This young emergent leaf is called an ensiform leaf or needle leaf. It is also called a spear leaf (Jones 1995). Inside the ensiform leaf are folded up leaflets, which open as the plant grows.

Leaflet insertion positions on rachis are asymmetrical (Nakamura et al. 2004a). The numbers of leaflets are different between the left and right sides of a rachis. In other words, leaflets on the right and left sides of a leaf are not jugate (Figure 3-7). Figure 3-8 shows leaflet insertion positions on a young open leaf. From the leaf base on the left side of the diagram, leaflets on the left side (L) and the right side (R) are numbered and their positions are marked on the lines. Every tenth leaflet is marked with a black dot; the diagram shows that the black dot positions on L and R do not correspond. There were 72 leaflets on the left side and 67 on the right side.

In sago palms, the insertion of a leaflet on the rachis has an inverted V shape (Figure 3-9). The Arecaceae family includes plants with induplicate (V-shaped, trough-shaped) leaflet insertions and those with reduplicate (inverted V-shaped, tent-shaped) leaflet insertions. The shape of the leaflet insertion is one of the basic markers for the classification of palms.

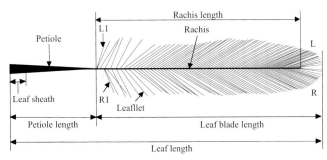

Figure 3-7 Pattern diagram of sago palm leaf and part names
Source: Goto and Nakamura (2004)

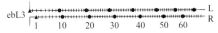

Figure 3-8 Leaflet attachment positions
Source: Goto and Nakamura (2004)

Figure 3-9 Edges of leaflets on rachis
Source: Goto and Nakamura (2004)

Sago palm leaflets range from linear lanceolate to narrow lanceolate. Lanceolate leaflets become pointed toward the apex and have a very thin elongated tip of various lengths and shapes.

When an emergent ensiform leaf opens, a cord-like tissue connects the tips of leaflets as if to trace the outline of the leaf (Figures 3-10, 3-11). In this state, the leaf looks like a simple leaf with many large slits between veins. As the leaf spreads, the cord-like tissue falls away (sometimes a cord over 1 m long is found on the ground), or breaks up, leaving individual leaflets fully independent. The thin elongated leaflet tip is a remnant of this cord-like tissue.

As this part has a highly variable form and does not contribute to leaf area, it is more convenient for discussion purposes to not include this part in the length of leaflet. For several leaflets at the base of the leaf, however, it is difficult to ascertain the position of the tip, as they are quite narrow to begin with and become even narrower toward the apex, naturally joining with the cord-like tissue. Leaflets are generally at their widest at around one third of the distance from the leaflet base.

Figure 3-10 Leaf immediately after opening
Leaflet tips are linked by a cord-like tissue.
Source: Goto and Nakamura (2004)

Figure 3-11 Leaf immediately after opening
Enlarged image of part of Figure 3-10.
Source: Nakamura et al. (2004a)

Typical individual sago palms growing in a sago plantation in Mukah, Sarawak, Malaysia, had 9 green leaves shortly after trunk formation (Nakamura et al. 2000). Except for the youngest ensiform leaf, the other 8 leaves had open leaflets. The youngest of the open leaves (12.4 m long) had a 4.6 m long petiole part and a 7.2 m long rachis part, with 69 leaflets on the left side, 65 leaflets on the right side and 1 bifurcated leaflet at the tip, a total of 135 leaflets. The longest of these leaflets measured 189 cm and the largest had a leaf area of 1,360 $cm^2$. The aggregate leaflet area was 12.1 $m^2$, including 6.28 $m^2$ on the left side and 5.82 $m^2$ on the right side. Measurements were carried out on another leaf, which had a total leaf area of 12.25 $m^2$. Based on these measurements, the estimated total leaf area of the 8 open leaves of this palm was approximately 100 $m^2$.

The accuracy of leaf area estimation was considered to be higher when it was derived from the integrated value of leaflet areas rather than the leaf outline based on the leaf development process or morphology (Nakamura et al. 2004a). Nakamura et al. (2005) conducted a study to find a way to estimate leaflet areas through detailed analysis of the relationship between leaf morphology and leaf area. It proposed a method to approximate the leaflet area using an ellipse with leaflet length as the major axis and leaflet width as the minor axis. In other words, the estimated leaflet area as an ellipse (Se) = ($\pi/4$) × L × W where L is leaflet length and W is the maximum width of leaflet. This method produces larger errors when estimating the areas of small leaflets arising at the leaf apex and base.

Omori et al. (2000a) calculated an α value by dividing leaflet area by the product of leaflet length and leaflet width and obtained 0.78–0.86 after examining many varieties and leaf positions. As the α value used in the elliptical approximation of leaflet area by Nakamura et al. (2005) was $\pi/4$, or 0.785, the α value obtained in the study by Omori et al. supports the validity of the elliptical approximation method.

Flach and Schuiling (1989) estimate the area of a leaf by using the following formula: leaf area = 2 × leaflet number on one side × longest leaflet length × width × coefficient (about 0.5) but they do not provide a definitive coefficient value. Nakamura et al. (2009a) examined a method to estimate leaf area based on the shape of leaf by converting leaflet shapes to rectangles of equivalent areas and arranging them on a rachis without any gaps between them in order to eliminate overlaps and gaps between leaflets (Figure 3-12). They found that the most accurate estimation was obtained when the proximal half of the leaf was calculated as a trapezium and the apical half was calculated as elliptic (Figure 3-13). The formula used in this leaf area (S) estimation is S = ab $\pi/8$ + ac/2 where a is rachis length, b is the total length of left and right leaflets at the midpoint of rachis after conversion to rectangles, and c is the total length of left and right leaflets at a point a/4 from the base after conversion to rectangles.

### 3.2.1 Phyllotaxis

Figure 3-14 is a pattern diagram of a cross section of partially overlapping petioles in a sago palm shortly after trunk formation. The leaves are younger toward the center where an ensiform leaf is found. In the diagram, the petioles are numbered from the youngest to the oldest with the ensiform leaf at the center being number 1.

Leaf arrangement is generally described from the leaf that opened earliest to the younger leaves and therefore in the reverse order to the numbering in the diagram. Accordingly, this palm has a counterclockwise arrangement, or so-called sinistrorse.

The way leaves are arranged on the trunk is called phyllotaxis. Sago palm reportedly has a phyllotaxis of 4/13 and a divergence angle of 110.77° (Jong 1991).

A divergence angle of approximately 110° means that 3 leaves make approximately 330° and therefore, the third leaf from the frontal leaf (counted as 0) appears to be attached slightly inside of the alignment (Figures 3-14, 3-15). The inward deviation of the third leaf is to the right in the case of dextrorse phyllotaxis and to the left in the case of sinistrorse phyllotaxis.

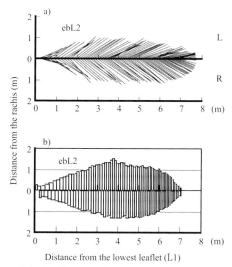

Figure 3-12 Leaf pattern diagram (a) and leaf image after all leaflets are converted into rectangles with equivalent areas (b)
Source: Nakamura et al. (2009a)

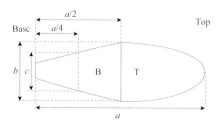

Figure 3-13 Schematic representation of leaf as a trapezium in the proximal half and a half ellipse in the apical half
Source: Nakamura et al. (2009a)

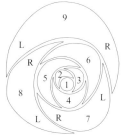

Figure 3-14 Cross section of petioles showing positional relations between leaves
Source: Nakamura et al. (2009a)

## 3.2.2 Leaf arrangement and leaflet attachment

In the case of sinistrorse leaf arrangement as shown in Figure 3-14, more leaflets are found at the extreme base of the leaf on the left side (L) than the right side (R) of rachis. There are also a greater number of leaflets overall on the L side as per the leaf shown in Figure 3-7. In the case of dextrorse leaf arrangement, the R side becomes the basal side for the leaflets found at the extreme base of the leaf and the R side of rachis has a greater number of leaflets.

## 3.2.3 Stomatal density

Omori et al. (2000b) conducted a detailed study of the stomatal density in leaves by leaf position, leaflet position and position on a leaflet (Riau, Indonesia). Figure 3-16 shows the distribution patterns of stomata on the adaxial surface (A) and abaxial surface (B) of a leaflet in a plant shortly after trunk formation. The patterns were transferred to an adhesive and microscopically observed. A comparison of stomatal densities between the topside and underside of a leaflet found markedly higher densities on the abaxial surface at 550–750 stomata per mm$^2$ compared with 100–250 on the adaxial surface. A comparison between the apical, middle and basal parts of a leaflet found that stomatal density tended to be smaller at the apical part. As the leaflet was thicker at the basal part and thin at the apical part, a significant positive correlation was demonstrated between leaf thickness and stomatal density on both the abaxial and adaxial surfaces. There was little difference found, however, in stomatal densities between different leaflet positions in a leaf (top, middle and base) and between different leaf positions in a plant.

Figure 3-15 Deviating phyllotaxis
Source: Goto and Nakamura (2004)

Omori et al. (2000b) studied stomatal densities in the middle part of leaflets to investigate the change of stomatal density by plant age using the middle leaflets of the middle leaves as samples (Sarawak, Malaysia). The stomatal density was about 400 stomata per $mm^2$ on the abaxial surface and about 50 stomata per $mm^2$ on the adaxial surface in a palm one year after sucker development. It was about 1,000 stomata per $mm^2$ on the abaxial surface and about 120 stomata per $mm^2$ on the adaxial surface in a palm in the fifth year of trunk formation. It changed very little thereafter on both surfaces. Yamamoto et al. (2006a) also reported that the stomatal density of a leaflet was markedly greater on the abaxial surface than the adaxial surface and tended to increase up until the stage of trunk formation on both surfaces. In other words, the stomatal density increases until the trunk formation stage and changes very little thereafter. The stoma length ranges from 10 to 16 µm on the adaxial surface and is smaller on the abaxial surface, ranging from 8 to 14 µm. The stoma length tends to become smaller with age.

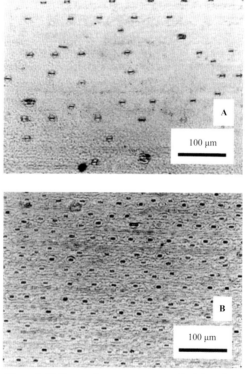

Figure 3-16 Stomatal arrangement on adaxial surface (A) and abaxial surface (B) of leaflet
Source: Omori et al. (2000b)

## *3.2.4 Internal morphology of leaf*

Ehara et al. (1995a) found that leaflet thickness ranged from 0.2 to 0.3 mm. A study by Yamamoto et al. (2007b) of 10 varieties in Indonesia found that the leaf thickness fell within the 0.3–0.4 mm range in all varieties. The mesophyll tissue, large vascular bundles and small vascular bundles are distributed in the space between 2 to 4 upper and lower layers of epidermis (Figure 3-17). There are 3–4 layers of palisade parenchyma on the adaxial side of the mesophyll and 8–10 layers of spongy parenchyma on the abaxial side with 3–6 small vascular bundles running between 2 adjoining large vascular bundles (Yamamoto 1998a). Around 17–21 large bundle sheath cells made of parenchyma are arranged into a single layer surrounding large vascular bundles although interrupted on the adaxial and abaxial sides. Small vascular bundles are surrounded by a layer of bundle sheath cells consisting of 7–10 parenchyma cells.

Nitta et al. (2004) collected leaves from a palm that was about 5 years after trunk formation, with 14 green leaves. They made detailed observations of the internal anatomy of leaflets at different positions on the leaf and at different leaf positions. Figure 3-18 shows cross-section images of leaflets at different attachment positions on a mid-level leaf (seventh from base) examined with a scanning electron microscope. The leaflet attached in the middle of the leaf had a thicker mesophyll tissue and greater leaf thickness compared with other positions due to a thicker spongy parenchyma and more layers. A comparison of cross sections of large and small vascular bundles in leaflets at middle and basal positions found little difference between them; the cross section of the large bundles was about 17,000 $\mu m^2$ and the small bundles ranged from 2,000 to 2,100 $\mu m^2$ in all of them. The distance between 2 adjoining bundles ranged from 860 to 870 µm in the case of large vascular bundles and was 165 µm in the case of small vascular bundles. Again there was little difference between different leaflet positions.

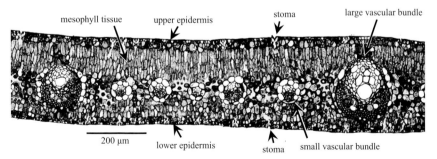

Figure 3-17 Internal structure of sago palm leaflet (cross section)
(Photo: Youji Nitta, original illustration)

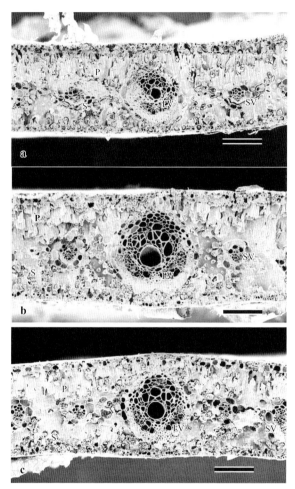

Figure 3-18 Cross sections of leaflets attached to apical (a), middle (b) and proximal (c) parts of mid-level leaf

LV: large vascular bundle, P: palisade parenchyma, S: spongy parenchyma, SV: small vascular bundle, image scaled to 100 μm.
Source: Nitta et al. (2004)

A comparison of internal leaf tissue characteristics at different leaf positions – high-level leaf (14th from base, Figure 3-19), low-level leaf (at extreme base, Figure 3-20) and mid-level leaf (Figure 3-18) – found that mesophyll and

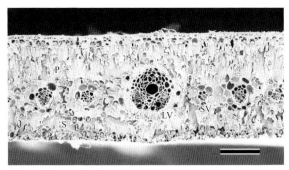

Figure 3-19 Cross section of leaflet attached to middle part of high-level leaf
Legend and scale as in Figure 3-18.
Source: Nitta et al. (2004)

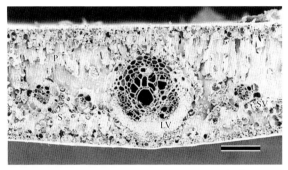

Figure 3-20 Cross section of leaflet attached to middle part of low-level leaf
Legend and scale as in Figure 3-18.
Source: Nitta et al. (2004)

leaf thickness were greater in high- and mid-level leaves than low-level leaf, mainly because their spongy parenchyma was thicker and had more layers. However, there was no difference between leaf positions with respect to the cross sections of large and small vascular bundles and the distance between bundles.

## 3.3 Stem (Trunk)

The trunk (stem) of the sago palm reaches 10 m or more in length and more than 50 cm in thickness over a period of 10 or more years after trunk formation. It can weigh in excess of 1,000 kg as it accumulates starch inside (Yamamoto 1998a). Its internal structure is an atactostele lacking the vascular cambium that is responsible for secondary auxetic growth.

## 3.3.1 Trunk configuration

As the leaf base attached to the trunk forms a leaf sheath that wraps around the trunk once, it leaves a node-like leaf base scar on the trunk after it withers and falls away. The removal of green leaves also leaves node-like leaf base scars on the trunk (Figure 3-21). It is therefore convenient to treat these leaf base scars as apparent nodes and the intervals between nodes as internodes for the purpose of describing sago palm growth and morphology.

Here, the node at which the apparently youngest leaf, i.e. the emerging ensiform leaf (ebL 1), is attached is called ebN 1, the node below it ebN 2, and subsequent nodes ebN 3, ebN 4 and so on. The internode immediately above ebN 1 is called ebIN 1, then ebIN 2, ebIN 3 and so on in a downward direction. Accordingly, the internode between ebN 1 and ebN 2 is ebIN 2.

A sago palm that is about 8 years after trunk formation with a trunk length of about 5 m was stripped of all leaves. The trunk is schematically represented in Figure 3-22. Internodes grow rapidly at the apical part of the trunk but growth stops at ebIN 4–5 and internodes below that level have very similar lengths (Nakamura et al. 2004b). Thickening appears to continue up to ebIN 15 or so and the trunk gradually thickens from the apex to this position. Trunk thickness remains almost constant in internodes below this position and becomes thicker in 7 or 8 basal internodes closest to the ground line.

## 3.3.2 Sucker

Branched sago palm stems, called suckers, exhibit peculiar behavior and are also used for propagation.

Most suckers emerge from their parent trunk near the ground line (Figure 3-23). They grow and form a thicket around the parent trunk or often creep sideways and form trunks away from their parent trunk. According to a survey in Mukah, Sarawak, Malaysia, the sideways-creeping suckers grow about 60 cm in a year. In many cases, they creep for 3–5 years over a distance of 2–3 m, then grow upright to form trunks (Figures 3-24, 3-25). In some cases, they may creep for nearly 5 m during a period of 7–8 years. Such plants may mature within 3–4 years of trunk formation. this early maturation means that only 3 m or so of their trunks can be harvested.

The creeping distance appears to be influenced by the ground conditions. Suckers appear to creep longer on harder ground surfaces (Mr. Smith, sago plantation owner in Mukah, personal correspondence). In fact, long creeping suckers were found more often in areas of hard mineral soils on river banks than in soft peat soil areas in Mukah.

In terms of external morphology, node-like features found on thick creeping suckers of sago palms growing in the mineral soils of Mukah were leaf scars and more suckers were emerging from some of these nodes. The creeping part was structurally the same as the trunk, but slightly flattened (Figure 3-26), and had accumulated starch inside. Based on these findings, thick, creeping

Figure 3-21 Leaf base scars after removal of green leaves

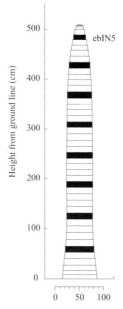

Figure 3-22 Schematic representation of sago palm trunk
(8 years after estimated trunk formation)
Source: Nakamura et al. (2004b), modified by author.

Figure 3-23 Emerged suckers

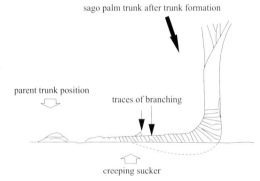

Figure 3-24 Schematic representation of sucker (grown about 3 m in nearly 5 years, about 1 year after trunk formation)
Source: Goto et al. (1998)

suckers were considered to be trunks which were growing horizontally instead of vertically although they had slightly shorter internodes. For these suckers, horizontal growth is perhaps no different from trunk formation and therefore they do not grow tall trunks after beginning to grow upright.

# MORPHOLOGY

Figure 3-25 Tip of a sucker continuing to creep sideways

Figure 3-26 Creeping sucker is being cut (left). Cross section of sucker stem (right).

### 3.3.3 Vascular bundle distribution in trunk cross sections

The cross section of a trunk can be divided into the outer part consisting of barks and the inner part consisting of pith or medulla (Figure 3-27). The pith has a dark brown periphery and a white central portion. Vascular bundles are scattered throughout the pith and visible to the naked eye.

Figure 3-28 graphs show the numeric distributions of vascular bundles in sago palms that are estimated to be 1, 4 and 7 years after trunk formation (Goto 1996). The outermost portion of each cross section in the graphs shows the number of vascular bundles per $cm^2$ in the zone 1 to 5 mm inside the pith periphery. This corresponds to the area inside the dark brown portion along the pith periphery.

In a cross section of a plant one year after trunk formation, the number of vascular bundles per unit area was particularly high in the pith periphery and decreased rapidly within several centimeters toward the center. Then the number decreased gradually until it increased slightly at the center. This basic pattern was observable in all cross sections. This arrangement appears to be similar to the one found in other arboreous plants of the family Arecaceae whose vascular bundles are more densely distributed in the marginal portion than at the center of trunk (Parthasarathy 1980). When a comparison was made between the 'low' cross sections of the sago palms 4 years and 7 years after trunk formation at the same height as the cross section of the sago palm 1 year after trunk formation (about 1 m above ground), they had a similar numbers of vascular bundles. When distances from the center were substituted with percentages of the pith diameter, the curves representing the relationships between distance from the center and the number of bundles found in the three cross sections were almost identical (Figure 3-29). It is conceivable that the trunk thickens at base while keeping the bundle distribution pattern unchanged even though it lacks auxetic growth involving cambium.

In the plant aged 4 years after trunk formation, the curves for 'middle' and 'high' overlapped (Figure 3-28). The bundle distribution patterns were similar in these parts, perhaps because it had not been long after their growth and their positions were relatively close (90 cm).

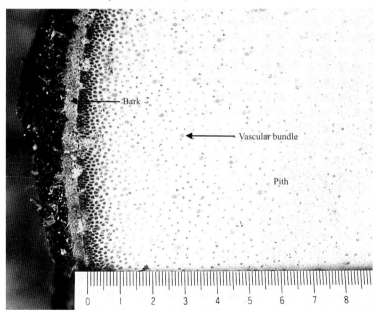

Figure 3-27 Trunk cross section at about 1 m above ground of sago plant about 8 years after trunk formation
Source: Yamamoto (1998a)

# Morphology

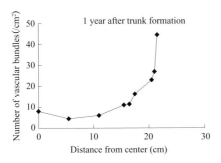

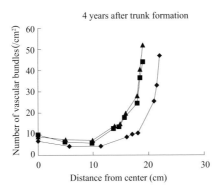

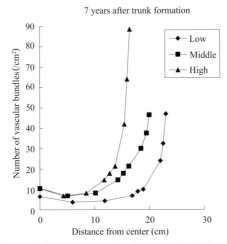

Figure 3-28 Vascular bundle distribution in sago palm trunk cross section
Low: basal part of trunk; Middle: middle part of trunk; High: upper part of trunk
Source: Goto (1996)

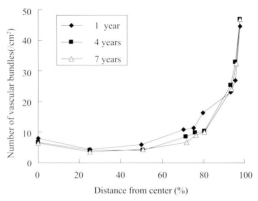

Figure 3-29 Number of vascular bundles in trunk cross section at about 1 m above ground (trunk radius being 1)

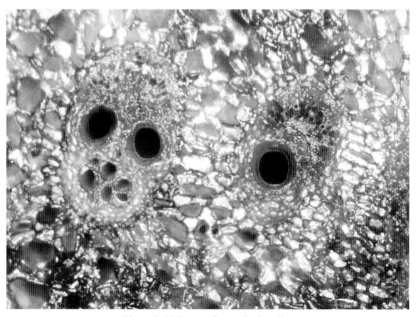

Figure 3-30 Types of vascular bundles
One type has a large vascular bundle (Type I) and the other type has 2 large vascular bundles (Type II). The image here is a cross section at a high level of a trunk probably 8 years old, about 3 cm inside of the periphery.
Source: Yamamoto (1998a)

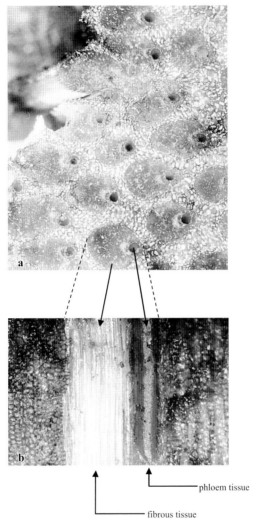

Figure 3-31 Fibrous tissues of vascular bundle: cross section (a) and longitudinal section (b) of peripheral vascular bundles
Arrows indicate corresponding tissues.
Source: Goto et al. (1994)

In the plant aged 7 years, the vascular bundle density was 'high' > 'middle' > 'low' at the same distance from the center. It was particularly high near the pith periphery in the 'high' cross section.

As a broad morphological characteristic, vascular bundles may have a single large vessel ('Type I' here for convenience) or 2 large vessels ('Type II') (Figure 3-30). Only Type II vascular bundles are found near the pith center, sparsely scattered throughout parenchyma. The proportion of Type I vascular bundles increases at around 3 cm inside the pith periphery, scattered rather sparsely across parenchyma.

Bundles near the pith periphery are filled with fibrous tissue, which is more developed laterally (toward the periphery), and phloem fibers (Figure 3-31). This fibrous tissue does not develop in bundles at the pith center portion. Phloem fibers that are found in bundles in the outer portion, i.e., about 4 cm inside of the periphery, are quite prominent and become thicker as they get closer to the pith periphery. Many of the bundles with developed fibrous tissue are Type I. Fibers found in the outermost part of the pith immediately inside of barks, especially in the zone 3 mm inside the periphery, are rigid and densely arranged. These fibers are considered to be responsible for maintaining the form of the palm trunk (Figure 3-32). This part does not appear to accumulate starch as an iodine-potassium iodine solution does not stain it, even in older plants.

### 3.3.4 Vascular bundle orientation

Figure 3-33 is a longitudinal section image of vascular bundles running from a leaf into the trunk through vascular bundles in the pith periphery. Vascular bundles from a single leaf enter the pith at several levels (Figure 3-34) and travel obliquely down toward the pith center. It has been observed in monocots, including genus *Rhapis*, that some leaf vascular bundles head toward the periphery again after entering into the pith center (Zimmermann and Tomlinson 1972). As the cross-section distribution patterns of vascular bundles in the pith were similar between sago palms and other Arecaceae plants (as mentioned earlier), it is likely that sago palm leaf vascular bundles also travel ultimately to the periphery. Such a structure is found even in quite young tissues (Figure 3-35), and leaf vascular bundles are already entering into the pith in immediate proximity to the youngest part where peripheral vascular bundles are visible to the naked eye. The oblique traveling angle varies from one bundle to another and the vascular bundle indicated by arrows in Figure 3-36 entered 5.5 cm inside from the periphery as it traveled down 18 cm. Consequently, the pith cross sections of relatively young plants show a regular arrangement of vascular bundle insertions from leaves near the periphery (Figure 3-27).

Thus, vascular bundles in the pith periphery portion and central portion are dissimilar from the point of branching (Figure 3-34, 3-35) and it appears that the main function of the pith periphery vascular bundles is to support the trunk while the interior vascular bundles enter directly from the leaves to transport nutrients to the inside of the pith.

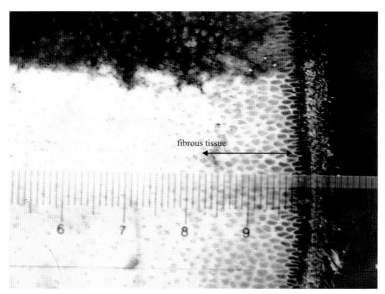

Figure 3-32 Fibrous tissue on the periphery of a trunk of approximately 12 years of age
Source: Watanabe et al. (2008)

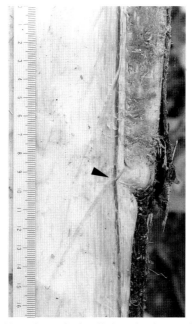

Figure 3-33 Leaf vascular bundle insertion into trunk (arrow)
Source: Watanabe et al. (2008)

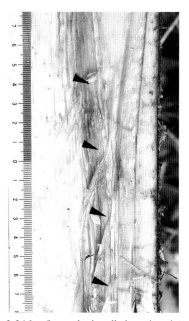

Figure 3-34 Leaf vascular bundle insertions into trunk
This is a relatively young leaf attachment. Vascular bundles of a leaf enter into trunk at several levels (arrows).
Source: Watanabe et al. (2008)

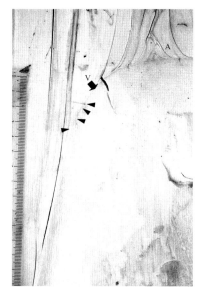

Figure 3-35 The youngest part of vascular bundles constituting fibrous tissue on pith periphery
V and thick arrow: The youngest part where peripheral vascular bundles are visible to the naked eye.
A: Growing point
Arrows: Leaf vascular bundle insertions into the pith
Source: Watanabe et al. (2008)

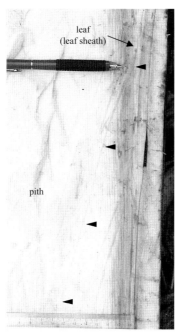

Figure 3-36 Tracing direction of leaf vascular bundle running into pith
It enters into the pith at the point of the pen (arrow) and descends obliquely toward the pith center.
Source: Watanabe et al. (2008)

### 3.3.5 Parenchyma in the trunk

Intercellular spaces are large in the ground parenchyma of sago palm pith (Nitta et al. 2005b). Each intercellular space is surrounded by 6 to 7 cells in the vicinity of the growing point (Figure 3-37), becoming larger as the surrounding cells elongate and thicken in the mature tissue below that point (Nitta et al. 2006; Warashina et al. 2007). Intercellular spaces and the surrounding cells form a network-like structure in the parenchyma. The percentage of intercellular spaces in the pith cross section is at the lowest level near the growing point, then increases in the part between the shoot apical meristem and 80 cm to the base, and becomes constant in the level beneath it (Warashina et al. 2007). However, the intercellular space ratio varies according to growing conditions; it is higher in plants growing in dry areas than those growing in swamps (Nitta et al. 2006). The ratio ranged from 36 to 45% depending on the variety of sago palm (Nitta et al. 2005a).

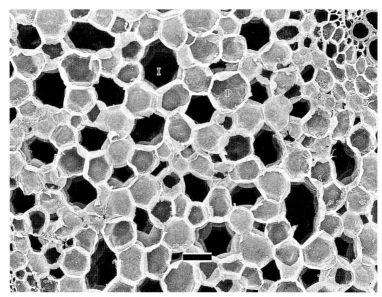

Figure 3-37 Scanning electron microscopic photograph of pith cross section of a sago palm estimated 3 years after trunk formation growing in a dry area
I: intercellular space, P: parenchyma, Crossbar length is 100 μm
(Photo: Youji Nitta)

An iodine-potassium iodine solution stains the parenchyma black, indicating a dense distribution of starch granules, but the peripheral parenchyma does not appear to accumulate starch as much as the central parenchyma as iodine staining is lighter in that portion. Amyloplasts accumulate inside the cells surrounding intercellular spaces (Figure 3-38) (Nitta et al. 2005b).

## 3.4 Flower, fruit and seed

The sago palm forms flower buds at the apical crown about 12 years after planting and takes a further 2 years or so to reach the flowering stage (Yamamoto 1998a). It requires at least 3 years from flower bud formation to fruit maturity (Flach 1997). Jong (1995a) reported that a single inflorescence (Figure 3-39) can branch to the third order and form 1,313–3,427 rachillae. Not all rachillae produce fruit but each rachilla sets about 2 fruits in many cases, which means that a single plant can produce several thousand fruits.

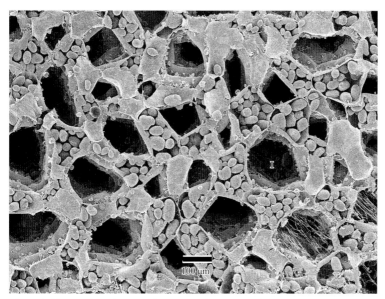

Figure 3-38 Scanning electron microscope photograph of pith cross section of sago palm variety Para Hongleu
I: intercellular space
(Photo: Youji Nitta)

## 3.4.1 Flower

Plants belonging to the genus *Metroxylon* produce perfect flower (hermaphrodite flower) and male flower (staminate flower), each having 3 sepals, 3 petals and 6 stamens (Tomlinson 1990). Figure 3-40 shows a hermaphrodite flower and a staminate flower of sago palms that are classified under the section *Metroxylon* (*Eumetroxylon*). The hermaphrodite flower has a gynoecium at its center surrounded by stamens. There is no size difference between the two types of flowers in sago palms but the staminate flower has a degenerated gynoecium called pistillode (Ehara et al. 2006a). In the genus *Metroxylon*, the timing of anther emergence and dehiscence after flowering is generally different between hermaphrodite flowers and staminate flowers and the former are cross-pollinated as their pollens are sterile. The pollens of both hermaphrodite and staminate flowers have 2 germination apertures on the horizontal short axis (Figure 3-41).

Figure 3-39 Inflorescence at the fruiting stage
A type of sago palm with sparse spines from Wewak, East Sepik, Papua New Guinea.

## 3.4.2 Fruit

The scaly exocarp on the outermost layer of the fruit has 18 vertical rows of scales. The number of scale rows varies from one species to another in the section *Coelococcus* and ranges from 21 to 31 rows with some intraspecific variation.

The proximal end of the fruit attaches to a rachilla (Figure 3-42). A mature fruit has a horizontal long axis of just over 5 cm, a short axis of just under 5 cm, a polar axis of about 5 cm and a fresh weight of 50–60 grams. Besides the section *Metroxylon*, which includes a single species, *M. sagu*, the genus *Metroxylon* include the section *Coelococcus* of which *M. amicarum* (H. Wendl.) Becc. (found in Chuuk, Pohnpei, and the Marshall Islands) and *M. warburgii* (Heim) Becc. (found in Vanuatu, Fiji, and Samoa) produce seeds with large diameters (about 9 cm and 6 cm in fruit horizontal long axis respectively) that are called palm ivory. These seeds are used as buttons and craft material (Dowe 1989; Ehara et al. 2003c; McClachey et al. 2006).

## 3.4.3 Seed

Sago palm seed is an albuminous seed with an endosperm. Its embryo is buried in the endosperm and covered by a cap-like organ called operculum (Figure 3-43). They are in turn covered with seed coat tissues called testa and pericarp. The double-layered testa consists of a thin inner seed coat (turns black at maturity) and a succulent sarcotesta outside of it. Pericarp also has two layers called mesocarp and exocarp.

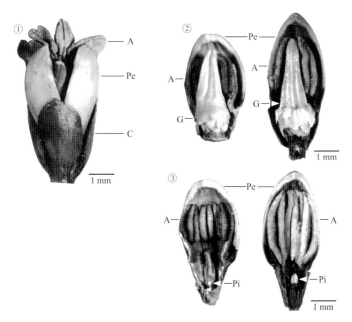

Figure 3-40 Exterior view of flower ① and internal views of hermaphrodite and staminate flowers ②, ③
C: calyx, Pe: petal, A: anther, G: gynoecium, Pi: pistillode
① spiny type from West Papua Province, Indonesia
② hermaphrodite flowers. Left: spiny type from West Papua Province. Right: spineless type from West Sumatra Province, Indonesia
③ staminate flowers. Left and right as in ②
Source: Ehara et al. (2006a)

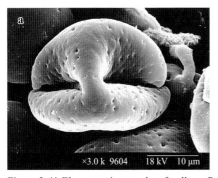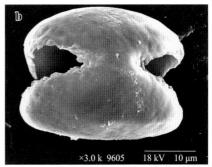

Figure 3-41 Electron micrographs of pollens. Pollens of hermaphrodite flower of spineless type (folk variety *Rumbio*) from Siberut Island, West Sumatra
a: Pollen grain with infolded aperture margins.
b: Pollen grain with aperture margins expanded (pollen grain in slightly oblique distal polar plane).
Source: Ehara et al. (2006a)

Figure 3-42 Fruit attachment
A spineless type from Batu Pahat, Johor, Malaysia. Left: Fruits are attached to a third-order branch (rachilla) on a second-order branch. Right: Fruit attached to the tip of a rachilla.
Source: Ehara (2006c)

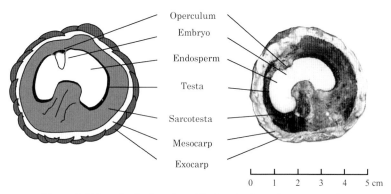

Figure 3-43 Longitudinal section of fruit and schematic representation
A spineless type from Batu Pahat, Johor, Malaysia.
Source: Ehara (2006c)

The endosperm is made of a reserve nutrient of cellulose and becomes extremely hard at maturity. This type of endosperm, which hardens like animal horn as it matures, is called horny endosperm (Rauh 1999). The degree of hardness of the sago palm seed is 400 kgf in a polar (longitudinal) direction (measured with a seed 2.7 cm in horizontal long axis and 2.1 cm in polar diameter). Compared with the degrees of hardness of the walnut or Japanese apricot kernel which are 20–50 kgf (horizontal long axis 3.5 cm, polar diameter 4.3 cm) and 30–80 kgf (horizontal long axis 1.8 cm, polar diameter 2.3 cm) respectively, it is evident that the sago palm seed is extremely hard.

Authors:
3.1: Youji Nitta
3.2: Satoshi Nakamura and Yusuke Goto
3.3: Manabu Watanabe, Satoshi Nakamura and Yusuke Goto
3.4: Hiroshi Ehara

# 4
# *Growth Characteristics*

## 4.1 Life cycle

In the first 4–5 years of the seed-to-seed life cycle of the sago palm, leaves continue to emerge from a slow growing stem and the plant body gradually enlarges. Leaves emerge near the ground surface in a rosette-like fashion. No trunk is formed during this period, which is called the rosette stage (Figure 4-1a). During the following trunk formation stage, the stem elongates and begins to form a trunk (Figure 4-1b). As the trunk grows, it accumulates a large amount of starch inside it. A flower bud eventually forms at the apex, growing into a large inflorescence that produces flowers and fruit. The plant withers and dies after the flowering-fruiting stage. Thus the sago palm is a monocarpic (hapaxanthic) plant which flowers and sets fruit only once in its life cycle. Figure 4-2 shows the life cycle of the sago palm under optimal ecological conditions (about 11 years) (Flach 1997). The plant generates many suckers (see 3.3.2) at the base of the trunk and forms a clump consisting of suckers and formed trunks at different growing stages.

The amount of starch accumulated in the trunk decreases rapidly in the process of inflorescence and fruit development (Jong 1995a). Therefore, sago palms are harvested in the period starting just before the flower bud formation and ending around the flowering stage when the maximum starch yield can be obtained. In the Malaysian state of Sarawak, the sago palm life cycle is divided into 12 growing stages from the rosette stage to the withering stage (Table 4-1); stages 7–9 correspond to the proper harvesting time (Yamamoto 1998a). Osozawa (1990) reports that the life-cycle is divided into 8 growing stages in the Indonesian province of South Sulawesi; here the proper harvesting time is considered to be between stage III (when a white bloom appears at the base of the petiole after trunk formation) to V (the trunk attains the final length and the petiole becomes white) (Yamamoto 1998a) which is slightly earlier than in Sarawak.

The sections below describe sago palm growing patterns for the rosette, trunk formation, and flowering-fruiting stages.

Figure 4-1 (a) Rosette stage; (b) Sago palm initiating trunk formation

Figure 4-2 Sago palm life cycle
Source: Flach (1996)

**Rosette stage**
The rosette stage is the period from transplantation to trunk formation. Sago palms are mostly propagated through sucker transplantation. Optimal densities of established sago palm stands are maintained by sucker removal (sucker control). To develop or expand a sago palm plantation suckers have to be cut and transplanted. The root establishing rate is improved if suckers are raised in nurseries to promote rooting prior to transplanting. The use of seedlings germinated from seeds may lead to divergence of character from the parent tree and their growth is 1–2 years slower than when grown from suckers.

Suckers develop from buds at the base of the leaf sheath. Their formation position has been studied from an external morphological viewpoint (Goto et al. 1998). The sago palm leaf at its trunk attachment forms a cylindrical sheath with its left and right edges fused together. The leaf attachment part of the trunk assumes a node-like appearance. The bud forms in the vicinity of the fused edges of the leaf that cradles it (Figure 4-3). This is on the opposite side of the so-called leaf axil position around the central shaft (Figure 4-4). It is not known whether bud formation takes place at every node as not all buds are visible to the naked eye.

At the rosette stage, the leaf length becomes progressively longer as each leaf spreads. The number of live leaves (the number of green leaves that form the tree crown) and the leaf size also increase to produce a greater leaf area. Trunk formation begins when a certain leaf size is attained and leaf size remains almost constant thereafter until flower bud formation. A study of identical plants at a plantation has found that the number of green leaves at the rosette stage reached 9–12 about 1 year after sucker transplanting and changed little thereafter until trunk formation (Yamamoto et al. 2005b)

**Trunk formation stage**
The size of sago palm leaf and trunk varies greatly depending on the ecological conditions of the growing location and from one plant to another. Generally speaking, though, the leaf length is about 10 m, the number of leaflets per leaf ranges from 140 to 180, and the number of green leaves at the crown ranges from 10 to 20 at the time of trunk formation. The leaf emergence rate after trunk formation is 0.5–1 leaves per month and the number of green leaves tends to increase as the plant develops. The rosette-stage shoots growing at the base of a trunked palm (parent tree) are put under shaded conditions as the parent tree grows larger.

The trunk surface has visible leaf scars. It is possible to estimate the amount of trunk elongation based on the leaf scar intervals, the leaf emergence rate and, to a certain extent, the ecological conditions during this period. In general, the leaf scar intervals are between 10 and 15 cm. Figure 4-5 shows a sago palm 5 years after trunk formation and a pattern diagram demonstrating the relationship between its trunk and leaf attachment. The trunk had 15 green leaves, including an ensiform leaf (ebL1). Each leaf was removed and its length measured. Each internode was also measured for length (leaf scar interval) and thickness (diameter of a virtual circle having the circumferential length of the middle part of the internode). The height of the attachment node of the outermost leaf (the lowest green leaf ebL 15) at the crown measured from the trunk base was about 161 cm and the distance from there to the trunk apex was about 181 cm.

Table 4-1 Sago palm growth stages in Sarawak, Malaysia

| Growth stage | Years after planting | Local name | Years after trunk formation | Growth stage characteristics |
|---|---|---|---|---|
| 1 | 1–5.5 | Sulur | 0 | Rosette stage, no visible trunk. |
| 2 | 5.5 | Angat burid | 0 | Trunk formation stage, transition from rosette stage to growth stage. A short trunk is visible on the ground surface near the base. |
| 3 | 7 | Upong muda | 1.5 | Trunk length 1–2 m approx. |
| 4 | 8 | Upong tua | 2.5 | Trunk length 2–5 m approx. (25 % of final trunk length) |
| 5 | 9 | Bibang | 3.5 | Trunk length 4–7 m approx. (50 % of final trunk length) |
| 6 | 10 | Pelawai | 4.5 | Trunk length 6–8 m approx. (75 % of final trunk length) |
| 7 | 11.5 | Pelawai manit | 6 | Trunk length 7–14 m approx. (100 % of final trunk length), small and upright leaves at the trunk apex. White bloom on petiole. |
| 8 | 12 | Bubul | 6.5 | Flower stalk formation stage. Terminal bud has a torpedo shape due to apical trunk elongation and bracteal leaf formation. |
| 9 | 12.5 | Angau muda | 7 | Flowering stage. First-, second-, and third-order rachillae develop and elongate on floral axes. Flowers before and after blooming are visible. |
| 10 | 13 | Angau muda | 7.5 | Young fruit stage. Fruit diameter 20–30 mm, seed (if formed) is still soft, many of the leaves are functioning soundly. |
| 11 | 14 | Angau tua | 8.5 | Fruit ripening stage. Fruit diameter 30–40 mm, seed (if formed) has dark brown testa and hard endosperm. Many of the leaves are at the withering stage. |
| 12 | 14.5 | Mugun | 9 | Withering stage. Most fruits drop. All leaves wither and die. |

Source: Jong (1995a)

In the most precise sense, the trunk height (or trunk length) extends from the trunk base to the growing point at the trunk apex. However, as the trunk apex of a sago palm covered with leaf sheaths is invisible and therefore unmeasurable (Figure 4-5a), and because only the visible part of the trunk is used as a log, the distance from the trunk base to the outermost leaf attachment node is generally taken as the trunk height (or trunk length). The palm height is basically the distance from the ground line to the uppermost position of the

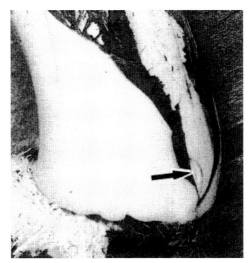

Figure 4-3 Differentiated bud inside the leaf margin
Source: Goto et al. (1998)

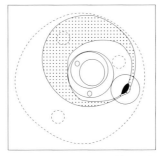

Figure 4-4 Position of bud differentiation
A differentiated bud cradled by the leaf sheath (shaded portion) initiates at the circled position in the diagram.
Source: Goto et al. (1998)

tree's appearance but as it is difficult to measure, a combined value of the trunk length and the leaf length is sometimes used as the tree length for convenience. The trunk grows from 0.6 to 2 m higher (longer) per year.

Starch accumulation begins from the lower part of the trunk during the early stage of trunk elongation and builds up to higher parts. Starch content differentials between different parts of the trunk decrease gradually as the plant develops. At around the proper time of harvesting (just before flower bud formation–flowering stage), the trunk length reaches 7–14 m, the diameter is approximately 40–60 cm, and the starch content reaches its maximum (see 7.1.3).

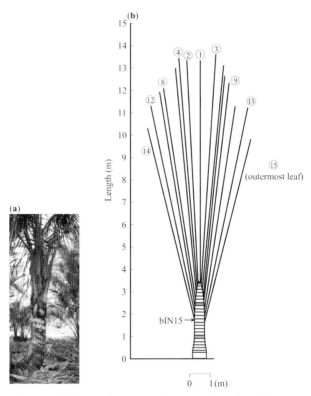

Figure 4-5 (a) Sago palm 5 years after trunk formation; (b) Pattern diagram showing the relationship between the trunk and leaf attachment positions
Source: Nakamura et al. (2004b)
The length of each leaf is represented by the straight line. Leaf positions are alternated from side to side. ① indicates ebL 1, ② ebL 2, ③ ebL 3 and so on.
bIN 15 refers to the internode between leaf attachment nodes bN 15 and bN 16.

Sago palm growth is generally best in mineral soils, followed by shallow peat soil and thick peat soil. It has been noted that few sago palms form trunks in peat soils which lack nutrients (Fong et al. 2005). Growth tends to be poor where groundwater levels are high (Jong 2006). Variations in the growing characteristics and starch productivity of different varieties are being studied (Yamamoto et al. 2000, 2004b, 2005b, 2006a, b, c, 2008b).

**Flowering-fruiting stage**
At least 10 years after seeding or sucker transplanting, a flower bud formation is initiated at the trunk apex. Once the flower bud forms, the length of emergent leaves becomes progressively shorter (Figure 4-6). The length of time up to

Figure 4-6 A large inflorescence emerged from this sago palm
Higher-positioned leaves near the inflorescence are small.

the flower bud formation stage varies depending on soil conditions, varieties, cultivars etc. It takes about 2 years from flower bud formation to flowering and another year to fruiting. The sago palm develops a terminal racemose inflorescence which branches to the third order. Staminate flowers and hermaphrodite flowers are set on third-order branches in pairs. The flowering period lasts about 2 months and it takes about 2 years for the inflorescence to develop to the fruit ripening stage (Kiew 1977). It is 19–23 months from pollination to fruit drop (Jong 1995a; Yamamoto 1998a).

## 4.2 Germination

The sago palm can propagate in two ways: vegetative propagation by adventitious buds arising from lower leaf axils or subterranean stems or suckers (tillers), and sexual propagation by seeds. For cultivation purposes, it is more common to transplant suckers, as seedlings take 1–1.5 years to grow to a size suitable for transplantation (Jong 1995a). The use of seedlings is slowly increasing, however, because it is not easy to secure sufficient numbers of suckers for transplanting in large-scale plantation projects (Jong 1995a; Ehara et al. 1998). However, sago palm seeds (fruit) are known to show low germination (Alang and Krishnability 1986; Flach 1983; Jaman 1985; Johnson and Raymong 11956; van Kraalingen 1984). According to Jong (1995a), the germination rate at a temperature range of 25–30 °C at the end of 6 weeks was around 5% when seeded as the whole fruit, about 10% when the husk (exocarp and mesocarp) was removed, and about 20% when sarcotesta was removed (samples from Sarawak, Malaysia, were used).

The section *Coelococcus* species found in the South Pacific belong to the same genus but they propagate by seeds alone as they do not produce suckers. Their seeds generally show high germination rates. *M. warburgii* in Vanuatu even produce viviparous seeds (Ehara et al. 2003d).

## 4.2.1 Determinants of germination

### 4.2.1.a Impacts of temperature

One of the external factors that are considered to influence germination is bed temperature. In an experiment conducted by Ehara et al. (1998), the fruit of spineless sago palms cultivated in Batu Pahat, Johor, Malaysia, were seeded into rockwool blocks (under dark conditions). The germination rate at the end of a 40-day period at 25 °C was not more than 20%. The rate improved to about 40% in 10–40 days when the temperature was subsequently increased to 30 °C. As the 40-day germination rate at 35 °C was about 20%, the optimum temperature for germination appears to be around 30 °C. Germination is delayed or inhibited at lower and higher temperatures.

### 4.2.1.b Internal factors

The sago palm seed is covered with 4 layers of seed husk tissues (see 3.4.3). The outermost exocarp consists of overlapping scales made of well-developed cuticles and the mesocarp and the sarcotesta are made of parenchyma. The seed coat surface is made of a suberized tissue and the surface of the endosperm

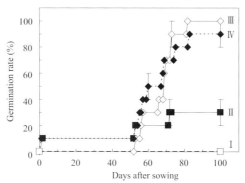

Figure 4-7 Changes in germination rate (soaked in water, 25 °C)
I: seed + exocarp & mesocarp + sarcotesta
II: seed + exocarp & mesocarp
III: seed + sarcotesta
IV: seed only
Source: Ehara et al. (2001)

inside it has a porous structure. The presence of these seed husk tissues is an internal factor affecting germination. The husk and the operculum physically hinder water and oxygen supply. Germination inhibitors (organic acids such as acetic acid and phenolics such as vanillic acid) contained in the husk, especially the exocarp and the mesocarp, have an adverse effect on embryo growth (Ehara et al. 2001; Komada et al. 1998).

In many plant species, treatment with growth regulating substances is known to be effective in improving germination viability or germination rate. Jong (1995a) reports that the treatment with $GA10^{-3}M$ (350 ppm) for 48 hours improved the germination rate in 12 days by 10% to around 40% and that the treatment with $IAA10^{-4}M$ (20 ppm) achieved a germination rate of 32%.

Figure 4-7 shows changes in the germination rate when husk tissues were removed from the fruit and seeds were immersed in beakers of water together with the removed tissues (Ehara et al. 2001).. There was no germination at all when the exocarp, mesocarp and sarcotesta were all present. However, the germination rate was 30% when only the exocarp and mesocarp were present, and it was practically unhindered in the presence of the sarcotesta alone. Thus the removal of seed husk tissues can lead to germination rates over 90%. As there was no significant difference in the germination rate between seeds free of husk tissues and seeds free of the operculum as well as husk tissues, the operculum set on top of the embryo is considered to assist in preserving water in aerially-exposed conditions. Therefore the removal of the exocarp, mesocarp and sarcotesta is considered effective as physical treatment prior to seeding.

## 4.2.2 Germination process

Germination under water immersion conditions at a temperature of 30 °C is usually between 10 and 40 days, but it can be as long as a year in rare cases (Ehara et al. 1998, 2001). Figure 4-8 illustrates the state of germination. Figure 4-9 shows a typical germination process. Although the number of days to germination varies from one plant to another, organs develop in an orderly manner after germination (Ehara et al. 1998).

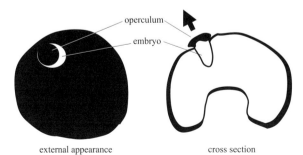

Figure 4-8 State of germination (pattern diagram)
Source: Ehara (2006c)

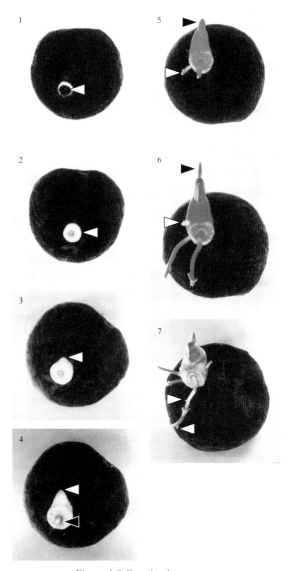

Figure 4-9 Germination process
1. △: operculum
2. △: coleorhiza-like organ
3. △: epiblast
4. △: epiblast, ▲: primary root
5. ▲: coleoptile, △: first adventitious root
6. ▲: first leaf, △: second adventitious root
7. △: branched root
Source: Ehara et al. (1998)

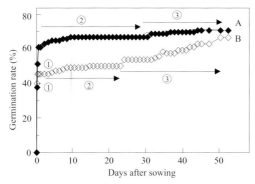

Figure 4-10 Changes in water uptake by germinated
seed (soaked in water, 25 °C)
A: Seed without exocarp, mesocarp and sarcotesta.
B: Seed without exocarp, mesocarp, sarcotesta and operculum.
①: first phase
②: second phase
③: third phase
Source: Ehara et al. (2001)

1. Germination is observable when the embryo elongates and pushes the operculum up.
2. The coleorhiza-like organ emerges about 6 days after germination.
3. The epiblast begins to elongate 2 or 3 days later.
4. The primary root emerges during the elongation of the epiblast (related to the timing of operculum displacement).
5. 2 or 3 days after the emergence of the primary root, the first adventitious root arises at a position above the base of the coleorhiza-like organ (at the base of the epiblast) and the coleoptile emerges along with it.
6. 2 or 3 days later, the second adventitious root appears and the first leaf emerges on the same day.
7. Around the same time as the elongation of the second adventitious root and the emergence of the third adventitious root, the branched root arises from the primary root.

Figure 4-10 indicates changes in water uptake by germinated seeds. The water uptake process by a seed is divided into 3 phases: first, the imbibition phase involves physical water absorption; second, the phase of constantly increasing water uptake is for metabolic preparation for germination; and third, the hydraulic growth phase involves the growth of the plumule and the radicle after germination. The sago palm seed has a hardness of 260 kilograms-force (of the seed coat; endosperm hardness is 560 kgf) which is markedly harder than the seed hardness of walnut and Japanese apricot (approx. 20–80 kgf). It is conceivable that the extreme hardness of the seed (endosperm) inhibits embryo expansion after the imbibition phase and is relevant to the long duration of the germination preparation phase.

## 4.3 Leaf formation and development

### 4.3.1 Leaf development

The sago palm leaf is pinnately compound, consisting of the rachis, petiole, and many leaflets attached to the rachis. The petiole base forms a leaf sheath that wraps around the trunk once. Leaf size changes according to the plant's growth. It produces progressively larger leaves during the rosette stage from seedling or sucker emergence to trunk formation; the leaves that develop during the trunk formation stage are the largest. It continues to produce leaves of near-maximum size for some time. The spatial position of leaves in the stand shifts upward as the trunk grows taller. The length of newly developed leaves shorten closer to the flower bud formation stage.

#### 4.3.1.a Before trunk formation

Flach (1977) reports that seedlings raised in good light conditions produce 2 leaves per month. As the plant developed 80–90 leaves by the trunk formation stage, it was estimated based on this leaf emergence rate that the duration from seedling to trunk formation would be around 3 years and 4–9 months. However, a study in South Sulawesi by Osozawa (1990) found that the leaf emergence rate in the early and later periods of the rosette stage was 0.88 and 0.52 leaves per month respectively. As light can be limited in sago palm plantations and other environmental conditions vary greatly, the leaf emergence rate of 2 per month reported by Flach (1977) appears likely to occur only under particularly good conditions. According to Nakamura et al. (2009b), the growth rate of ensiform leaves in a plant in its third year after sucker transplanting was 7.0–8.0 cm per day.

A study on the number of live leaves in the crown during the rosette stage, i.e., the number of green leaves (including ensiform leaf) that are considered to contribute to the photosynthesis process, followed the growth of identical trees at a sago palm plantation in Tebing Tinggi Island, Riau, Indonesia (Yamamoto et al. 2005a). It reports that the number of live leaves reached 9–12 leaves about one year after sucker transplanting and changed little thereafter.

During the rosette stage, the sago palm develops progressively larger leaves as it grows (Flach 1977). Figure 4-11 shows examples of leaves at the rosette stage. These leaves were among the 10 green leaves, including an unopened ensiform leaf, of a rosette-stage tree about 9 m in height growing at a sago plantation in Mukah, Sarawak, Malaysia. In the illustration, leaflets are represented by line segments, which correctly reflect the attachment position and angle on the rachis and the length of each leaflet (Watanabe et al. 2004). The position of the emerging youngest ensiform leaf is labeled as ebL 1 and the leaves thereafter are labeled as ebL 2, ebL 3 and so on. The leaf blade length and the petiole length were 430 cm and 257 cm for ebL 6 and 545 cm and 330 cm for ebL 3. In other words, the leaf at a higher position was larger in blade length and petiole length. The number of leaflets and the leaf area were 106 leaves (55 on the left and 51 on the right) and 4.81 m$^2$ for ebL 6 and 113 leaves (58 on the left and 55 on the right) and 6.39 m$^2$ for ebL 3. Again, the leaf at a higher position had more leaflets and leaf area. The largest leaflet was 121 cm long and 7.4 cm wide on ebL 6 and 147 cm long and 8.5 cm wide on ebL 3. The constituent leaflets were also larger on the leaf at a higher position, i.e. one that developed later.

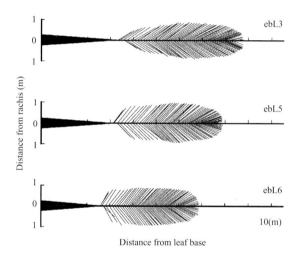

Figure 4-11 Sago palm leaf appearance at the rosette stage
Source: Prepared by author based on Watanabe et al. (2004).

### 4.3.1.b Trunk formation stage

Leaves develop to approximately the same size during the trunk formation stage. As an example, Figure 4-12 shows the leaf blade length and the petiole length (including the leaf sheath) of a tree about 2 years after trunk formation with 11 green leaves growing in a sago palm plantation in Mukah (Nakamura et al. 2004a). The diagram includes leaves from ebL 2 to ebL 11; ebL 1 is not included because it was an unopened ensiform leaf. The leaf blade length ranged from 6.0 m (ebL 6) to 7.2 m (ebL 11) and the petiole length ranged from 1.8 m (ebL 3) to 3.1 m (ebL 10). Figure 4-13 is a pattern diagram of the open leaves from ebL 2 to ebL 10 of the same tree (Nakamura et al. 2004a).

All of the characteristics except the width of the petiole base reflect the actual measurements as in Figure 4-11. The leaf blade of ebL 2 was just starting to open from the apical leaflets and the leaflets of ebL 3 were almost fully open. The leaflet attachment angle on the rachis was larger on the proximal side and smaller near the leaf tip except ebL 2.

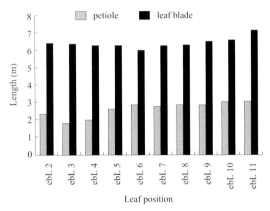

Figure 4-12 Leaf blade length and petiole length at the trunk formation stage
Source: Nakamura et al. (2004a)

The most common leaf length at the trunk formation stage was around 10 m according to preceding sago palm studies. While some found little variation between varieties (Omori et al. 2000c), others reported intervarietal differences (Yamamoto et al. 2002a, 2007a). It was reported that the leaf length was shorter in sago palms growing near the coast than those growing inland within the same area (Yamamoto et al. 2004a). From the flower bud formation stage, higher-positioned leaves tend to be shorter and have smaller numbers of green leaves and leaflets per leaf (Yamamoto et al. 2002a, 2006a).

Although it is possible to hypothesize that the leaf emergence rate at the trunk formation stage would vary greatly with individual plants and environmental conditions just as at the rosette stage prior to it, the variation range is yet to be ascertained. For example, it is 1 leaf per month according to Flach (1977) while a survey in Sulawesi Province, Indonesia, by Yamamoto et al. (2000) found 8.4–10.3 leaves per year and another survey in Papua Province, Indonesia, found 5.8–7.0 leaves per year (Yamamoto et al. 2006b), which equate to 0.48–0.86 leaves per month. Leaf emergence rates of more than 1 leaf per month have also been reported. For example, a study in Sarawak, Malaysia, by Yamaguchi et al. (1997) discovered 12.0 leaves per year on trees growing in the presence of thick peat soil (increased to 17.0–19.2 leaves per year 3–4 years after trunk formation) and 13.4–15.5 leaves per year on trees growing in the presence of thin peat soil. Thus it is unclear how the leaf emergence rate changes from the rosette stage to the trunk formation stage.

## 4.3.2 Leaf formation

The pre-emergent sago palm leaf has a firmly folded structure and its formation process remains largely unexplained. Of the family Arecaceae plants with pinnate compound leaves, Dengler et al. (1982) reported on the morphogenetic process in the early stage of leaflet formation in *Chrysalidocarpus lutescens*. While a similar morphogenetic process is expected to take place in sago palm leaves, which are also pinnately compound, it warrants further studies as *Chrysalidocarpus lutescens* is morphologically different from the sago palm in some aspects, such as its clearly distinguishable petiole and leaf sheath.

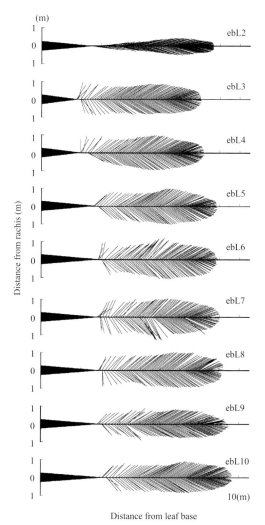

Figure 4-13 Sago palm leaf appearance at the trunk formation stage
Source: Prepared by author based on Nakamura et al. (2004a)

Here, the positions of externally invisible pre-emergent leaves are labeled in relation to the youngest emergent leaf, the ensiform leaf ebL 1. The leaf position immediately above (inside) ebL 1 is uL 1, and the leaf positions further up (inside) are labeled (acropetally) uL 2, uL 3 and so on.

The pre-emergent leaves of very young sago palms were observed with suckers used for transplantation. Figure 4-14a shows a sucker cultivated for propagation in a sago palm plantation in Mukah. The sucker was fixed in FAA and the young leaves inside it were examined. The sucker's leaves had been trimmed off in preparation for transplantation but counting from the outermost (lowest) leaf, the sixth leaf was the youngest externally visible leaf (ebL1) which was either ensiform or had some open leaflets. Figure 4-14b shows the eighth leaf in position uL 2. The leaf length was about 7 mm, of which the petiole length was about 4.5 mm. The tip of the ninth leaf (uL 3) was visible in the petiole part under the leaf blade; its leaf length was about 3.5 mm.

Figure 4-14 (a) A prepared sucker with leaves trimmed off; (b) Young leaves inside
Source: Nakamura et al. (2008)

The leaf formation process in a trunked plant was observed using a tree about 10 years after trunk formation, having 17 green leaves (Figure 4-15). It was found that a pre-emergent leaf elongates its leaf blade during the period from the emergence of the third leaf below it to its emergence and the leaf emerges when its petiole elongates rapidly after that period. More specifically, the duration of rapid leaf blade elongation was from uL 4 (8.0 cm) to ebL 1 (720 cm) whereas the largest change of the petiole length from uL 1 (9.0 cm) to ebL 1 (131 cm) suggested a rapid elongation when it emerged as an ensiform leaf. The ratio of the leaf blade length to the leaf length was 64–76% from uL 8 to uL 4 while it was extremely large at uL 3, uL 2 and uL 1 at 97–99%. It was 85% in the ensiform leaf at ebL 1 and 83% in the open leaf at ebL 2.

Figure 4-16 shows the lengths of mostly pre-emergent leaves of sago palms grown in Mukah 2 years, 4 years and 8 years after trunk formation having 11, 15 and 17 green leaves respectively. The vertical axis has a logarithmic scale. The leaf length by leaf position gets shorter in a linear fashion acropetally from around uL 4 to uL 12. The relationship between the leaf position and the leaf length suggests that a differentiated leaf elongates exponentially at a constant rate until the third or fourth leaf below it emerges, then elongates rapidly to emerge from the trunk.

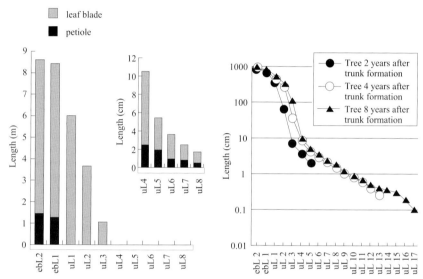

Figure 4-15 Leaf blade length and petiole length in a tree 8 years after trunk formation

Figure 4-16 Leaf length in trees approx. 2, 4 and 8 years after trunk formation

When paraffin sections were prepared and observed from the above trees 4 and 8 years after trunk formation, the leaf positions of the youngest leaf primodia were uL 18 and uL 19. These numbers were 3 and 2 leaves more than the 15 and 17 emerged and opened green leaves of the respective trees. Flach and Schuiling (1989) reported that the number of differentiated pre-emergent leaves was 24, which was the same as the number of green leaves attached to the trunk, in the early stage of trunk formation. Further studies are anticipated on the details of leaf formation by tree age, including the relationship between the leaf emergence rate and the leaf differentiation rate and whether the number of visible leaves can be used as an indicator of the number of differentiated leaves.

## 4.4 Trunk formation and elongation/thickening

The sago palm stem continues to produce leaves for several years after seed germination or sucker transplantation without forming a substantial trunk. In other words, it grows in a rosette-like fashion. After a certain period, however, the stem begins to elongate upward and forms a trunk. This is called trunk formation. After trunk formation, the prominent trunk continues to elongate until the flower bud formation stage.

It is useful for the analysis of trunk elongation and thickening to consider the trunk as a series of internodes stacked on top of one another (see 3.3.1). Node-like leaf base scars are left on the sago palm trunk once leaves have withered and fallen away. These are regarded as apparent nodes and the sections between them as internodes. Analyzing the growth of these internodes makes it easier to understand the trunk formation process.

As the node is where a leaf is attached to the trunk, it can be considered that as soon as a new leaf emerges, a growing internode becomes almost an extension of the one below itself. Based on the continuous growth of internodes, each internode takes the place of the one immediately below itself every time a new leaf emerges, resulting in trunk growth. Using this relation between the leaf emergence and the internode growth, it is possible to infer a growth pattern of internode elongation and thickening over time by investigating plant samples at a given point of time.

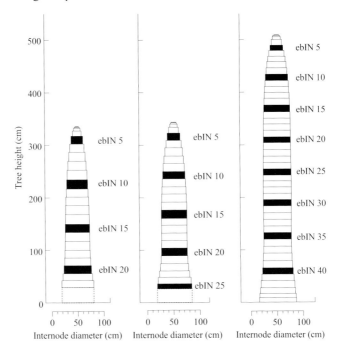

Figure 4-17 Pattern diagram of trunk
Tree A, Tree B and Tree C from left to right.
Source: Nakamura et al. (2004b)

In this section, the node to which the ensiform leaf (ebL1) is attached immediately after emergence and before opening is called ebN 1 and the subsequent nodes below it are successively called ebN 2, ebN 3 and so on. The internode immediately above ebN 1 is called ebIN 1 and the internodes below it are called ebIN 2, ebIN 3 and so on (see 3.3.1). The nodes that are younger than ebN 1 are acropetally labeled uN 1 immediately above ebN 1, then uN 2, uN 3 and so on. The internodes are also acropetally labeled uIN 1 immediately above ebIN 1, then uIN 2, uIN 3 and so on. The leaf position immediately above ebL 1 is called uL 1, followed by uL 2, uL 3 and so on.

This overview of the relationship between internode elongation and thickening is based on a study of trunked sago palms grown in Mukah, Sarawak, Malaysia (Nakamura et al 2004b). The trunk length of the trees included in the study was about 3.5 m in the trees 4 years after trunk formation (trees A and B) and about 5 m in the tree 8 years after trunk formation (tree C) (Figure 4-17).

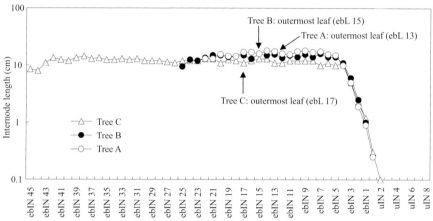

Figure 4-18 Length of each internode (exponential scale on vertical axis)
Source: Nakamura et al. (2004b)

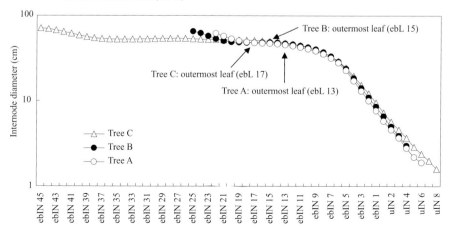

Figure 4-19 Diameter of each internode (exponential scale on vertical axis)
Source: Nakamura et al. (2004b)

Figure 4-18 shows the length of each internode in each tree. The vertical axis has an exponential scale. As similar patterns were exhibited by all 3 trees, it is likely that the marked internode elongation began at uIN 1 and continued to ebIN 5. From uIN 1 to ebIN 4 in particular, internodes appear to have elongated exponentially at a constant rate. Internodes below ebIN 5 had an almost uniform length, suggesting that elongation had finished in these internode positions. In other words, the range from uIN 1 to ebIN 4 appears to be the trunk elongation zone. Differences in the internode length below ebIN 5

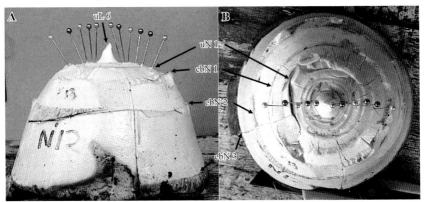

Figure 4-20 Shape of trunk apex
(A) side view, (B) overhead view.
Source: Nakamura et al. (2004b)

Figure 4-21 Node and internode positions on sucker longitudinal section
Source: Nakamura et al. (2008)

were confined to a few cm except for a few internodes at the base, which were slightly shorter. The length of post-elongation internode, excluding the trunk base internodes, was about 16 cm in tree A, about 15 cm in tree B, and about 12 cm in tree C.

However, the internode diameter appears to have enlarged in a wide range from a newly differentiated internode to around ebIN 15 (Figure 4-19). This range includes the trunk elongation zone from uIN 1 to ebIN 4. These trees exhibited exponential thickening, especially in the zone from the newly differentiated internode to around ebIN 6. As the internodes above the node of ensiform leaf attachment undergo thickening with little elongation, the apical part of the trunk forms a dome shape with the flattened tip (Figure 4-20).

As the tree grows, its leaves wither and fall away successively from the low-level node to the high-level node. The outermost (lowest) leaf position was ebIN 13 in tree A, ebIN 15 in tree B, and ebIN 17 in tree C (Figure 4-19). Very little internode thickening occurred in the internodes 2 or 3 levels below the node of the outermost leaf attachment. Accordingly, thickening appears to have ended at around the internodes above and below ebIN 15, which was the node of the outermost leaf attachment.

As internode elongation ends at ebIN 5 and thickening ends at ebIN 15, it is estimated that a particular internode continues to thicken while 10 more leaves develop after the end of internode elongation regardless of the number of years after trunk formation.

Nakamura et al. (2008) studied stem elongation and thickening before trunk formation using suckers in the earliest stage of growth being used as saplings for cultivation. Like the trunk, the longitudinal section of the sucker's stem (Figure 4-21) had no visible morphological features to mark nodes or internodes. When notional nodes are marked based on leaf attachment positions

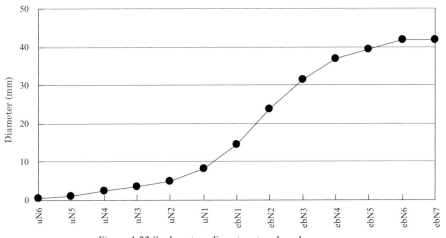

Figure 4-22 Sucker stem diameter at each node
Source: Nakamura et al. (2008)

(Figure 4-21), the nodal planes from ebN 7 to ebN 4 were almost parallel whereas those at higher positions were thicker toward the bottom and curved upward. The length of internodes at ebIN 4 and below (measured at a right angle to the nodal plane) was approximately 5 mm. The internode length was mostly unmeasurable for leaf positions younger than the emerging leaf (ebL 1).

The leaf in Figure 4-21 (uL 1) was found to contain 5 young leaves (up to uL 6) inside it. The diameter of each node up to uN 6 (Figure 4-22) suggests a pattern according to which the stem in the vicinity of the attachment of an emerging and spreading leaf thickens rapidly. Although the number of indernodes involved is smaller, the vigorous growth of stem mass near the node of the opened leaf attachment is analogous to trunk growth after trunk formation.

## 4.5 Root formation and elongation

The sago palm is a monocot with a fibrous root system consisting of many adventitious roots arising from suckers and stems (trunks) and lateral roots branched from them (Nitta and Matsuda 2005). As the sago palm is generally propagated by suckering, its root system is formed by adventitious roots without normal seminal roots.

### *4.5.1 Root development and elongation from suckers*
To promote root development after transplantation, suckers are often cultivated on rafts over waterways prior to transplanting. Root development and leaf emergence begin about 1 month after the commencement of cultivation. The starch content in the pith begins to diminish at the same time (see 6.3.1 and Figure 6-14). In sago palm plantations, suckers are usually cultivated for a period of 2 to 3 months before transplanting. During this time the leaves and roots begin to grow rapidly and the accumulated starch in the pith is consumed (Omori et al. 2002). After the third month, the leaves and roots continue to grow without changing the pith starch content. It is therefore assumed that the plant switches to autrophic photosynthesis in leaves. After 3 months of cultivation, the mean number of developed roots in suckers was 18. Those suckers that had produced only 10 or fewer roots had low survival rates 6 months after transplantation. The number of roots is closely related to the weight and leaf number of suckers: larger suckers with more leaves develop more roots and hence exhibit higher root establishing rates. Good root taking is important as it ensures fast early-stage growth after transplanting and shorter periods leading up to trunk formation (Yamamoto et al. 2005a).

### *4.5.2 Root system formation*
The root system of the sago palm is responsible for supporting the enormous plant body and supplying large amounts of water and nutrients. It is therefore

important to understand the overall picture of the root system in establishing cultivation management. However, the sago palm's root system is very difficult to study as it has a huge above-ground body and in many cases is grown in swamps with high water table levels (Nitta and Matsuda 2005).

According to Kasuya (1996) and Omori et al. (2002), sago palm roots reach a depth of over 1 m in the first year after transplantation and continue to elongate at a rate of 1–2 m per year thereafter. Elongation rates of 1 m in 5 months and 1.8 m in 12 months have been reported (Jourdan and Rey 1997). A similar early-stage growth rate is found in oil palms. Miyazaki et al. (2003, 2008) measured the sago palm root system at different tree ages from 1 to 11 years after sucker development by digging trenches for 1 and 2 m horizontally and 0 to 90 cm vertically from the trunk base. They found that the dry weight of the roots increased with tree age and that the increase was especially marked after trunk formation (Figure 4-23). The dry weight was highest directly below the tree trunk (0 m from the trunk) and in a soil horizon 0–30 cm below ground at all tree ages (Figure 4-24). The weight ratio of the roots directly below the trunk increased with tree age. In the 11-year-old tree, 70–80% of the roots were concentrated directly below the trunk (8 years after trunk formation) (Figure 4-24a). The root weight ratio tended to increase in a shallow soil horizon 0–30 cm below ground with tree age prior to trunk formation whereas it tended to increase in a deeper soil horizon 30–90 cm below ground after trunk formation. These findings suggest that the roots elongate with tree age both horizontally and vertically before trunk formation but elongation in a vertical direction becomes more prominent after trunk formation. Kasuya (1996) studied the root mass of an approximately 8-year-old sago palm to a depth of 40 cm in soil and found that the roots up to 5 mm in diameter were predominant in a shallow soil horizon 0–20 cm below ground while thicker roots over 5 mm in diameter were predominant in a deeper soil horizon 20–40 cm below ground (Figure 4-25). Thick roots have a well-developed aeration function through lysigenous aeranchyma formation (Nitta et al. 2002a; Nitta and Matsuda 2005), and are thus capable of penetrating the water-logged deep soil layer. The thick roots that develop after trunk formation are adventitious roots emerging from the surface of the trunk which penetrate deep into the soil directly below the trunk (Miyazaki et al. 2008). Both Kasuya (1996) and Miyazaki et al. (2003) found a vigorous root system growing in a shallow soil horizon directly below the trunk. This information is useful for soil management, including fertilizer application and irrigation. In oil palms, in contrast, roots elongate mostly in a vertical direction in the early growth stage (0–1 year after germination) and horizontal elongation gradually becomes predominant after that (Jourdan and Rey 1997). The root system of one 20-year-old palm was found to have spread 6 m vertically and 25 m horizontally. Further studies are needed on matters such as interspecific variations to determine the differences in the root elongation direction and other characteristics between oil palms and sago palms.

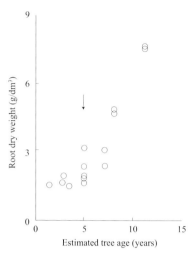

Figure 4-23 Changes in root dry weight by tree age (Kendari, Indonesia)
The arrow indicates the trunk formation period.
Source: Miyazaki et al. (2008)

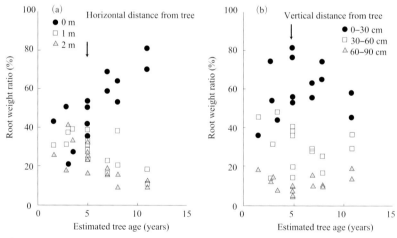

Figure 4-24 Changes in root weight ratio between horizontal (a) and vertical (b) directions by age (Kendari, Indonesia)
Root weight ratio = Root dry weight by direction / Total root dry weight
The arrow indicates the trunk formation period. Horizontal distance values in (a) are total values at 0–90 cm in (b). Vertical distance values in (b) are total values of values at 3 positions in (a).
Source: Miyazaki et al. (2008)

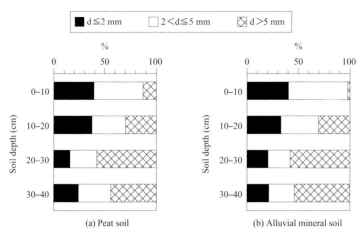

Figure 4-25 Root ratio by diameter of sago palm in peat soil and alluvial mineral soil
d: diameter
Source: Kasuya (1996)

### 4.5.3 Impact of soil environment

Root elongation in the sago palm is greatly influenced by the soil environment. Omori et al. (2002) compared root growth in sago palms after culturing them for 1 year in the soil with a low water table 40–50 cm below ground and a high water table 0–30 cm below ground. They reported smaller root dry weights and shorter root lengths at higher groundwater levels although no clear differences were found in above-ground growth. They also observed that more roots developed near the ground surface and elongated horizontally at higher groundwater levels. Kasuya (1996) reported lower root mass and slower growth in peat soils consisting of organic matters and less nutritive salts compared with alluvial mineral soils. While the root mass decreased at greater distances from the tree trunk center in alluvial mineral soils, it was reported that this tendency was not observed in peat soils. More samples must be studied to verify these findings, as others have reported that denser and larger root systems are formed in peat soils than in mineral soils (Tie et al. 1987).

## 4.6 Reproductive growth

### 4.6.1 Transition from vegetative growth to reproductive growth

When the trunk grows and matures to a sufficient degree, inflorescence begins to initiate from the growing point at the trunk apex and the reproductive growth phase commences. It is unclear if the transition to reproductive growth is triggered by the trunk length, the number of developed leaves, or the trunk growth duration. Observations of sago palm flowering over the last two decades suggest that flower bud initiation is not necessarily caused by a single factor even though each factor seems important. It appears that flower bud initiation begins when trunk growth reaches a certain level of physiological maturity. Tree growth is expected to vary significantly from one individual to another depending on climatic and nutrition conditions.

According to a study on the number of leaves up to the transition to reproductive growth in many folk varieties (simply called 'varieties' hereafter), the number of leaf scars left on the trunk surface of flowering trees varied greatly from 60 to 120. This study was conducted near Lake Sentani in Papua, Indonesia, (Yamamoto et al. 2004b) and found that Rondo had the lowest number of leaf scars and Para had the highest.

When the distance between the trunk base and the sheath base of the lowest green leaf is measured and treated as 'trunk length', the trunk length is strongly influenced by environmental factors such as shading, soil nutrients and water stress. However, the varietal characteristics appear to a certain extent, even if the trees are grown under similar conditions. In the cases of the flowering plants near Lake Sentani, the trunk length was the shortest (about 5–6 m) in Rondo and the longest (16 m) in Yepba. Trees with trunks around 14 m were also observed among a few other species (Yamamoto et al. 2005b). At a sago palm plantation where uniform species were densely planted in peat soils in Tebing Tinggi Island, Riau, Indonesia, the earliest-flowering spiny sago palm group had trunk lengths ranging from about 8 to 10 m. However, there was significant variation as a small number of tree trunks were only 4–5 m long.

## 4.6.2 Flower bud development at the stem apex

The early development of the floral axis in the trunk apical meristem is concealed by unopened leaves and unobservable from outside. The trees that are considered by farmers to be nearing harvest maturity generally have 6–12 unopened leaves below the flower bud. As the inflorescence will not emerge until most of them have opened, it is typically 6–12 months from this stage to inflorescence emergence, assuming each leaf takes 1 month to open.

Small leaves gradually develop over a period of more than 6 months around the time of transition from vegetative growth to reproductive growth. Eventually, small bracteal leaves without distinct leaflets and an inflorescence (Figure 4-26) emerge from the apex. The trunk apex assumes a thick cylindrical appearance during this period.

Immediately above the small bracteal leaf attachments, the first-order floral axes (ax1) of the inflorescence develop rapidly, then the second-order floral axes (ax2) branch out from the first-order floral axes, followed by the third-order floral axes (ax3) soon afterward (Figure 4-27). The third-order floral axes have a finger-like shape on which flower buds form and are sometimes called rachillae (Thomlinson 1971) or spikes (Beccari 1918). Flower buds usually form on third-order floral axes in pairs. Each pair of flower buds are

# Growth Characteristics

Figure 4-26 Inflorescence at sago palm trunk apex

covered by a cup-shaped scaly leaf called bracteole (Beccari 1918) (Figure 4-28). Tomlinson (1971) separated this into inner and outer bracteoles. The scaly globose fruit sets on the bracteole.

The whole inflorescence, including the first-, second- and third-order floral axes and flower buds, develop within 6–8 month of the commencement of inflorescence formation.

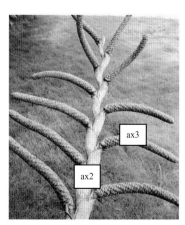

Figure 4-27 Second-order floral axis (ax2) and third-order floral axis (ax3)

Figure 4-28 Cup-shaped bracteoles covering flower buds on third-order floral axis (ax3)

## 4.6.3 Flower bud numbers

A study of 5 spineless sago palms in Sarawak (Jong 1995a) found 15–19 first-order floral axes, 173–344 second-order floral axes and 1,313–3,427 third-order floral axes on a single inflorescence. In the early growth stage, flower buds form in pairs on third-order floral axes. One of the buds is a male (staminate) flower and the other is a perfect (hermaphrodite) flower. In the subsequent growth stage up to flowering, stunting of flower buds occurs in a regular fashion. In other words, one of the paired flower buds in the bracteole stops growing and drops off. It appears that this event happens so that sufficient space is created for the growth of the other flower. Stunting can occur to either flower of the pair: the staminate flower or the hermaphrodite flower or the proximal flower or the distal flower within the bracteole. It is very rare to find a pair of flower buds in the bracteole at the end of flower bud formation. In most cases, the bracteole has only one flower or no flower. More hermaphrodite flowers survive than staminate flowers and the ratio of hermaphrodite flowers among blossoming flowers is high during the flowering period.

Tomlinson (1971) estimated that the number of flowers of *M. vitiense* was 917,280 of which one half were hermaphrodite flowers and that the number would double in larger trees. Jong (1995a) estimated that 121–273 flower buds formed on each third-order floral axis and the number of flower buds per tree just before flowering would be 276,000–864,000 (Table 4-2). This estimate is supported by Utami (1986) and Kiew (1977).

Table 4-2 Total number of ax1, ax2 & ax3 and estimated number of flower buds

| Tree | ax1 | ax2 | ax3 | Flower buds / ax3 | Flower buds / tree |
|---|---|---|---|---|---|
| 1 | 18 | 278 | 3,165 | 273 ± 86 | 864,000 |
| 2 | 15 | 173 | 1,313 | 210 ± 65 | 276,000 |
| 3 | 16 | 177 | 1,443 | 216 ± 80 | 312,000 |
| 4 | 17 | 255 | 2,848 | 242 ± 73 | 689,000 |
| 5 | 19 | 334 | 3,427 | 121 ± 67 | 418,000 |

Spineless sago palms shortly before flowering in Sarawak
Source: Jong (1995)

## 4.6.4 Types of flowering

Jong (1995a) conducted a study of flowering using 9 sago palm trees in flower in Sarawak, Malaysia. He observed two types of flowering. One type involved the coexistence of staminate flowers and hermaphrodite flowers on a single tree, which confirmed that the sago palm was a monoecious plant that has male and female flowers on a single inflorescence. The normal hermaphrodite flower has long filaments with large bright orange or yellow anthers.

The other type also involved the coexistence of staminate and hermaphrodite flowers but normal and abnormal flowers were observed among the latter. The abnormal hermaphrodite flowers had short and underdeveloped stamens, i.e., pale anthers on short filaments. These abnormal hermaphrodite flowers often suffered from growth insufficiency and incomplete flowering.

### 4.6.5 Flowering

Staminate flowers bloom 3–4 weeks earlier than hermaphrodite flowers. In general, the blossoming of both staminate and hermaphrodite flowers progress from the proximal first-order floral axes toward the apical first-order floral axes. The time difference between the start of flowering on the first first-order floral axis and the start of flowering on the last first-order floral axis is 3 weeks for staminate flowers and 2 weeks or so for hermaphrodite flowers. On many first-order floral axes, the flowering density starts low, increases to 50% in around 2 weeks, and decreases after that.

### 4.6.6 Flowering process and pollination mechanism

In a tree with a mixture of normal staminate flowers and normal and abnormal hermaphrodite flowers, the timing of flowering is anomalistic. A majority of abnormal hermaphrodite flowers open before or concurrently with the opening of staminate flowers just as normal hermaphrodite flowers do. Staminate flowers and hermaphrodite flowers may open concurrently in some trees. They are therefore not strictly protandrous flowers in such cases.

#### 4.6.6.a Staminate flowers

The flowering of staminate flowers starts at 10:30 and peaks between 11:00 and 12:00. On a third-order floral axis with staminate flowers in full bloom, it looks as if only 5–10% of all flower buds are open sporadically. The blooming staminate flowers are seen to start wilting by 15:00 and a majority of anthers wither on wilted filaments in open staminate flowers by 17:00.

Figure 4-29 Staminate flowers in full bloom

It takes 20–30 minutes for a staminate flower to open completely. Sepals split open along fusion lines. Anthers emerge from the open flower, extend upward completely and separate from the center of the flower. These anthers are pale purple in color and gradually begin to dehisce before filaments extend completely and project over sepals (Figure 4-29). The emerged anthers gradually change their color and turn orange–yellow in an hour.

Nectar begins to secrete within an hour of opening and increases gradually in the next 30–60 minutes (12:00–12:30) to fill the void at the center of the open flower (Figure 4-30). Excess nectar overflows when the flowers are completely open but most of it is consumed by insects.

#### 4.6.6.b Hermaphrodite flowers

Hermaphrodite flowers open almost at the same time as staminate flowers do. They begin to open just past 10:00. Stamens and pistils emerge completely from open flowers in 30 minutes (Figure 4-31). Anthers are slightly paler purple than in staminate flowers but they dehisce in a similar manner. A majority of active flowers open between 11:00 and 12:00. No regularity is found in the order of blooming on second- and third-order floral axes. Sometimes more than 50% of flower buds open on third-order floral axes in the same morning but the opening of 20–30% of the flowers is more common. Nectar secretion starts at 11:30 and gradually increases to fill the center of the flower. Nectar overflows out of flowers in 30 minutes.

The secreted nectar is all consumed by the gathering insects within 2 or 3 hours. Nectar secretion continues. After a swarm of insects departed at 14:15, nectar was seen spilling over again at around 15:30. Anthers begin to wilt and curl up by around 14:30 and a majority of them wither by 16:30.

### *4.6.7 Pollination and pollinators*

#### 4.6.7.a Pollen properties

When the anther is removed from an open staminate flower under a dissecting microscope, it dehisces within a few seconds and presents moist and glossy pollen. The pollen grains are purplish orange oval spheres which stick to the edges of the pollen sac in a mass. Moistness and viscosity are the properties of pollen that are useful for pollination by insects.

A fresh pollen measures about 45 μm in length and about 30 μm in width. The anthers and pollen of the hermaphrodite flower immediately after opening are almost identical with those of the staminate flower except that they have a lighter color.

#### 4.6.7.b Pollinating insects

The main flower visiting insects include stingless bees such as *Trigona itama* and *Trigona apicalis*, honey bee *Apis dorsata* and a type of wasp *Vespa tropica*.

Figure 4-30 Nectar production by blooming flowers

Figure 4-31 Stamens and pistils of hermaphrodite flowers in full bloom

*Trigona* bees visit the flowers between 7:00 and 18:00. The visitor number is estimated to increase from several hundreds in the early morning to 3,000–4,000 at the peak of blooming of hermaphrodite flowers around noon. The visitor number stays the same until it begins to decrease at around 16:00. These bees mostly appear to collect pollen into pollen baskets on their hind legs.

*Apis dorsata* visit flowering trees in a swarm of 2,000–3,000 bees. The observation of a tree over 3 days found a distinctive visiting time of honey bees, starting around 12:00 when nectar was secreting in abundance and ended around 14:30. The bees move from one flower to the next while spending only a couple of seconds on each flower to suck nectar. They do not appear to collect pollen.

*Vespa tropica* visit in a swarm of less than 100 bees which were seen sucking nectar as well as eating pollen. The visitor numbers appears to correlate with the flowering density and the peak flowering period, but they arrive later and leave earlier than *Trigona*.

When *Trigona* and *Apis* were observed under a dissecting microscope, pollen grains were found on the head, legs and wings of all of the captured bees. This demonstrates that these insects transport pollen by successively visiting different flowers that are open on the same or different trees. Only *Trigona* were found to carry pollen in pollen baskets.

### 4.6.8 Pollination and fruiting

The study of 9 sago palm trees produced some findings about pollination and the subsequent development of fruit in addition to the aforementioned observations.

First, 4 trees were self-pollinated using the pollen from staminate and hermaphrodite flowers. The entire second-order floral axes were covered after pollination. All 4 trees set normal fruit after self-pollination (Figure 4-32) but they were seedless (Figure 4-33). These 4 trees were cross-pollinated using the pollen gathered from other flowering trees. This experiment also produced seedless fruit.

The remaining 5 trees were examined to see if their ripe fruits had seeds. The number of fruit set on 4 of the trees ranged from 2,174 to 5,875 all of which were seedless. Although many insects visited their flowers during the flowering period, none of their fruit set a seed. Only one of the 5 trees produced 343 fruits that had a seed with the horseshoe-shaped endosperm (Figure 4-34) out of 6,675 fruits. It amounted to 5.1% of the total number of fruits.

One of the trees in the pollination experiment had 3 second-order floral axes covered in bags from flower bud formation to fruit ripening. This tree was visited by thousands of *Trigona* and *Apis* during the flowering season. The covered second-order floral axes set 64 mature fruits, none of which had a seed. Another set of 3 second-order floral axes were similarly covered in bags, which were opened for 3 days during the flowering season to allow insects to visit before being closed again. They set 75 fruits of which 4 had seeds. This indicates that about 5% of the fruit were successfully pollinated by insects. This suggests that the proportion of the pollen carried by insects was small, or the pollen used in the cross-pollination experiment was either incompatible or nonfunctional for some reason.

In the aforementioned study using 5 sago palm trees, the duration of fruit development from fruit setting to fruit drop was observed. It was 19–23 months from flowering to fruit drop. The pattern of fruit size increase involved rapid

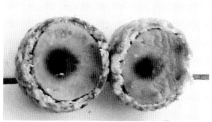

Figure 4-33 Cross section of seedless fruit

Figure 4-32 Fresh mature fruit of sago palm

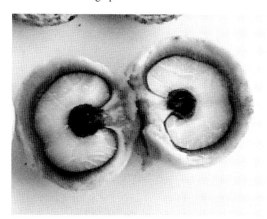

Figure 4-34 Cross section of fruit with seed having horseshoe-shaped endosperm

early growth which moderated toward ripening. The width (diameter) of the ripe fruit ranged from 35 to 44 mm. The number of ripe fruits was 8.6–29.6 per second-order floral axis and 0.8–2.3 per third-order floral axis.

Authors:
4.1 and 4.3: Satoshi Nakamura, Manabu Watanabe and Yusuke Goto
4.2: Hiroshi Ehara
4.4: Manabu Watanabe, Satoshi Nakamura and Yusuke Goto
4.5: Akira Miyazaki
4.6: F. S. Jong

# 5
*Physiology*

Sago palm grows under an extremely wide-range of environmental conditions from peat soil to mineral (inorganic) soil, from brackish water to fresh water and so on. While its growth rate varies depending on soil conditions, it can attain a very vigorous growth rate of 70 t (dry matter) per hectare per year in a favorable environment (Flach 1983). This chapter provides an overview of various physiological properties that are relevant to the growth of the sago palm. Section 5.1 discusses the speed of water absorption and transpiration, which largely determines the nutrient absorption ability of the sago palm. Section 5.2 describes photosynthesis, which is the plant's ability to produce organic matter. Section 5.3 looks at dry matter production in the sago palm from the viewpoints of leaf area, various above-ground plant parts, and starch yield. Section 5.4 explains the water stress adaptation mechanism of the sago palm, which is capable of growing in a wide spectrum of hydric environments. Section 5.5 considers the acidity adaptation mechanism of the sago palm in terms of adaptation to acidity itself and adaptation to the toxicity of aluminum that becomes solubilized under acidic conditions at pH 5 or lower as the sago palm growing soils, ranging from peat soils to mineral soils, often become acidic. Section 5.6 describes the seawater adaptation mechanism. Section 5.7 discusses nitrogen fixing bacteria, which supply nitrogen to sago palms that grow in infertile soils.

## 5.1 Water absorption and transpiration rates

Research on the water physiology of the sago palm has been slow. This is because sago palms that are distributed across tropical low lands of relatively high rainfall are not so susceptible to water stress and because it is difficult to measure the water absorption and transpiration rates of an arboreous plant such as the sago palm. The sago palm is expected, however, to have good salt tolerance as it is able to grow in brackish waters. Physiological studies on salt tolerance using seedlings are under way. The transpiration rate in sago palm seedlings has been measured as part of these physiological studies.

Ehara et al. (2008a) cultivated 8- or 9-leaf stage sago palm seedlings in pots and measured transpiration by the gravimetric method for about 30 days in

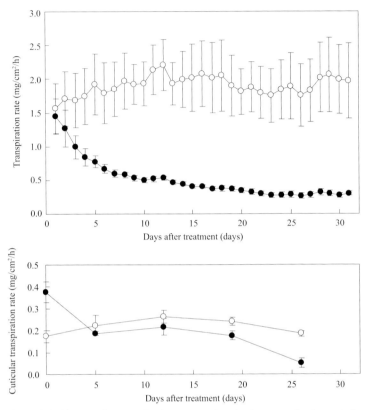

Figure 5-1 Changes in transpiration rate and cuticular transpiration rate in sago palm seedlings treated with sodium chloride (black dot) and untreated (white dot)

NaCl was added to Kimura B solution to make a 342 mM (equivalent of 2 %) solution (pH 5.5) for treatment for 30 days. Data are 5-day moving averages and plotted as mean ± standard deviation from 3 replicates.
Source: Ehara et al. (2008a)

order to understand the mechanisms of salt stress resistance. The transpiration rate in sago palm seedlings in the control plot not treated with sodium chloride (NaCl) was approximately 1.6–2.2 mg/cm$^2$/h (Figure 5-1). The transpiration rate in the plot treated with 342 mM of NaCl (equivalent to 2% NaCl) gradually declined from 1.5 mg/cm$^2$/h on the first day of treatment to 0.3 mg/cm$^2$/h on the 30th day. It is likely that the plant under salt stress reduces transpiration by closing stomata in order to avoid body water loss. The transpiration rate per palm per day in the control plot was 55 g as the 8- or 9-leaf stage seedlings used in the experiment had a leaf area of about 1,200 cm$^2$. Transpiration can be divided into stomatal transpiration through stomata and cuticular transpiration from the cuticular layer. Assuming that transpiration at night when stomata are closed represents cuticular transpiration, the cuticular transpiration rate was about 0.2 mg/cm$^2$/h, approximately 10% of the daytime transpiration rate. The cuticular layer is considered to develop as the foliage matures, but it is still unclear how cuticular transpiration changes with leaf development. As the leaf stomatal density increases with palm age, especially rapidly up to the trunk formation stage (Omori et al. 2000c), the transpiration rate per unit leaf area is likely to change with palm age.

The sago palm distribution zone is mostly confined within latitudes about 10 degrees north and south of the equator. Natural stands are found in coastal fresh water swamps and river floodplains up to an altitude of 300 m or so (Takamura 1995). Normal types of soils are considered to be more favorable for sago palm growth than clay or swampy soils (Schuiling and Flach 1985). Sago palms have recently been found to be able to grow in Morogoro, Tanzania, with an annual precipitation of 900 mm and 5 months of dry season from May to September (Tarimo at al. 2006). A better understanding of sago palm's physiological response to soil water is desirable so that suitable sago palm cultivation sites can be ascertained or cultivation techniques for high starch yield can be established.

## 5.2 Photosynthesis

The sago palm presents $C_3$ photosynthetic properties with low apparent photosynthetic rates (13–15 mg $CO_2/dm^2/h$), a low light saturation point (750 µmol/m$^2$/s), a high $CO_2$ compensation point (44 ppm), and a medium range of optimum temperature for the maximum photosynthesis (26.0–27.4 °C) (Table 5-1, Uchida et al. 1990). The leaf internal anatomy shows characteristics of a typical $C_3$ plant, which has distinct palisade tissue and spongy tissue without the Kranz anatomy, a characteristic of a $C_4$ plant (Nitta et al. 2005a) (see 3.2). As the maximum photosynthetic rate and the light utilization efficiency (quantum yield) are not significantly influenced by shaded conditions (80%), the leaves seem to be well adapted as shade leaves (Uchida et al. 1990). It has been reported that the leaves are thick (0.2–0.4 mm) and hence the chlorophyll content is markedly high (60–80 in SPAD value) (Yamamoto et al. 2006a, 2007b). Reaching the maximum photosynthetic rate in low light canopy conditions appears to be the basic ability to produce high starch production.

The leaf photosynthetic rate in the sago palm reaches a maximum 37–45 days after unfolding of the leaves and maintains more than 50% of the maximum photosynthetic rate about 70 days after unfolding of the leaves (Figure 5-2). This suggests slower development and deterioration of photosynthetic ability than annual crops (Uchida et al. 1990). This is due to low and stable nitrogen (N) concentration in leaves with less N remobilization (Matsumoto et al. 1998). It has been reported that the lifespan of the sago palm leaf is 8–13 months (Yamamoto 1998a) or about 2 years (Flach and Schuiling 1989) which is much longer than that of the upper leaves of annual crops such as rice and wheat (about 60 days). Leaf longevity is an important property which compensates

Table 5-1 Photosynthetic properties of sago palm (seedling-stage)

| Property | Unshaded | 80% shaded[1] |
|---|---|---|
| Maximum photosynthetic rate (mg $CO_2/dm^2/h$) | 13.5 ± 2.8 | 12.8 ± 2.0 |
| Light saturation point ($\mu mol/m^2/s$) | 741 ± 86.2 | 613 ± 13.7 |
| Light compensation point ($\mu mol/m^2/s$) | 5.3 ± 3.0 | 4.6 ± 1.8 |
| Quantum yield (mmol $CO_2$ mol/ photon) | 54.7 ± 10.1 | 71.3 ± 27.5 |
| Optimum temperature (°C) | 27.4 | 26.0 |
| $CO_2$ compensation point (ppm) | 43.6 ± 6.6 | 43.8 ± 6.6 |

[1] Grown under 80% shaded conditions for 12 months prior to measurement.
Source: Uchida et al. (1990)

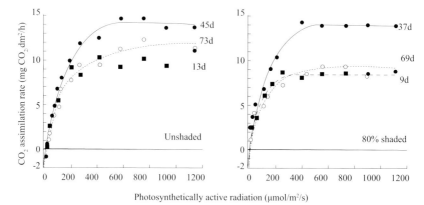

Figure 5-2 Light-photosynthesis curve of sago palms grown under different light conditions (seedling stage)
Source: Uchida et al. (1990)

for the low photosynthetic rate in the sago palms grown in infertile soils. The cumulative amount of carbon assimilation, obtained by multiplying the photosynthetic rate by leaf lifespan, was similar to or higher than that of wheat (Matsumoto et al. 1998).

The maximum photosynthetic rate (mg $CO_2/dm^2/h$) in the sago palm is expected to change with palm age. In the case of sago palms grown in pots under glasshouse, it was 13–15 at the seedling stage (Uchida et al. 1990) or 8–12 in a young sago palm (Flach 1977). On the other hand, it was up to 25–27 in the trunked sago palms grown in the field, which was higher than in the sago palms before trunk formation (16–18) (Miyazaki et al. 2007).

These variations in the photosynthetic rate are caused by increases in photosynthesis-related quantitative characters with palm age. For instance, the sago palm leaf becomes thicker with age, which is associated with the increase in the chlorophyll content (SPAD value) (Figure 5-3, Yamamoto et al. 2006a). It was reported that the transpiration rate (g $H_2O/dm^2/h$) was 0.67 at the seedling stage (Uchida et al. 1990), but 2.3 after trunk formation (Miyazaki et al. 2007). As a background reason for this, it was found that the stomatal density increased with palm age from the seedling stage to the trunk formation stage and that it was markedly higher at the trunk formation stage than in any annual crops (Figure 5-4; adaxial surface 120/mm$^2$, abaxial surface 1,000/mm$^2$) (Omori et al. 2000b).

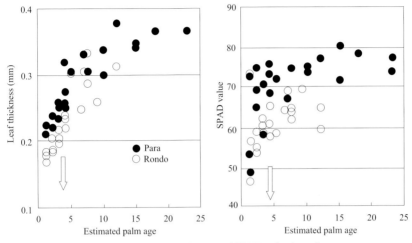

Figure 5-3 Changes in leaf thickness and SPAD value by palm age
Measurements were taken at the center of middle leaflets of middle leaves.
Arrows in the diagrams indicate the trunk formation stage.
Source: Yamamoto et al. (2006a)

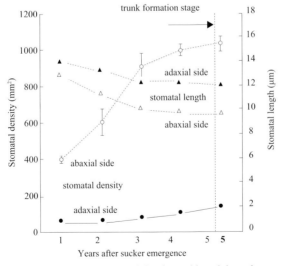

Figure 5-4 Changes in stomatal density and length by palm age
Stomatal density and length were measured at the center of middle leaflets of middle leaves.
Source: Omori et al. (2000b)

These changes in photosynthesis-related characteristics with palm age are considered to influence photosynthesis as the controlling factors. This means that at the seedling stage the transpiration rate and the stomatal density were low, therefore the extent of rate-limiting by the stomata was substantial (30% or more), suggesting that photosynthesis was mainly controlled by the stomatal factor (Uchida et al. 1990). As the photosynthetic rate after trunk formation demonstrated a strong correlation with mesophyll conductance rather than stomatal conductance, it was suggested that the carbon fixation ability in leaves had a major influence on photosynthesis (Miyazaki et al. 2007). Further studies on photosynthetic properties at various growth stages are needed to clarify these differences.

The photosynthetic rate among folk varieties was compared on the shores of Lake Sentani in Jayapura (Table 5-2). While no clear difference in the apparent photosynthetic rate was found between Para (high yielding) and Rondo (early maturing) (Miyazaki et al. 2007), Para had thicker leaves with higher chlorophyll contents than Rondo (Miyazaki et al. 2007, Yamamoto et al. 2006a). As the photosynthetic rate is a widely variable characteristic that fluctuates every moment according to the environment, difference among folk varieties also deserves more detailed investigation.

Table 5-2 Difference among folk varieties in photosynthetic rate, stomatal conductance, mesophyll conductance, SPAD value and leaf thickness (3–5 year-old sago palms, before trunk formation, Jayapura)

| Folk variety | Leaf position[1] | Photosynthetic rate (mg $CO_2/dm^2/h$) | Stomatal conductance (cm/s) | Mesophyll conductance (cm/s) | SPAD value | Leaf thickness (mm) |
|---|---|---|---|---|---|---|
| Rondo | 4th | 21.6a | 0.274a | 0.123a | 58.4cd | 0.329cde |
| Rondo | 5th | 19.2a | 0.275a | 0.101abc | 58.3cd | 0.338bcd |
| Para | 4th | 19.8a | 0.281a | 0.108ab | 77.6a | 0.369ab |
| Para | 7th | 13.4b | 0.177b | 0.072bc | 76.5a | 0.396a |
| Yepha | 5th | 12.8b | 0.165b | 0.059c | 65.5bc | 0.368abc |
| Yepha | 6th | - | - | - | 56.9d | 0.300def |
| Yepha | 7th | 17.4ab | 0.255ab | 0.081abc | 61.7bcd | 0.269f |
| Follo | 4th | 16.5ab | 0.215ab | 0.076abc | 69.8b | 0.341bc |
| Follo | 8th | - | - | - | 66.2b | 0.298ef |
| Average | | 17.2 | 0.235 | 0.089 | 65.6 | 0.334 |
| Coefficient of variation (%) | | 19.1 | 20.8 | 25.4 | 11.8 | 12.1 |

Values in a column followed by different letters are significantly different at 5% level between folk varieties.
[1] Leaf position counted from the uppermost leaf.
Source: Miyazaki et al. (2007)

## 5.3 Dry matter production

The starch harvested from sago palm is of course a product of photosynthesis in the leaf. A better understanding of variation in dry matter production by the sago palm according to the type (variety) or environmental conditions is important in understanding and improving its starch productivity. Dry matter production in sago palm is determined by the leaf area per plant or per unit area as well as the mean photosynthesis rate per unit leaf area. And the average photosynthesis rate per unit area is determined by the leaf photosynthetic capacity and the amount of light received by each leaf. Among these characteristics involved in dry matter production by sago palm, the photosynthetic capacity has been described in Section 5.2. However, very few studies have been undertaken with regard to the dry matter production characteristics of the sago palm so far. Yamamoto et al. (2002a, b) used 2 types of sago palms with vastly different starch productivity (Molat and Rotan, see 7.2.1.a) to study the growth traits of individual sago palms of different ages between the trunk formation and harvesting stages (year 7 for Molat and year 5.5–7 for Rotan) in Kendari, Southeast Sulawesi, Indonesia. The following discussion is based on this study.

### 5.3.1 Leaf area

The leaf area per plant of Molat tended to increase from about 126 $m^2$ soon after trunk formation to 400–450 $m^2$ at the proper time for harvesting (year 7 after trunk formation). In contrast, the leaf area per plant of Rotan changed little from trunk formation to harvesting and fluctuated in the 100–200 $m^2$ range. The leaf area per plant is obtained by multiplying the number of leaves by the mean leaf area. While the leaf area per plant showed very significant positive correlations with both the number of leaves and the mean leaf area in Molat, it had a significant correlation with the number of leaves alone in Rotan. The number of leaves per plant was 10–15 up to around year 5 after trunk formation as lower leaves were being harvested for atap, and 20-21 leaves in older palms. Rotan leaves were not used for atap and the number of leaves was 22 in flowering plants and 10–15 in plants of other ages. The mean leaf area was greater in Molat than Rotan; it was up to 20–22 $m^2$ in the former and about 10–13 $m^2$ in the latter. Their mean leaf area showed very significant positive correlations with both the number of leaflets and the mean leaflet area but the correlation was stronger with the mean leaflet area than the number of leaflets in Molat. And the leaf area per leaf was closely associated with the mean leaflet area in both varieties. The leaf length was 8–13 m and the rachis length was 6–9 m in both varieties but Molat tended to surpass Rotan in both parameters. The number of leaflets attached to rachis was 150–180 in Molat and 130–160 in Rotan. The maximum leaflet length and width were 160–180 cm and 10–14 cm in Molat and 150–160 cm and 8–10 cm in Rotan; again Molat surpassed Rotan in both measurements. It became clear based on these results that the aforementioned intervarietal variation in the mean leaf area depends on variation in both the number of leaflets per leaf and the area of each leaflet.

Flach and Schuiling (1991) studied the characteristics of sago palm leaf in Batu Pahat in Johor and Mukah in Sarawak, Malaysia, and Seram Island, Indonesia, and found marked variation in the leaf area per leaf at 4 $m^2$, 9 $m^2$ and 20 $m^2$ in respective sample populations. The study demonstrated that the regional variation in the leaf area per leaf arose from variation in the number of leaflets per leaf and the leaflet area, and that variation in the mean leaflet area was attributable to variation in the maximum leaflet length and/or the leaflet width. It is estimated that this regional variation in the sago palm leaf area was influenced by the sago palm type (variety) and the growing environment.

### 5.3.2 Total above-ground biomass fresh weight and dry matter weight

Figure 5-5 shows changes in the total above-ground biomass fresh weight of individual Molat and Rotan palms after trunk formation through to the harvest stage (Yamamoto et al. 2002b). While the total above-ground biomass, not only fresh weight, but also dry weight, increased exponentially in both varieties, intervarietal differences grew larger with years after trunk formation as the magnitude of the change in fresh weight and dry matter weight was 600–3,050 kg and 100–1,030 kg respectively in Molat and 490–1,120 kg and 90–440 kg in Rotan. The difference in total above-ground weight at the harvesting stage between the two varieties was 2.8 times in fresh weight and 2.7 times in dry weight. Changes in total above-ground dry matter ratio based on fresh weight and dry matter weight remained small from shortly after trunk formation to year 4 after trunk formation at below 20% for Molat and around 20% for Rotan and rapidly increased thereafter to reach about 35% at the harvesting stage (Table 5-3).

### 5.3.3 Dry weight by above-ground part

Looking at changes in the dry weight of the leaves and the trunk separately, the leaf dry weight increased exponentially from 80 kg to 290 kg in Molat while Rotan showed a relatively moderate increase from 60 kg to 110 kg. The trunk dry weight increased exponentially in both varieties from 30 kg to 750 kg in Molat and from 25 kg to 340 kg in Rotan. Molat was 2.7 times heavier than Rotan at the harvesting stage. Change in the leaf dry matter ratio with palm age after trunk formation was smaller than that in the trunk dry matter ratio and increased within the 25–35% range (Table 5-3). The trunk dry matter ratio in Molat remained at the relatively low level of just over 10% until around year 4 with minor change and increased rapidly thereafter. By contrast, the trunk dry matter ratio in Rotan tended to start increasing shortly after trunk formation (Table 5-3). The difference in the pattern of increase in the trunk

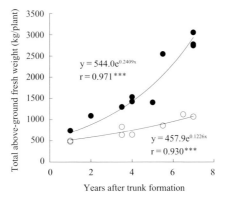

Figure 5-5 Relationship between years after trunk formation and total above-ground fresh weight
● Molat, ○ Rotan
***: 0.1% level of significance
Source: Yamamoto et al. (2002b)

Table 5-3 Changes in dry matter ratio (%) in above-ground parts by years after trunk formation

| Variety | Years after trunk formation[1] | Leaves Total | Trunk | Above-ground |
|---|---|---|---|---|
| Molat | 1 | 25.5 | 9.4 | 17.4 |
|  | 1 | 25.7 | 10.4 | 18.8 |
|  | 1 | 25.9 | 11.1 | 19.0 |
|  | 2 | 27.0 | 12.5 | 19.3 |
|  | 3.5 | 25.9 | 12.7 | 18.0 |
|  | 3.5 | 27.4 | 11.4 | 16.6 |
|  | 4 | 27.5 | 10.6 | 16.3 |
|  | 4 | 27.2 | 13.4 | 18.4 |
|  | 5 | 30.3 | 27.5 | 28.2 |
|  | 5.5 | 31.3 | 26.6 | 28.2 |
|  | 7 | 31.2 | 34.7 | 33.7 |
|  | 7 | 32.1 | 35.3 | 34.4 |
|  | 7 | 31.4 | 37.1 | 35.5 |
| Rotan | 1 | 24.9 | 10.7 | 18.1 |
|  | 1 | 27.6 | 11.9 | 19.5 |
|  | 1 | 28.7 | 13.0 | 20.8 |
|  | 3.5 | 28.2 | 16.3 | 20.2 |
|  | 3.5 | 30.1 | 22.8 | 25.5 |
|  | 4 | 28.5 | 17.1 | 20.9 |
|  | 5.5 | 29.5 | 33.0 | 31.9 |
|  | 6.5 | 31.5 | 34.7 | 33.7 |
|  | 7[2] | 36.3 | 42.2 | 40.8 |

1) Years after trunk formation  2) Flowering stage
Source: Yamamoto et al. (2002b)

dry matter ratio between these varieties is likely to be linked to the fact that pith starch accumulation begins earlier in Rotan than in Molat. The trunk dry matter ratio at the harvesting stage was 35–40% in both varieties with only minor difference between them (except that Rotan in year 7 showed a slightly higher ratio at the young fruit stage).

When the trunk dry weight was separated between the bark part and the pith part, it increased exponentially after trunk formation in both parts of both varieties. The pith dry weight changed from 20 to 600 kg in Molat and 20 to 270 kg in Rotan. The difference between the two varieties increased with years after trunk formation; the mean weight at the harvesting stage was 581 kg in Molat and 213 kg in Rotan, a difference of 2.7 times. The bark dry weight changed in the range of 10–150 kg in Molat and 5–65 kg in Rotan. The dry matter ratio in the bark part up to year 4 after trunk formation was 20–35%, which was higher than in the pith part but the difference between the two parts became smaller thereafter. The dry matter ratio at the harvesting stage was around 35% in the pith part and 35–40% in the bark part.

## 5.3.4 Dry weight percentage of above-ground parts

Based on the above dry weight measurements in each plant part, changes in them can be summarized as follows (Yamamoto et al. 2002b) (Table 5-4).

Table 5-4 Changes in dry weight percentage (%) in above-ground parts by years after trunk formation

| Variety | Years after trunk formation | Leaves Total | Trunk | Inflorescence | Above-ground |
|---|---|---|---|---|---|
| Molat | 1 | 73 | 27 | 0 | 100 |
|  | 1 | 75 | 25 | 0 | 100 |
|  | 1 | 73 | 27 | 0 | 100 |
|  | 2 | 65 | 35 | 0 | 100 |
|  | 3.5 | 58 | 42 | 0 | 100 |
|  | 3.5 | 54 | 46 | 0 | 100 |
|  | 4 | 57 | 43 | 0 | 100 |
|  | 4 | 54 | 46 | 0 | 100 |
|  | 5 | 29 | 71 | 0 | 100 |
|  | 5.5 | 37 | 63 | 0 | 100 |
|  | 7 | 28 | 72 | 0 | 100 |
|  | 7 | 28 | 72 | 0 | 100 |
|  | 7 | 24 | 76 | 0 | 100 |
| Rotan | 1 | 72 | 28 | 0 | 100 |
|  | 1 | 69 | 31 | 0 | 100 |
|  | 1 | 69 | 31 | 0 | 100 |
|  | 3.5 | 45 | 55 | 0 | 100 |
|  | 3.5 | 43 | 57 | 0 | 100 |
|  | 4 | 46 | 54 | 0 | 100 |
|  | 5.5 | 29 | 71 | 0 | 100 |
|  | 6.5 | 28 | 72 | 0 | 100 |
|  | 7[1] | 20 | 77 | 3 | 100 |

1) Flowering stage
Source: Yamamoto et al. (2002b)

Firstly, difference between the two varieties was small in the percentage of the leaf dry weight and the trunk dry weight to the total above-ground dry weight. They changed antithetically from the leaf weight percentage of 75–70% and the trunk weight percentage of 25–30% shortly after trunk formation to 20–30% and 70–80% respectively at the harvesting stage.

The dry weight ratios of leaflet, rachis and petiole (including leaf sheath) to the total leaf weight changed relatively little after trunk formation. They were 30–40% in the bark part and 60–70% in the pith part up to year 4 after trunk formation and increased in the pith part thereafter to reach about 20% in the bark part and about 80% in the pith part. These results suggest that the dry weight of the harvestable pith part accounts for about 60% (56–64%) of the total above-ground dry weight at the harvesting stage.

Flach and Schuiling (1991) report that the weight percentage of leaflet, rachis and petiole (including leaf sheath) of the harvesting-stage leaves of Tuni, a sago palm variety found in Seram Island, Indonesia, are 40.6%, 28.1% and 31.3% which are consistent with the aforementioned results. They also report that the percentage of the total leaf dry weight to the harvesting-stage total above-ground dry weight (521.2 kg) of the sago palm growing in Sarawak, Malaysia, was 22.6%, the trunk dry weight ratio was 77.4%, and the dry weight ratios of leaflet, rachis and petiole (including leaf sheath) to the total leaf dry weight were 33.2%, 33.0% and 33.8% while the dry weight percentage of bark and pith to the trunk dry weight were 13.3% and 86.7% respectively. Kaneko et al. (1996) studied a 13-year-old (supposedly at the harvesting stage) sago palm (total above-ground dry weight of 387 kg) growing in shallow peat soils of Sarawak, Malaysia, and found that the dry matter weight ratios of leaf, bark and pith were 26%, 23% and 51% respectively. Yatsugi (1987b), however, reported that the percentage of the bark weight and the pith weight to the trunk weight were 20% and 80%.

The percentage of the trunk dry weight and the leaf dry weight to the total above-ground dry weight at the harvesting stage and the dry weight percentage of leaflet, rachis and petiole (including leaf sheath) to the total leaf dry weight reported in these studies are mostly consistent except that the percentage of the bark dry weight to the trunk dry weight in the data from Sarawak in Flach and Schuiling (1991) was lower than in the other studies.

### 5.3.5 Starch yield and dry matter production properties

Starch yield of sago palm is calculated by the same formula used for other crops, that is, Starch yield = total plant dry weight × starch yield / total dry weight.

The right side of the equation is further expressed as: total plant dry weight × pith dry weight / total plant dry weight × starch yield / pith dry weight. This means that the total plant dry weight, the ratio of the pith weight to the total plant dry weight, and the starch content per pith dry weight are highly relevant to starch yield determination. The total dry weight consists of the above-ground dry weight and the root dry weight but since it takes a great deal of time and labor to measure the root dry weight, the total plant dry weight has been substituted with the above-ground dry weight in presenting the starch yield-related dry matter production properties of the two varieties in Table 5-5. The table shows that the difference in starch yield between the two varieties stemmed from differences in their above-ground dry matter production. No difference was found in the ratio of dry matter distribution to the pith in which starch accumulated or the starch content per pith dry weight (see 7.2). The above-ground dry matter production was greater in Molat than Rotan. Having a greater leaf area appears to be closely associated with dry matter productivity.

Table 5-5 Varietal differences in dry matter production properties related to sago palm starch production

| Variety | Total above-ground dry weight (kg/palm) | Pith weight percentage[1] (%) | Pith starch content (%) |
|---|---|---|---|
| Molat | 1,000 | 60 | 60–70 |
| Rotan | 400 | 60 | 70 |

1) Ratio to total above-ground dry weight

## 5.4 Mechanisms of adaptation to brackish water

Plants are subjected to stress from excessive water and from the salt in brackish waters. This section describes the mechanisms of adaptation to water stress and salt stress separately.

### 5.4.1 Mechanisms of adaptation to water stress

Plants require water as an essential material for all metabolic reactions as well as for proper maintenance of their plant bodies. Sago palm, a $C_3$ plant, consumes 50–100 mol of water through transpiration in order to fix 1 mol of carbon dioxide through photosynthesis. Accordingly, its roots need to supply the amount of water consumed in transpiration. However, it is thought that plants typically suffer from oxygen starvation if they are immersed in water and suffer oxygen damage when they are exposed to air again.

Brackish water areas in the sago palm growing zone have high ground water levels either permanently or seasonally. Only plants that have some form of moisture tolerance mechanism (water stress exclusion mechanism) (sometimes called hydrophytes (Shimoda 2005) or helophytes) (Figure 5-6) can grow there. The main mode of water transport in plant cells is by osmosis through a semi-permeable membrane. More active modes of transport are unlikely. Water potential in plant cells change in response to the external moisture level. Plant cells increase their intracellular pressure in response to a rapid rise in water potential in order to prevent the potential difference in

Floating-leaved vegetation in a pond surrounded by a forest: *Brasenia schreberi, Nymphaea tetragona, Potamogeton fryeri* etc.

A pond densely covered with the floating leaves of *Trapa japonica*.

Figure 5-6 Helophytes
Source: Shimoda (2005)

and out of the cells from widening and thus to avoid cell rupture by restricting excessive water absorption (Kato 2002). Cell wall and vacuole, which are characteristic plant tissues, prevent cells from being ruptured by excessive water.

When plants are starved of oxygen due to excess water, they experience hypoxia and try to supplement oxygen. Many helophytes have well-developed lysigenous aerenchyma (Figure 5-7) in the leaf, stem and root. Air drawn in through the leaf stoma and oxygen generated by photosynthesis are transported through the lysigenous aerenchyma to reach the underground part and used in oxygen-utilizing reactions. Part of the oxygen transported from the aboveground part to the underground part is used in root cells; the rest is released from the root to maintain the oxidative environment around the root. While Yamamoto (1997) grouped sago palm roots into thick roots (diameter 7–11 mm) and fine roots (diameter 4–7 mm) for reasons of expediency, he reported that both types of roots had a well-developed lysigenous aerenchyma. It is no exaggeration to state that this lysigenous aerenchyma provides the sago palm's

water stress adaptation mechanisms.
The sago palm avoids excessive water stress by developing the lysigenous

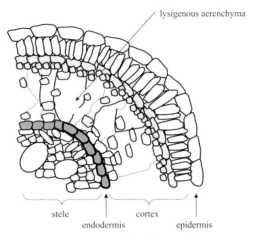

Figure 5-7 Cross section of sago palm root

aerenchyma in the root, which transports oxygen from the above-ground part to the underground part (Yamamoto 1998a). As a result, the surface of the sago palm root is brown in color. Fe (II) ions in the soil solution are oxidized by oxygen released from the root and deposit on the surface of the root where oxidized iron compounds such as goethite α-FeOOH and lepidocrocite γ-FeOOH can be observed.

## 5.4.2 Mechanisms of adaptation to salt stress

There are two pathways for transporting the ions absorbed from plant roots: the symplast pathway through parenchyma cells in the root and the apoplast pathway via cell walls and intercellular spaces (Matoh 1991). The symplastic ion transport is regulated by ion channels in the cell membrane. The apoplastic ion transport relies on mass flow and diffusion; plants therefore cannot actively control ion migration via the apoplast pathway.

Many land plants suffer from low growth at a salt concentration of 0.1% and marked growth suppression when it exceeds 0.3% (Matoh 2002). Salt stress is divided into two types: the inhibition of water absorption due to osmotic pressure generated by salt (osmotic stress) and excess damage caused by the specific physiological effects of the salt-constituting ions (ionic stress). Ionic stress is generally regarded as the primary factor of salinity stress.

Takahashi (1991) lists 7 mechanisms by which plants adapt to salt stress:

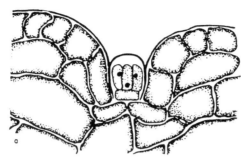

Figure 5-8 Salt gland in *Avicennia marina*
Source: Waisel (1972), Takahashi (1991)

1. Excluding sodium ions using barriers in the roots to inhibit sodium ion absorption.
2. Inhibiting the transportation of absorbed sodium ions by storing them in its roots, stems or petioles (Matoh 2000).
3. Translocating sodium ions that have been transported to the above-ground part back to the roots and expelling them.
4. Storing sodium ions in bladder hairs differentiated from the epidermal cells, eliminating the sodium ion-filled bladder hairs out of the body, creating new bladder hairs, and repeating this cycle as necessary to avoid salt stress.
5. Differentiating salt glands (Figure 5-8) on the leaf surface and expelling sodium ions out of the body through them.
6. Storing sodium ions in vacuoles of mesophyll cells to segregate sodium ions from the cytoplasm and then using them to create cellular turgor pressure.
7. Increasing cytoplasmic concentrations by synthesizing compatible solutes such as potassium, sucrose, amino acids and betaines (Figure 5-9) (Wada 1999) thereby creating a higher osmotic pressure (Matoh et al. 1987).

Mechanisms 1, 2 and 3 to limit cytoplasmic sodium ion concentration constitute a series of adaptive mechanisms of the plant against salt stress. Mechanisms 4, 5 and 6 are specific mechanisms against salt stress that have developed in halophytic plants. As the sago palm is highly tolerant to salinity stress, it is likely to be equipped with mechanisms 4, 5, 6 and 7.

Halophytes are higher plants that are able to flower and fruit in a sodium chloride solution of 100 mM or more. Sodium ions are absorbed via the low-affinity potassium channel in the cell membrane (Munns 2001). Halophytes accumulate sodium ions in the leaf and increase osmotic pressure in order to absorb water from a solution of high sodium salt concentration. As excessive sodium ions cause ionic stress in the cytoplasm, halophytes keep sodium ions segregated in the vacuole and accumulate compatible solutes to adjust osmotic pressure.

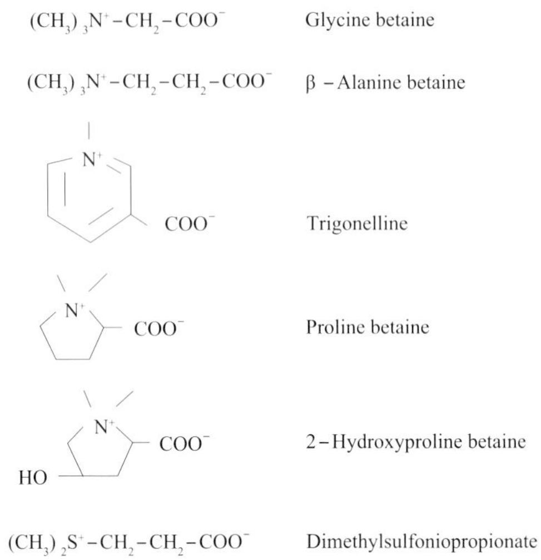

$(CH_3)_3N^+-CH_2-COO^-$   Glycine betaine

$(CH_3)_3N^+-CH_2-CH_2-COO^-$   β –Alanine betaine

[pyridinium ring with COO⁻]   Trigonelline

[pyrrolidinium ring with COO⁻]   Proline betaine

[hydroxy-pyrrolidinium ring with COO⁻, HO-]   2–Hydroxyproline betaine

$(CH_3)_2S^+-CH_2-CH_2-COO^-$   Dimethylsulfoniopropionate

Figure 5-9 Betaines

Studies on the salinity tolerance of plants are under way, including research into sodium proton antiporter activity (involved in the ability to transport sodium ions to the vacuole), proton ATPase activity, the sodium reabsorption-transport ability in the vascular parenchyma, and the ability to accumulate compatible solutes (Girjia et al. 2002). Cloning of tonoplastic sodium proton antiporter protein genes and cytoplasmic potassium channel protein genes are being undertaken (Matoh 2002).

The mechanisms by which the sago palm has adapted to salt stress are yet to be fully understood. Flach et al. (1977) studied salinity tolerance in sago palm seedlings and concluded that growth is not adversely affected by seawater incursion at a frequency of up to once a fortnight. Ehara et al. (2003b, 2006b) reported that sago palm roots exclude sodium ions using barriers to inhibit sodium ion absorption.

In reed grass, it has been found that the site of sodium ion control (about 1 cm thick) is located at the base of the stem rather than in the roots (Maruyama et al. 2008). The reed controls its sodium ion concentration by transporting sodium ions into starch granules at this site as well as into the vacuoles. Further studies are needed to examine whether a similar mechanism operates in the sago palm.

Yoneta et al. (2006) surmised that the sago palm maintains a high osmotic pressure in the cytoplasm primarily by using potassium ions as the osmotic pressure regulator (compatible solute). They concluded that only small amounts of other compatible solutes such as proline and glycine betaine are produced. Many salinity tolerant plants synthesize betaines and keep them in the cytoplasm. The choline monooxygenase (CMO) gene and the betaine aldehyde dehydrogenase (BADH) gene are clearly involved in betaine synthesis. CMO synthesizes betaine aldehyde from choline. CMO was successfully obtained from spinach by partial purification (Burnet et al. 1995) but genomic cloning is yet to be carried out. BADH converts betaine aldehyde to betaine. Purified BADH has already been obtained and its genes cloned successfully.

## 5.5 Mechanisms of adaptation to low pH

### 5.5.1 Mechanisms of adaptation to hydrogen ion stress

All biomembranes have cytoplasmic membrane proteins which are responsible for ion transportation (Figure 5-10). Among these transportation proteins, the ones that are capable of creating a high-energy state within a membrane by transporting ions using either physical or chemical forms of energy and functions as primary active transport is called a pump (Figure 5-10). A type of membrane protein that binds with ions at specific sites before transporting them is called a carrier. Some carriers perform active transportation while others provide passive transportation. A carrier that transports ions against its electrochemical potential gradient is considered to be a secondary active transport. A membrane protein that selectively allows passage of certain ions is called a channel, which transports ions passively. There are other types of primary active transportation called transporters.

The ion pump (proton pump, proton ATPase) inside a plant cell membrane translocates hydrogen ions (protons) out of the cell. This pump is thought to have a molecular weight of about 100,000 and functions to maintain a weak alkalinity inside the cell and acidity outside the cell. At the same time, an electrochemical potential gradient across the cell membrane from outside to inside is created and utilized in the transportation of other ions.

The inhibition of plant growth caused by hydrogen ion increases is called acid stress (hydrogen ion stress).

### 5.5.2 Mechanisms of adaptation to aluminum stress

When soil pH falls, aluminum becomes solubilized at pH 5.0 or lower and may reach toxic concentrations at pH 4.0 or lower. Under acidic conditions, plant growth inhibition is caused by two things: increased hydrogen ion levels and solubilized aluminum ions.

Plants with high aluminum tolerance have a root apical cell plasma membrane with an excellent aluminum exclusion mechanism. Aluminum stress induces the synthesis of β-1,3-glucan (callose) outside of the root apical cell plasma membrane, which inhibits the migration of several substances in the apoplast and the symplast between adjacent cells by depositing itself in the apoplast and around the plasmodesma (Wagatsuma 2002). The Al-induced synthesis of β-1,3-glucan does not occur readily in highly aluminum tolerant plants. The aluminum tolerance in plants is also regulated by citric acid, oxalic acid and malic acid released by the roots. The release of these organic acids is in turn induced by aluminum stress (Ma et al. 2001; Furukawa and Ma 2006)

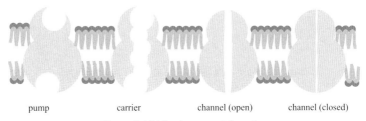

pump        carrier        channel (open)        channel (closed)
Figure 5-10 Membrane protein action

(Figure 5-11). Although the extent of the aluminum absorption inhibition by the organic acids released from roots was unclear, as the organic acid concentration in the vicinity of roots was markedly lower than the aluminum concentration, the organic acid concentration on the surfaces of roots appears to be high enough to explain the aluminum tolerance (Koyama 2002).

Aluminum tolerance in plants appears to be controlled by separate mechanisms which contribute to the process to different degrees. These mechanisms can be divided into three groups: (1) aluminum detoxification outside of the cell; (2) binding to the surfaces of cell walls and membranes; (3) detoxification and protection inside the cell (Figure 5-11).

### 5.5.2.a Extracellular aluminum detoxification

One type of extracellular aluminum detoxification mechanism in plants involves the release of organic acids from the roots (Ma et al. 2001). Oxalic acid, citric acid and malic acid form stable complexes with aluminum. The release of these organic acids is induced by aluminum stress. Aluminum stress is completely controlled when the ratios of citric acid, oxalic acid and malic acid to aluminum are 1:1, 3:1 and 10:1 respectively (Wagatsuma 2002). The formed complexes are low in toxicity and poorly absorbed by the roots (Figure 5-12).

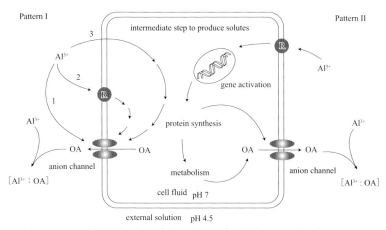

Figure 5-11 Model of aluminum-induced organic acid ion secretion from roots
R: Al ion-specific receptor
OA: Some of the substances shown in Oxalic Acid Pattern I have been identified but those in Pattern II are suspected substances.
Source: Ma et al. (2001)

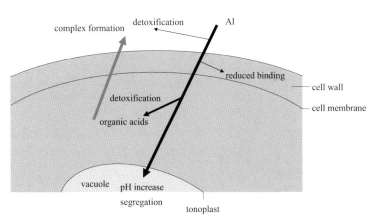

Figure 5-12 Plant's aluminum stress tolerance mechanism

### 5.5.2.b Binding to the cell wall and membrane surfaces

Aluminum makes contact with the surface of the cell wall or cell membrane and gradually migrates to the symplast, thereby lowering ion absorption and migration as well as disturbing intracellular communication. The cell membranes of Al-tolerant species can maintain their cell membrane functions as their low zeta potentials mean that aluminum ions bind to the cell membranes poorly.

### 5.5.2.c Intracellular detoxification and protection

Intracellular defense against interferences by aluminum involves enclosing the aluminum in the vacuoles. Aluminum-accumulating plants such as buckwheat keep aluminum in the vacuoles in the form of a complex with an aluminum to oxalic acid ratio of 1:3 (Ma et al. 2001). The isolation of aluminum-induced gene clusters is currently under way. De la Funte et al. (1997) created transgenic tobacco and papaya by introducing microbially-derived citrate synthesase. They discovered that these plants synthesized large amounts of citric acid and expressed aluminum tolerance. Ezaki et al. (2000) screened plant-derived aluminum-induced genes for aluminum tolerance. They reported that *AtBCB*, *parB*, *NtPox* and *NtGDI1* genes could confer aluminum tolerance, suggesting the existence of tolerance mechanisms other than organic acid chelation. Iuchi et al. (2007) discovered a mutant of *Arabidopsis thaliana* called *stop1* which was unable to elongate its roots under acidic conditions. They concluded that the *STOP1* gene played a critical role in the expression of molecular mechanisms for acid tolerance. Figure 5-13 demonstrates the way both pH and aluminum ions cause growth inhibition in *stop1* mutant (Iuchi et al. 2007).

It is not known whether the low pH adaptation mechanism in sago palm involves *STOP1* gene expression. Further research is needed in this regard.

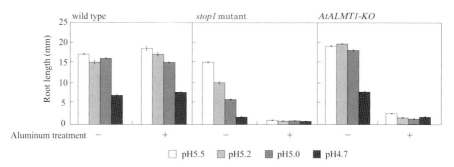

Figure 5-13 Influences of pH and aluminum on *stop1* variant
Source: Iuchi et al. (2007)

## 5.6 Mechanisms of adaptation to seawater

Sago palm properties such as salinity tolerance are important in expanding cultivation areas and in avoiding competition with other crops (Sato 1993). There is thus a need for a clear understanding of intervarietal or interspecific differences in salt tolerance. The sago palm growing environment is very diverse, including both brackish water and fresh water areas (Shimoda and Power 1990). Its ability to grow in low land swamps affected by seawater suggests that it possesses a degree of salinity tolerance. The genus *Metroxylon* is generally found growing in the backland behind mangroves belonging to the genera *Rhizophora* and *Bruguiera* or nipa palms (*Nypa fruticans*). Yamamoto et al. (2003a) surveyed the growth status of sago palms at various distances from the coast and reported that the growth of sago palms found within 20 m of the seashore was markedly poor compared to those situated at 100 m or more from the seashore. However, little is known about the level of salt stress they are exposed to under such natural conditions or the extent of growth inhibition by such salt stress.

Ehara et al. (2006b) treated 17-leaf or 19-leaf sago palm seedlings with a 86 mM NaCl solution for 30 days. They found that sago palm seedlings kept the sodium concentration in leaflets at low levels by accumulating sodium in the roots or lower petioles. Another experiment in which 8-leaf sago palms were treated with 0, 86, 171 and 342 mM NaCl solutions (equivalent of 0, 0.5, 1.0 and 2.0%) for 30 days found that although the sodium concentration in the roots and petioles increased at higher treatment concentrations, there was no clear difference in the upper leaflet sodium concentration between the treated plots (Ehara et al. 2006b) (Figure 5-14). These results suggest that the sago palm possesses some avoidance mechanisms to control the translocation of sodium ions to the leaflets and keep the leaflet sodium concentration low by storing sodium in the roots and petioles. Many crops are known to absorb excessive amounts of sodium under salt stress, but this generally then inhibits the absorption of potassium, which is an essential nutrient. In the sago palm, however, the absorption and translocation of potassium are not reduced by sodium accumulation in roots and petioles under salt stress; rather they tend to increase. The leaflet sodium concentration at 342 mM differed little from other treated plots but the transpiration rate was lower. As the decreased transpiration rate means stomatal closure, the photosynthesis rate is likely to have been lowered as well. Since sago palm seedlings died on the third day of treatment with 400 mM NaCl in an experiment conducted by Yoneta et al. (2003), the salt concentration of around 342 mM NaCl may be the upper limit for sago palm survival.

It is generally important for mesophytes to inhibit sodium accumulation, particularly in the leaf blade, whereas highly salt-tolerant halophytes maintain low cytoplasmic sodium concentrations by storing sodium ions in the vacuoles when they reach the leaf blade cells. In the coconut palm, which appears to be as salt tolerant as the sago palm, the leaf blade sodium content was found to be 4.56 mg per gram dry weight in those growing on the seashore (Magat and Margate 1988). This value is more than 7 times higher than 0.60 mg per gram dry weight in sago palms similarly growing on the seashore (Naito et al. unpublished). The leaf blade potassium content of 9.73 mg per gram dry

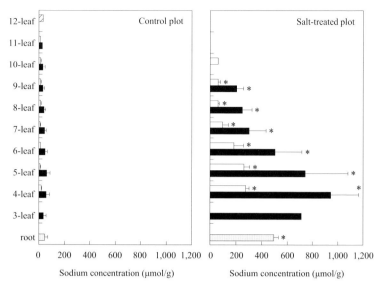

Figure 5-14 Sodium concentration in sago palm (*M. sagu*) parts treated with 342 mM NaCl
White bar: sodium concentration in leaf blade; black bar: sodium concentration in petiole
* indicates significant difference at the 5% level in the t-test between the control plot and the treatment plot.
Source: Ehara et al. (2006b, 2008a)

weight in coconut was also higher than 6.73 mg per gram dry weight in sago palm. These findings point to the possibility that the coconut palm absorbs some amount of salts in the rhizosphere and use them as osmotic regulators whereas the sago palm maintains its growth by inhibiting sodium absorption. Besides these inorganic ions, osmolytes such as glycine betaine and proline are known to regulate osmosis but they are not considered to act as major osmotic regulators in the sago palm. As the leaf blade potassium concentration tends to increase in the sago palm under salt stress (Ehara et al. 2008a), potassium appears to be an osmotic regulator, albeit to a lesser extent than in the coconut palm.

Salt stress response varies between species within the genus *Metroxylon*. While sodium accumulates in the roots and the lower petioles more than the upper petioles in *M. sagu*, *M. vitiense* exhibits a lesser capacity to accumulate sodium in the roots and the petioles (of lower leaves in particular); sodium accumulation is small in the roots and the sodium concentration rises in the petioles of all leaves (Ehara et al. 2008b). In *M. warburgii*, the sodium content in the roots is small while it tends to increase in the petioles of upper leaves under salt stress (Ehara et al. 2007) (Figure 5-15). This shows that it has limited capabilities to exclude sodium at the roots and store sodium in the roots and the petioles. Consequently, *M. warburgii* is prone to increased sodium concentrations in the leaf blades which inhibit the opening of new leaves or cause lower leaf withering or discoloration. It follows that the sodium exclusion function in the roots and the sodium storage capacity in the roots and the lower leaves have a great bearing on interspecific differences in salinity tolerance of the genus *Metroxylon*.

Except for the aforementioned halophytic plants which accumulate absorbed sodium ions in the vacuoles and actively use them in osmotic regulation, the sodium exclusion function at the roots is critically important in the maintenance of crop growth under salt stress. In rice, the exclusion of sodium by the root is thought to take place at the endodermis. When sodium concentrations were measured at the stele and the rest of the root in order to determine the distribution of sodium in the roots of sago palm under salt stress, they were found to be higher at the outer part of the root than at the stele. X-ray microanalysis using a scanning electron microscope also found

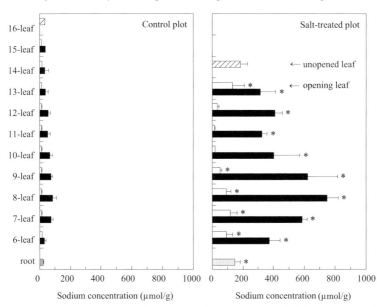

Figure 5-15 Sodium concentration in *M. warburgii* parts treated with 342 mM NaCl
Bars and * signify the same as in Figure 5-14.
Source: Ehara et al. (2007)

that the sodium concentration was low at the stele and high at the innermost part of the cortex adjacent to the endodermis (Figure 5-16). This suggests that sodium exclusion is performed at part of the endodermis in the genus *Metroxylon* as well. *M. vitiense* with a lesser sodium exclusion function and a higher translocation of sodium to the above-ground part has a lesser degree of thickening of the surface adjacent to the pericycle of the endodermal cell compared with *M. sagu*. It appears that the level of development of this part has an important implication for sodium exclusion.

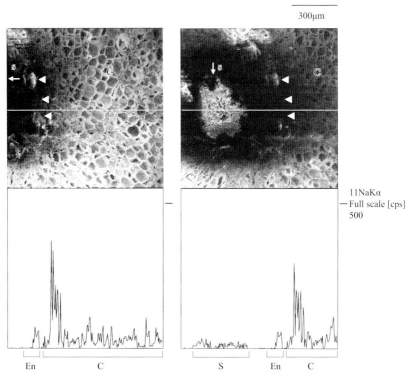

Figure 5-16 Cross section of salt-treated sago palm (*M. sagu*) root and sodium distribution by energy dispersive X-ray spectroscopic analysis
The bottom charts show the distribution of sodium along the red line in the top images.
C: cortex, S: stele, En: endodermis, △: outer edge of endodermis
Source: Ehara et al. (2006b)

## 5.7 Nitrogen fixing bacteria

Nitrogen fixing bacteria convert molecular nitrogen abundantly present in the atmosphere into ammonia nitrogen. Symbiotic nitrogen fixing bacteria are found on leguminous crops such as soybeans and peanuts and leguminous trees such as *Acacia* growing in the tropical and subtropical regions. Filamentous bacteria belonging to the genus *Frankia* live in symbiosis with tree species such as *Alnus japonica* and *Alnus firma* and are also known to fix nitrogen.

Non-leguminous crops such as rice and sugarcane are not known to harbor symbiotic nitrogen fixing bacteria but various kinds of free-living nitrogen fixing bacteria establish themselves around or inside these plants and perform nitrogen fixation. A comparison of wild and cultivated species of rice (Elbeltagy et al. 2001) found that nitrogen fixing bacteria belonging to the genus *Herbaspirillum* colonized the stems and leaves of wild rice species extensively and fix nitrogen whereas their densities and nitrogen fixing activity were low in cultivated species. Various other bacteria belonging to such genera as *Azospirillum* and *Enterobacter* have been isolated from a variety of crops, including rice, sugarcane, corn and sweet potato. Nitrogen fixing bacteria generally fix nitrogen actively when the nitrogen nutrition status of the host plant is poor and reduce their nitrogen fixing activity when nitrogen nutrition is sufficient. Sago palm, which is a tropical non-leguminous plant, is usually cultivated without fertilization (Okazaki et al. 2005). It can grow in peat soils as well as in mineral soils. As sago palm can thrive under infertile conditions, it is expected to possess some mechanisms to obtain nutrients, especially nitrogen, effectively under such conditions. Shrestha et al. (2006) studied nitrogen fixing bacteria in various parts of the sago palm.

Table 5-6 Acetylene reducing activity in different parts of sago palm collected from the Philippines (nitrogen fixing activity)

| Collection date | Sago palm part | Acetylene reducing activity (nmol $C_2H_4$/g/d) |
|---|---|---|
| March 2003 | root | 960, 490, 250, 210, 102, 89, 36, 10, 4.5 |
|  | leaf sheath | 66 |
|  | leaflet | 0 |
| July 2004 | root | 26, 13, 2.8, 84, 33 |
|  | leaf sheath | 1.6, 25, 28 |
|  | pith | 1.1, 140, 110 |
|  | bark | 42 |
| March 2005 | root | 26 |
|  | starch | 210 |
|  | leaf sheath | 66 |
|  | pith | 590 |
| August 2005 | root | 110 |
|  | starch | 330 |
|  | pith | 21 |
|  | leaf sheath | 33 |
| April 2006 | root | 440, 310 |
|  | leaf sheath | 12 |
|  | pith | 590 |
|  | starch | 79 |
| July 2006 | root | 270, 92, 25 |
|  | starch | 140 |

Different numbers are the mean values of 2–4 samples from different sago palms.
Source: Shrestha et al. (2006)

Various parts (root, leaf sheath, leaflet, bark, pith and starch) of sago palm were sampled from the Philippine islands of Leyte, Cebu, Panay and Aklan for testing the nitrogen fixing activity. The measurement of nitrogen fixing activity with the acetylene reduction assay detected nitrogen fixing activity – indicating the presence of nitrogen fixing bacteria – in all samples except leaflets (Table 5-6; Shrestha et al. 2006). Starch and roots showed relatively high values. When the surfaces of sago palm roots were sterilized with a 70% ethanol solution, the samples still showed almost the same level of nitrogen fixing activity as in non-sterilized roots. This suggested that the nitrogen fixing activity of sago palm roots was generated by nitrogen fixing bacteria colonizing inside the roots. Nitrogen fixing activity in starch, in contrast, was largely lost after a thorough washing. This suggested that nitrogen fixing bacteria were loosely attached to starch grains. As nitrogen fixing activity was found to be highly variable with all samples, it was likely that nitrogen fixing bacteria were unevenly distributed over the various parts of the sago palm.

Samples showing high nitrogen fixing activity were then selected for bacterial isolation. Nitrogen fixing bacteria were isolated by aerobic cultivation using a culture medium containing no nitrogen source. The activity of isolates was monitored using a soft agar medium containing no nitrogen source. All isolated bacteria with high nitrogen fixing activity were analyzed for 16S rDNA gene sequences and their phylogenetic positions were estimated. It was found that nitrogen fixing bacteria in the sago palm belonged to a wide variety of species, including *Klebsiella pneumoniae*, *K. oxytoca*, *Pantoea agglomerans*, *Enterobacter cloacae*, genus *Burkholderia*, and *Bacillus megaterium* (Figure 5-17). The degree of similarity between the 16S rDNA sequences of the strains isolated at different times (March and July 2003, July 2004, and March 2005) was 99–100% for *K. pneumoniae* relatives, 98.4–100% for *E. cloacae*, and 100% identical for *K. oxytoca* and *P. agglomerans*. The result pointed to the possibility that the isolates colonized the sago palm ubiquitously.

Various factors influence the ability of these isolated nitrogen fixing bacteria to fix nitrogen in the plant body. One factor is the carbon source. The synthesis of nitrogenase, an atmospheric nitrogen fixing enzyme, requires a lot of energy. Symbiotic nitrogen fixing bacteria receive some photosynthetic products as carbon sources from the host plant. The rhizosphere has relatively abundant carbon sources as photosynthetic products leak out of the plant roots. It is therefore likely that many free-living nitrogen fixing bacteria can fix nitrogen there. This is one of the reasons why sago palm roots exhibited high activity. Another potential energy source is the presence of several polymeric compounds in the plant body. These include cellulose (a cell-wall constituent), pectin and lignin (in intercellular spaces), and starch (accumulated in the pith) in the case of sago palm. Although nitrogen fixing bacteria do not generally have enzymes to break down polymeric compounds, they can utilize these compounds as carbon sources if they can cooperate with polymer degrading bacteria.

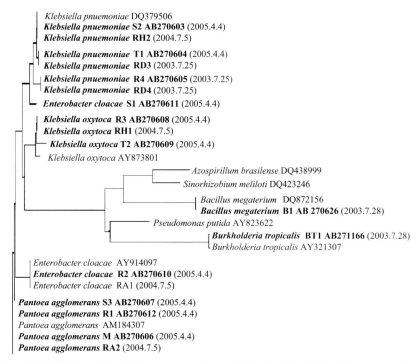

Figure 5-17 Phylogenetic relations among nitrogen fixing bacteria isolated from sago palms based on 16S rDNA gene sequences

Strains with sampling dates in brackets are the isolates from sago palms. AB, AY and DQ followed by numbers are accession numbers.
Strains in bold were isolated as part of the present study with isolation dates in brackets.
Source: Shrestha et al. (2007)

A second factor is oxygen partial pressure. There are many types of nitrogen fixing bacteria, including aerobic, strict anaerobic and facultative anaerobic, but nitrogenase itself becomes inhibited at high oxygen concentrations in each type. In this regard, wetland sago palms provide a suitable environment for nitrogen fixation, growing as they do in oxygen-poor soils. Dominant bacteria and various polymer degrading bacteria were isolated from various parts of the sago palm. The impacts of interactions among these microbes and nitrogen fixing bacteria on nitrogen fixing activity were evaluated.

The effect of interaction among microbes can be either inhibitory or friendly. The nitrogen fixing activity was around 2.1–31.4 nM $C_2H_4$ per culture per hour when nitrogen fixing bacteria alone were cultured but it increased markedly to 17.4–281.8 nM $C_2H_4$ per culture per hour when they were co-cultured with non-nitrogen fixing bacteria (Shrestha et al. 2007). Measurement of the activity of nitrogen fixing bacteria under low oxygen conditions found a marked increase in activity. It was therefore considered that the mechanism of increase in the nitrogen fixing activity in co-culture involved increase in nitrogenase activity as oxygen partial pressure around nitrogen fixing bacteria decreased due to oxygen consumption by indigenous bacteria. In other words, any microbes have the potential to enhance nitrogen fixing activity by consuming oxygen using some form of carbon source in the neighborhood of nitrogen fixing bacteria.

When starch that was abundantly present in trunked sago palms was used as a carbon source, nitrogen fixing bacteria showed little nitrogen fixing activity

Table 5-7 Nitrogen fixing activity enhancement by co-culture of nitrogen fixing bacteria and starch degrading bacteria

| Nitrogen fixing bacteria strain | Acetylene reducing activity (nmol $C_2H_4$/culture/h) | |
| --- | --- | --- |
| | Nitrogen fixing bacteria alone | Nitrogen fixing bacteria & *B. megaterium* BT1 |
| *Enterobacter cloacae* S1 | 0.1 | 270 ± 15 |
| *Klebsiella pnuemoniae* S2 | 1.5 | 258 ± 7 |
| *Pantoea agglomerans* S3 | 0 | 154 ± 5 |

Cultured in a nitrogen-free medium with starch as a carbon source.
Source: Shrestha et al. (2007)

Table 5-8 Estimated biologically fixed nitrogen in mature sago palm stand

| | Nitrogen fixing activity by part (nmol $C_2H_4$/g/d) | Mean weight per palm (kg) | Fixed nitrogen per palm per year (kg) | Fixed nitrogen per hectare per year (kg) |
| --- | --- | --- | --- | --- |
| Root | 221 | 68 | 0.04 | |
| Pith | 213 | 721 | 0.3 | 210 (923 sago palms/ha) |
| Leaf sheath | 35.4 | 97 | 0.01 | |

Source: Estimates based on Jong (1995b), Kaneko et al. (1996), Josue and Okazaki (1998), Miyazaki et al. (2003) and Yamamoto et al. (2005c).

on their own but their nitrogen fixing activity increased markedly when they were co-cultured with a starch degrading bacterium *Bacillus* sp. B1 (Table 5-7). The activity of nitrogen fixing bacteria also increased markedly when they were co-cultured with a hemicellulose degrading bacterium *Agrobacterium* sp. HMC1 or *Flexibacter* sp. HMC2 in a hemicellulose medium. As these starch and hemicellulose degrading bacteria were isolated from the sago palm, it was conceivable that these degrading bacteria and nitrogen fixing bacteria co-acted with each other to fix nitrogen using starch and hemicellulose localized throughout the plant body as carbon sources. As nitrogen fixing activity intensifies, intracellular nitrogen levels in nitrogen fixing bacteria increase. While it appears unlikely that nitrogen thus produced can be utilized directly by other bacteria or the plant body, nitrogen fixing bacteria become beneficial for the sago palm after death as their cell components will turn into simple organic matter and eventually transform into mineral nitrogen that can be utilized by other microorganisms and the plant body. However, since co-culture of pectin degrading bacteria and nitrogen fixing bacteria did not result in increased nitrogen fixing activity, the presence of polymeric compounds and their degrading bacteria would not necessarily enhance nitrogen fixing activity.

While the estimation of fixed nitrogen based on acetylene reducing activity contains some uncertainties, the amount of fixed nitrogen in a sago palm stand was calculated at approximately 200 kg per hectare per year based on the mean acetylene reducing activity values in the sago palm root, pith and leaf sheath measured by ourselves, the weight of each part found in literature, and the planting density of mature sago palms per unit area (Table 5-8).

The above findings suggest that nitrogen fixing bacteria colonize various parts of the sago palm and actively fix nitrogen. In the process they enhance their nitrogen fixing activity by using polymeric compounds such as starch and hemicellulose in the sago palm as nutrient sources as well as by cooperating with other bacteria that degrade such materials.


Authors:
5.1: Hitoshi Naito
5.2: Akira Miyazaki
5.3: Yoshinori Yamamoto
5.4 and 5.5: Masanori Okazaki
5.6: Hitoshi Naito and Hiroshi Ehara
5.7: Koki Toyota

# 6
# *Cultivation and Management*

## 6.1 The current state of sago palm harvesting and cultivation methods

### *6.1.1 Cultivation at small-scale farms*

The sago palm cultivation method used by small-scale farms varies from one area to another and depending on the purpose, i.e., whether they are grown as food for personal consumption, for commercial sale, or for socio-cultural purposes. In Papua New Guinea, Irian Jaya, western Indonesia, the Philippines and Thailand where sago palms are primarily used for personal consumption, socio-cultural purposes (see Chapter 10), small-scale sale and roof thatching, they are cultivated in the vicinity of villages and planted to define boundaries between individual homes or fields. The number of cultivated palms tends to be small and their varieties (folk varieties) are limited to those which are capable of supplying suckers or are of cultural importance.

In the Indonesian province of Riau and the Malaysian state of Sarawak where sago palms are cultivated as a cash crop, in contrast, each farm operates a relatively large growing area. These range from 1 ha for small farms to several hundred ha for large farms.

#### 6.1.1.a Field preparation

When small farms open fresh ground for the cultivation of sago palms, they weed and clear the planting area, either partially or entirely. Where the entire area is cleared, the field is burned after the removal of weeds and trees. For partial clearing of a field, rows are prepared where sago palm suckers are to be planted. Weeding is carried out in such a way that the surrounding vegetation is not adversely affected. Suckers are planted at intervals of 6–10 m in a square or rectangular planting pattern.

#### 6.1.1.b Transplanting in the field

Many farms select high quality suckers and transplant them directly in the field without any prior preparation in a nursery. A small percentage of farms place them on rafts in shallow drainage canals to raise suckers for a period of 1 to a few months prior to transplantation.

Figure 6-1 Sucker placed in a transplant hole
Two sticks are crossed to support the sucker.

Holes are dug to a depth of 20–50 cm in the prepared transplanting site (where field preparation has been completed). Suckers are placed in these holes the root end down. Two sticks are placed in a cross shape over the rhizome part to support the sucker (Figure 6-1).

In Malaysia, a single sucker is planted in each hole. Very few farms replace suckers that have died after transplantation. Transplanted fields only receive minimal management such as weeding at early stages.

In Riau, Indonesia, it is common practice to plant 2 or 3 suckers in each hole. This is useful for improving the successful rate of transplantation as suckers are rarely replaced at early stages. As with small Malaysian farms, sago palm fields in Riau receive minimal management with no fertilizer application.

## 6.1.2 Small-scale farms in Riau, Indonesia

Tebing Tinggi Island, which is a district situated in Riau Province, Indonesia, is the center of commercial sago production. A large part of sago starch produced by the local flour mills in this area is consumed domestically.

There are 30 sago palm farms on the island ranging from 0.6 to 62 ha in size with an average of 11 ha. The results of a study on their cultivation practice are described below (Jong 2001).

As mentioned earlier, the farms transplant 2–3 sago palm suckers in each hole in order to minimize post-transplantation management and sucker losses. On average, 87 sago palm suckers are transplanted in each hectare of land but the density of the mature sago palm fields among the surveyed farms varied widely from 5.6 to 53 palms/ha with an average of 26 palms. The fields managed by sago palm processing factory owners had an average planting density of 45/ha.

Eighty-six percent of the surveyed small farms pre-sold their sago palms before maturity ('pajak' in the vernacular). These pre-sales are made when the farms require cash for urgent expenses, including house construction, weddings and celebratory occasions for example. The farms sell some or all of their sago palms as the need arises. Sago palms are sold at the earliest 3 years after transplanting but usually 8 years after transplanting.

The selling price of an immature sago palm is determined not only by its growth condition, which affects starch production, but also by its age and the distance from transport waterways. The sellers continue to grow the pre-sold sago palms at their farms for the buyers with no charge for subsequent husbandry.

Figure 6-2 NTFP sago plantation

It generally takes between 11 and 15 years from transplanting to maturity, with an average of 12 years. Assuming that the starch yield per sago palm is 250 kg, the annual sago starch yield is 6 t/ha/yr on small-scale farms and 11 t/ha/yr on the farms of starch factory owners. Nevertheless, considerable amounts of starch are lost in the form of unextracted starch during the conventional starch refining process at the existing sago starch factories. This is due to factors such as sago log deterioration in storage, coarse milling and low starch extraction rates.

According to a survey of starch factories in the area conducted in 2007, 175 kg of dry starch (moisture content: 13%) were produced from sago palm trunks weighing 1 t in fresh weight. However, sago palms of Tebing Tinggi Island are thought to be able to produce more starch. The same 1 t fresh weight trunks yielded 210 kg of dry starch when they were milled and extracted thoroughly and sieved through an 80-μm screen (Jong, unpublished).

It is estimated that about 100,000 t of dry starch are produced by many small-scale starch factories on Tebing Tinggi Island each year. This means that about 570,000 t of mature sago logs are harvested in this area annually.

### 6.1.3 Intensive sago palm cultivation at plantations

Intensive sago palm cultivation at plantations broadly means that sago palms are being cultivated in a large growing area, generally 1,000 ha or more, with the use of appropriate planting densities and cultivation techniques (Figure 6-2). While new plantations are being developed, the existing sago palm plantations are limited to a small number being operated in Malaysia and Indonesia.

The first large-scale sago plantation was a 7,700-ha site developed by Estate Pelita in the Malaysian state of Sarawak during the 1980s. It employed a modern cultivation system. It was expanded to 17,700 ha several years later (Sahamat 2007). No detailed information about this plantation is available publicly.

In Indonesia, the first license for commercial sago palm plantation development was issued to PT. National Timber and Forest Product (NTFP). The initial plan entailed the development of 20,000 ha in Tebing Tinggi Island, Riau. Transplantation commenced in 1996. A total area of 12,000 ha had been transplanted by 2007. Further transplanting is scheduled to take place in 2008. The first batch of sago palms that was transplanted in an area of 2,000 ha in 1997 has reached the harvesting stage.

Two privately-owned plantations are developed on Tebing Tinggi Island in Riau and Sumatra Island. The size ranges from 1,000 ha to 3,000 ha. They use semi-intensive cultivation systems.

Intensive cultivation systems that are used at plantations are discussed below based on the sago palm plantation operated by PT. NTFP as an example.

Figure 6-3 Multi-purpose canal at NTFP plantation

Figure 6-4 Single-track railway laid at the plantation
This is used to transport fertilizer and other material.

### 6.1.3.a Infrastructure development

Infrastructure is of critical importance to large-scale sago palm plantations. Infrastructure includes a means of transport, buildings and a means of communication. The establishment of a transport network is important for work management, the supply of materials to the field and the shipping-out of a harvest. At sago palm plantations developed in low land swamps, irrigation canals (Figure 6-3), railways and roads (Figure 6-4) are used in combination for practicality and cost effectiveness. The railway is laid in a strategically important area as the central means of transport that can be used throughout the year. The road is built in an area that has been drained and compacted for the construction of the canal. The canal also plays an important role in transporting materials into the field and transporting the harvest out of the field, as well as fire control, groundwater table control and as a nursery for raising sago palm suckers.

Managers and workers who are involved in plantation management typically live on site in order to minimize travel time. The plantation at PT. NTFP is divided into lots of 2,000 ha each. A camp is set up in each lot to provide a base from which to carry out field work and management. An assistant manager is the head of each camp, which independently carries out day-to-day field operations. A workers' dwelling is located on every 200 ha from which 10 or so workers perform routine tasks.

Figure 6-5 Mechanically weeded transplanting row and suckers immediately after transplantation

### 6.1.3.b Groundwater level management

Maintaining the groundwater table at an appropriate level is as important as infrastructure development for sago palm growth. In the peat soil area, the groundwater level is controlled to stay within 25 cm below the ground surface in order to avoid land subsidence and irreversible damage from excessive drainage. Water canals are usually constructed about a year before commencing transplantation (Figure 6-3).

PT. NTFP has adopted a closed system of water canals at its plantation, punctuated by floodgates in appropriate positions. The groundwater level is maintained at about 18 cm below ground on average throughout the year. The movement of groundwater decreases and stagnates (oxidation-reduction potential of -15 mV) in the dry season with few deluging rainfalls. This is important for minimizing the leaching losses of applied fertilizer.

The first step in field preparation is to determine the planting site (transplanting rows) based on the ease of work and access. Weeds and large shade trees growing in a 2–3 m strip of land along each transplanting row are manually or mechanically removed (Figure 6-5).

After transplanting, vegetation between the rows is gradually cleared as part of post-transplanting management because it might shade the transplanted suckers, blocking the sunlight. Consequently, the original vegetation is altered gradually as sago palms grow.

### 6.1.3.c Procurement, grading and cultivation of suckers

Suckers are procured from several suppliers under contract. The quality of the procured suckers varies greatly. Individual suckers differ in size, shape, freshness (elapsed time since separation from the parent palm), water content, temperature and physical damage. Causes of sucker quality deterioration include the following.

1. The contractor took a long time to collect suckers or traveled a long distance to deliver them.
2. The low quality of available suckers (immature, too large or too small, and improper shape as too much of the rhizome, roots or rachis were cut off).
3. Suckers are damaged by sago beetles or other insects, infested with insect eggs or other pathogens.
4. Suckers are too dry from overexposure to the sun or too hot from lack of ventilation due to poor management during transportation.

Accordingly, suckers are graded upon delivery to the plantation in order to ensure high survival rates after transplanting and contractors are advised not to supply low quality suckers.

To achieve a high post-transplanting survival rate, suckers must be fresh, only a short time (usually a few days) after separation from the parent palm.

The rhizome is still moist and no pathogens or sago beetles are found on the cut surfaces. The rachis and leaf sheath are deep green in color and not dry. A physiologically mature sucker has a long rhizome, which is often cut and separated from the parent palm at a narrow, hard section (neck).

The raising of suckers in the nursery prior to transplanting is discussed in Section 6.3.1 below.

#### 6.1.3.d Field preparation and transplanting

As environmental impacts need to be considered in field preparation, burning off vegetation is prohibited. There is no need to completely clear an entire field (complete removal of vegetation) but sucker transplanting rows must be cleared thoroughly. It is recommended that vegetation in transplanting rows should be mechanically cleared in the case of large-scale transplantations (Figure 6-5).

Transplanting rows need to be marked manually prior to clearing the vegetation. Transplanting rows are set at intervals of 10 or 12 m. The use of heavy equipment weighing around 15 t (excavator) is advisable for the removal of vegetation and the clearing of transplanting rows. The removed vegetation should be set aside on either side of each row and a strip of cleared ground about 3 m wide is created. If the ground is root-bound and the water table is low, the excavator should be used to dig a trench in the middle of each transplanting row to a depth of 10–20 cm. This is a preparatory measure to achieve the desired depth of transplanting holes.

Transplanting positions are marked by hand at intervals of 8 m (where rows are 12 m apart) or 10 m (where rows are 10 m apart) on the cleared transplanting rows. Transplanting holes are dug with a spade down to immediately above a water table or until moist soil appears. This task can be performed by the excavator during the removal of vegetation. The size of the holes depends on the size of suckers but it is usually sufficient as long as the rhizomes can fit in them.

Sucker planting and rooting are described in Section 6.3 and post-planting management, harvesting and transportation are discussed in Sections 6.4 and 6.5 below.

## 6.2 Propagation

### *6.2.1 Propagation by suckering*

Sago palm (*M. sagu* Rottb.) can reproduce by way of either vegetative propagation with axillary buds initiating from lower leaf axils or adventitious buds initiating from the underground part of the trunk (stem) or by way of seed propagation. In general, farmers use suckers for new plantings as seedlings germinated from seeds need to be raised in a nursery for 12 to 18 months (Jong 1995a). Sago palms in natural stands propagate by both suckering and seeding.

Sago palm is a monocarpic (or hapaxanthic) plant that dies after flowering and fruiting. As both seedlings and suckers begin to produce suckers at a relatively young age (as early as 2 years after germination in the case of seedling), however, the clump itself continues to live after the original parent palm withers and dies. In addition to the perception that suckers require a shorter time in the nursery and hence reach harvesting stage sooner, suckers are widely used for new plantings due to a belief that they express morphological and physiological characteristics that are identical to those of the parent palm.

According to Yamamoto (1998a), suckers may be directly planted in the field when there is a sufficient amount of precipitation (300 mm/month) but they are generally placed on rafts floating on water and nursed for 3–5 months until they begin to produce new roots before being planted in the field. Shading is essential during the nursing period in many cases. According to sago palm growers in Sarawak, Malaysia, suckers (within 12 months of initiation; 1.5–2.0 m in length) of harvesting-stage sago palms that have initiated from higher positions on the trunk tend to root better in the nursery and therefore grow better after planting in the field.

While larger suckers, within limits, reportedly show better growth than seedlings after planting (Jong 1995a), they also pose problems in terms of the difficulty of transporting and handling, require more nursery space, and incur higher planting costs (Yamamoto 1998a). There have been some studies about the optimal sucker size for planting. Jong (1995a) suggests 7–10 cm in diameter and 2–5 kg in weight. Similarly Flach (1984) suggests 10–15 cm in diameter and about 2.5 kg in weight. Approximately 10 cm diameter and just over 2 kg in weight seems to be generally optimal for ease of handling, but there also appears to be a requirement for a certain size in order to ensure a necessary amount of nutrient stored in the rhizome.

A post-planting survival rate of around 90% can be achieved when suckers are planted within a week or so from picking under shaded conditions or within 3 days without shade (Jong 1995a) as mentioned below (see 6.3.1.a). The survival rate declines as the storage period lengthens. The survival rate after 3 weeks in storage was reportedly around 60% when shaded and 30% or less when unshaded (no such drop in the survival rate occurs when suckers are nursed). The planting depth should not be too deep, but it is desirable to bury between one-half and the whole subterranean stem in the ground (Jong 1995a). Applying some bactericide or fungicide to cut surfaces is beneficial for increasing the survival rate of suckers.

### 6.2.2 Propagation by seedling

Sago palm (*M. sagu* Rottb.) is capable of both vegetative propagation and seed propagation. Vegetative propagation using suckers is more common but it is not easy to procure a large number of suckers at once. For this reason a wider use of seedlings as transplanting stock has been anticipated in recent years.

However, sago palm seeds (inside the fruit) tend to have a low germination rate (Johnson and Raymond 1956; Flach 1984; van Kraalingen 1984; Jaman 1985; Alang and Krishnability 1986) as well as significant variation in the length of time from sowing to germination (Ehara et al. 1998, 2003d). The germination rate can be improved by removing seed coat tissues such as exocarp, mesocarp and sarcotesta to promote imbibition and eliminate germination inhibitors time (Ehara et al. 2001) (Figure 6-6). As the growth of each organ of the seedling progresses at relatively a constant rate after germination (Ehara et al. 1998), enhanced germination can lead to the production of uniform seedling stock. The germination rate is maximized at a temperature of 30°C and declines at temperatures both lower and higher than that (Ehara et al. 1998).

Figure 6-6 Germinated seedlings grown in coconut palm fiber medium
2 weeks after germination, at the Sokoine University of Agriculture, Tanzania, March 2005.
Source: Ehara et al. (2006c)

Propagation by suckers is more common in sago palm plantations because seedlings need 12 to 18 months to grow to a size suitable for transplantation (Jong 1995a). When sago palms were grown from fruits or seedlings at the Sokoine University of Agriculture, Morogoro, Tanzania, the seedlings reached a plant body height of 130–160 cm during a nursing period of 15 months from germination with a leaf emergence rate of about 1.5 leaves/month (Ehara et al. 2003c) (Figure 6-7). Once transplanted in the field at the university, their body height increased by more than 200 cm in the following 20 months (Figure

6-8). When sago palms are cultivated in Japan, growth stagnates in winter even if heating is provided; the average leaf emergence rate throughout the year is about 1 leaf/month; and seedlings take 18–20 months from germination to reach a body height of 130 cm (Ehara, unpublished). Sago palm seedlings begin to produce suckers as early as in their second year after germination but the timing varies greatly from one seedling to another; it can be as late as three full years after germination for some seedlings to initiate their first sucker. The sago palms that were introduced into Tanzania grew 2–6 suckers by the third year (Ehara et al. 2006c). The duration from germination to trunk formation through the seedling and rosette stages is about 4 years.

In contrast, the section *Coelococcus* plants that are found in the South Pacific propagate by seed only. They do not produce suckers even though they belong to the same genus. *M. amicarum* from Micronesia, *M. warburgii* from Vanuatu and *M. vitiense* from Fiji show high seed germination rates. *M. warburgii* in particular produces viviparous seeds (Ehara et al. 2003d). Post-germination growth in terms of plant body height elongation is about 35 cm in 5 months and about 90 cm in 12 months in the case of *M. amicarum*, 60–70 cm in 5 months in *M. warburgii*, and about 30 cm in 5 months in *M. vitiense*. The average leaf emergence rate is about 1 leaf/month (all in a greenhouse, heated in winter, in Japan) (Ehara, unpublished).

Figure 6-7 Seedlings nursed in pots
The rightmost palm was aged 15 months and the rest were aged 10 months after germination, at the Sokoine University of Agriculture, Tanzania, August 2006.
Source: Ehara et al. (2006c)

Figure 6-8 Palm transplanted in the field 20 months earlier after a nursing period of 10 months in a pot
At the Sokoine University of Agriculture, Tanzania, November 2006.
Source: Ehara et al. (2006c).

## 6.3 Planting and establishment

Important issues concerning sago palm cultivation by sucker propagation include the establishment rate (survival rate) after planting and early growth performance. To improve the establishment rate and promote early growth after planting, collected suckers are raised in a nursery prior to planting.

### 6.3.1 Sucker nursing

Except when collected suckers are planted directly in the field at small-scale sago palm farms, they are nursed for 1–5 months in creeks or swamps near the intended field until they initiate new roots and leaves prior to planting (Figure 6-9). Nursing collected suckers is important in confirming viability when large numbers of suckers are collected and stored for prolonged periods for large-scale cultivation. To nurse suckers in creeks, rafts made of sago palm rachises and petioles are used (Figure 6-10).

Figure 6-9 Nursing collected suckers in a creek (Top) and a swamp (Bottom)

Figure 6-10 Sucker nursery rafts made of sago palm rachises and petioles
About 3 m long and 75 cm wide.

Figure 6-11 Trimmed and prepared suckers

### 6.3.1.a Sucker preparation

*Rhizome cutting*
To facilitate handling and minimize transpiration from leaf surfaces, folded leaves are cut to 30–50 cm lengths from the base. Only the young unfolded (spear) leaf is left on the sucker (Figure 6-11). The rhizome is sometimes cut for ease of transportation. When suckers with different rhizome lengths (10–30 cm at 5 cm intervals) were planted 1 day and 14 days after collecting, no clear difference in the survival rate was found among those that were planted 1 day after collecting but suckers with longer rhizomes showed higher survival rates among those that were planted 14 days after collecting (Jong 1995a). It is likely that suckers with more nutrition reserves in the rhizome pith survived the prolonged storage better.

*Root trimming*
Collected suckers have numerous roots, which take up more space during transport and make handling more difficult at the nursery. For these reasons, sucker roots are cut to certain lengths after collecting. The length of the remaining roots (1–10 cm) was found to have no impact on the survival rate of suckers that were placed in the nursery immediately after collecting (Jong 1995a). Sucker roots are commonly trimmed to a length of 2–3 cm.

*Chemical treatment of sucker rhizome cut surfaces*
Suckers are cut from the parent palm at a point close to their attachment. They sometimes become infected by fungi (*Penicillium, Mucor, Aspergillus*) in the cut surfaces and die in the nursery. The survival rate is improved when the cut surfaces are immersed in a fungicide solution of 1% benlate or 1% carbofuran or a mixture of the two before nursing. Chemical treatment with these solutions is expected to be effective especially where and when fungal incidence is high.

*Sucker storage period*
It is important to commence nursing immediately after sucker collecting to achieve high survival rates in the nursery. If suckers have to be stored, those which are kept under shade, e.g., fully covered with sago palm leaves, show about an 80% survival rate after 2 weeks or so, but the survival rate drops rapidly after the third day in storage when they are kept unshaded (Table 6-1) (Jong 1995a).

### 6.3.1.b Sucker characteristics

The timing of emergence and the number of new leaves and roots during the nursing period depend on the sucker characteristics. Factors such as the shape and size (weight, rhizome diameter) of the sucker, and the age and growing environment of the parent palm are considered to be relevant characteristics, but only a few studies have been conducted on this subject.

Table 6-1 The survival rate of sago palm suckers stored for different periods under shaded and unshaded conditions

| Storage duration (days) | Survival rate (%) | |
| --- | --- | --- |
| | Shaded [1] | Unshaded |
| 0 | 86.0 b | 93.0 d |
| 3 | 93.3 b | 90.0 d |
| 7 | 93.3 b | 56.3 bc |
| 14 | 80.0 b | 60.3 cd |
| 21 | 56.7 a | 26.7 ab |
| 28 | 60.0 a | 13.3 a |
| Standard error | 6.9 | 14.1 |
| Coefficient of variation (%) | 10.8 | 30.0 |

1) Shaded by sago palm leaves (50%)
The figures followed by different letters differ significantly at p=0.05 by DHRT.
Source: Jong (1995a)

Figure 6-12 Forms of collected suckers

*Sucker shape*
The shapes of the collected suckers after their above-ground part is trimmed to a length of 30–50 cm can be broadly divided into two types: a J shape and an L shape (Figure 6-12). The L-shaped suckers are preferred as they can be set on nursery rafts and stabilized in planting holes more easily.

*Sucker size*
The age (years after emergence) of the suckers collected for propagation varies from area to area but they are generally about 1–3 years old. Although suckers grow at an increasing rate as they age, the rate of growth varies greatly

depending on environmental conditions. In suckers raised on creek rafts for 2–4 months, sucker weight (1–5 kg) made no difference to the length of new leaves or new roots but the heavier the sucker, the greater the number of new roots (Figure 6-13) (Irawan et al. 2005). Thus, heavier suckers perform favorably in terms of new root emergence in the nursery, establishment after planting and early growth but are problematic in terms of transport, handling, nursery space and labor requirements for planting. As mentioned earlier, a diameter of about 10 cm and a weight of 2–5 kg are considered to be the most desirable size of suckers for plantation.

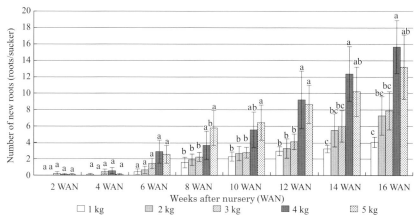

Figure 6-13 Effect of sucker weight on the number of new roots during the nursery period
Different letters indicate significant difference at the 5% level
Source: Irawan et al. (2005)

*Age of the parent palm*
It is common practice among sago palm growers to collect suckers for propagation from mature palms nearing the flowering stage rather than young palms. However, when suckers taken from young parent palms of 1–2 years after trunk formation and mature parent palms at the flowering stage (about 10 years after trunk formation) were raised on rafts for 2.5 months, both groups showed a survival rate of 70–80%. This absence of significant differences suggests that the age of the parent palm has little influence. This might be because little difference is found between the suckers taken from the parent palms of different ages in the nonstructural carbohydrate (total sugar and starch) content as well as the mineral content in the rhizome pith (Irawan et al. 2009b).

*Soil type of the parent palm habitat*
Sago palms are found growing in a wide range of soil types from mineral soil to deep peat soil. When suckers taken from young parent palms (1–2

years after trunk formation) and mature parent palms (nearing the flowering stage) growing in mineral soils and deep peat soils (about 3 m thick) were raised on rafts for 2.5 months, no significant difference was found in their survival rates. This suggests that the soil type has only a minor influence on the sucker characteristics for propagation. No clear difference was found in the nonstructural carbohydrate content in the rhizome pith (Irawan et al. 2009b).

### 6.3.1.c Sucker nurseries

*Planting depth*
When suckers are nursed in low land swamp soils, holes need to be deep enough to accommodate at least half and sometimes the entire rhizome in order to maintain high survival rates. Placing the rhizome on the surface of the soil or burying it too deep decreases the survival rate (Table 6-2) (Jong 1995a).

Table 6-2 The survival rate of sago palm suckers planted at different depths

| Planting depth | Survival rate (%) |
|---|---|
| Rhizome left on soil surface | 76.7 ab |
| Rhizome half buried on lower side | 90.0 ab |
| Rhizome fully buried just below the surface | 90.0 ab |
| Rhizome buried 8 cm below the surface | 70.0 b |
| Standard error | 7.4 |
| Coefficient of variation (%) | 11.4 |

The figures followed by different letters differ significantly at p=0.05 by DMRT
Source: Jong (1995a)

*Shading after planting*
The survival rate of suckers nursed in the soil varies markedly according to moisture conditions. During the dry season (August) (precipitation: 87 mm/month), the survival rate was enhanced by shading the suckers in the nursery (about 50% shading with sago palm leaves) (Table 6-3). Conversely, the survival rate was reduced by shading during the wet season (December) (precipitation: 391 mm/month) (Jong 1995a). Rotting leaves and growth points was found in the shaded suckers during the wet season. Thus, the effect of shading in the nursery differs greatly depending on the amount of precipitation. It appears, however, that shading can effectively prevent the suckers from drying in the dry season when precipitation and humidity are generally low.

Table 6-3 Effect of shading on the survival of suckers during wet and dry seasons

| Shading condition | Survival rate (%) | |
|---|---|---|
| | Dry season | Wet season |
| Shaded[1] | 83.3 a | 66.7 b |
| Unshaded | 66.7 a | 90.0 a |
| Standard error | 29.0 | 6.7 |
| Coefficient of variation (%) | 23.6 | 5.2 |

1) Shaded by sago palm leaves (50%)
The figures followed by different letters differ significantly at p=0.05 by DMRT
Source: Jong (1995a)

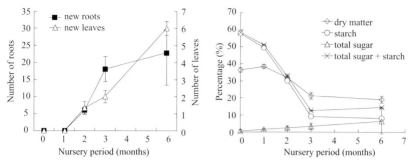

Figure 6-14 Changes in the number of new roots and leaves (left), dry matter, total sugar and starch percentage (right) during nursing period on rafts (Riau, Indonesia, 1999)
Mean values for a sample of 9 suckers at the 0, $1^{st}$, $2^{nd}$ and $3^{rd}$ months and 4 suckers at the $6^{th}$ month. Vertical bars on symbols in the graphs indicate standard errors.
Source: Omori et al. (2002)

### 6.3.1.d The emergence of new leaves and roots during the nursing period

Figure 6-14 shows the emergence rates of new leaves and roots as well as changes in the total sugar and starch contents of suckers that were nursed on rafts floating on a creek for 6 months (Omori et al. 2002). The emergence of new leaves and roots was noticed in the second month after the commencement of nursing. The numbers of new leaves and new roots in the third month were 2 and 18 respectively. The numbers of leaves and roots continued to increase until the sixth month of nursing.

The starch content in the sucker rhizome pith was high at around 60% when nursing commenced, then decreased to about 10% as new leaves and roots emerged but continued to grow through to the third month. The total sugar content tended to increase. This demonstrates that starch was used to form the substance that supported the emerging and growing leaves and roots. Despite the further increase in the number of new leaves and roots after the third month, however, the total sugar and starch contents actually increased. This shows that the growth of new leaves and roots lead the suckers into the autotrophism stage.

## 6.3.2 Sucker planting and establishment

When high water tables provide high moisture levels at the planting site, or continuing rainfall is expected after planting (i.e., during the wet season), suckers may be planted directly, bypassing the nursery period. In general, however, they are raised in a nursery for 3 to 5 months as described above in order to promote rooting after planting. Those with 2–3 new leaves and some new roots are then selected for planting.

Figure 6-15 A planted sucker
3 sticks were used to stabilize the sucker.

### 6.3.2.a Planting method

Holes are dug at designated planting positions according to the size of individual suckers (approximately 30–40 cm long, 30 cm wide and 30–40 cm deep). Suckers are placed in the holes with their growing points facing the same direction and covered with topsoil (Figure 6-15). If the water table is high, the holes are made shallower so that the growing points are not submerged. Planted suckers are sometimes pinned down with sticks to prevent movement (Figure 6-1). Once planted, suckers begin to grow in a month or so. A 2–3 m-long pole is erected on the ground to mark each planting position.

About 1 month after planting, workers begin patrolling the planted site to replace dead suckers and adjust plant posture.

The post-planting establishment rate of suckers that grew new leaves and roots in a nursery is largely governed by soil moisture conditions. An establishment rate of 80–90% can be expected when water tables and soil moisture levels are high immediately after planting. The establishment rate can be as low as 10–20% when groundwater tables and soil moisture levels are low. The wet season rainfall patterns in Southeast Asia have become irregular in recent years and some cases of markedly low establishment rates due to lack of post-planting rainfall have been reported (Omori 2001). In order to promote establishment after planting, some large-scale plantation sites keep their water tables high for a period of time after planting by regulating the water level in their canals.

### 6.3.2.b Early growth after establishment

In Riau, Indonesia, suckers were raised on a floating raft nursery for 3 months and planted in deep peat soil (3 m-thick peat layer) at a site of low water tables (annual average of 40–50 cm) and a site of high water tables (annual average of 0–30 cm) for 12 months for growth comparison. No difference was found in the growth conditions of the above-ground part (above-ground length: about 1.6 m; leaves: 16; fresh weight: 4 kg) between the two groups. By that time, the pith starch content had recovered to about 50% from a low level shown immediately after rooting. Although there was little difference in the number of roots at around 80 roots, the suckers growing at the low water table site were superior in the maximum root length, total root length, average root length and root weight. This indicated that groundwater levels have a major impact on early root growth after establishment (Table 6-4). These differences in the early growth of roots after establishment probably have a significant influence on the subsequent growth of the above-ground part.

Hence, it is clear that while a high water table is required for establishment after planting, it is important for good early growth to lower the water table once the suckers establish in order to promote root growth.

Table 6-4 The effects of groundwater level on root growth of sago palm suckers grown for one year after planting

| Groundwater level | Total number of roots | Total root length (m) | Mean root length (m) | Mean maximum root length (m) | Root dry matter weight (g) |
|---|---|---|---|---|---|
| Low (40–50 cm) (n=5) | 83 ± 32 | 49 ± 21 | 58 ± 14 | 170 ± 54 | 107 ± 42 |
| High (0–30 cm) (n=4) | 79 ± 34 | 27 ± 13 | 34 ± 5 | 95 ± 18 | 60 ± 31 |

Mean ± standard error
Source: Omori et al. (2002)

Table 6-5 Sago palm growth 2 years after planting

| Ketro variety (n=13) | | | Anum variety (n=11) | | | Makapun variety (n=6) | | |
|---|---|---|---|---|---|---|---|---|
| Palm No. | Height (m) | Number of leaves | Palm No. | Height (m) | Number of leaves | Palm No. | Height (m) | Number of leaves |
| 1 | - | - | 1 | 2.0 | 9 | 1 | 3.1 | 10 |
| 2 | 3.1 | 8 | 2 | 3.9 | 16 | 2 | 4.0 | 15 |
| 3 | 2.5 | 6 | 3 | 3.8 | 26 | 3 | - | - |
| 4 | 3.0 | 15 | 4 | 2.5 | 7 | 4 | 3.9 | 33 |
| 5 | 3.3 | 16 | 5 | 5.0 | 36 | 5 | 3.9 | 14 |
| 6 | 1.8 | 6 | 6 | 3.1 | 12 | 6 | 4.0 | 10 |
| 7 | 2.2 | 8 | 7 | 3.1 | 20 | 7 | - | - |
| 8 | 2.0 | 13 | 8 | 2.6 | 13 | 8 | - | - |
| 9 | 3.2 | 17 | 9 | 3.8 | 26 | 9 | - | - |
| 10 | 3.1 | 20 | 10 | 2.6 | 28 | 10 | - | - |
| 11 | 1.6 | 8 | 11 | 2.4 | 7 | 11 | - | - |
| 12 | 1.6 | 4 | 12 | - | - | 12 | - | - |
| 13 | 1.8 | 4 | 13 | - | - | 13 | - | - |
| Mean ± S. D. | 2.4 ± 0.7 | 10 ± 5.5 | Mean ± S. D. | 3.2 ± 0.9 | 18 ± 9.7 | Mean ± S. D. | 3.8 ± 0.47 | 16 ± 9.6 |

Source: Japan-PNG Goodwill Society (1984)

Table 6-5 shows the growth performance of 3 sago palm varieties (folk varieties) in the second year after planting in the Sepik River area, Papua New Guinea (Japan-PNG Goodwill Society 1984). There are significant variations in the above-ground height (1.6–5.0 m) and the number of leaves (4–36). While these variations are no doubt partially attributable to intervarietal variations, differences in the establishment performance of individual suckers and the soil environment of their planting sites are considered to have also been significant factors.

## 6.4 Post-planting management

### 6.4.1 Sucker preparation (see 6.1.3.d)

Although the sago palm can propagate by seeds, it is more common to adopt asexual propagation by suckers for new plantings of high-yielding varieties

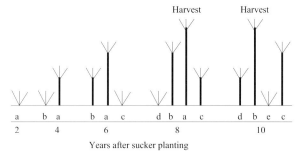

Figure 6-16 Sago palm sucker management method
Source: Sato et al. (1979)

(Watanabe 1984). The planted suckers produce many new suckers in the course of their growth. Consequently, once the original sucker matures and its trunk is harvested, its younger suckers can be used for continuous sago palm cultivation. In order to make this type of cultivation possible, each clump needs to be managed to contain trunked palms of different ages (Shimoda and Power 1992b). Without proper sucker management, the sago palm stand becomes too dense and can be harmed by competition between a parent palm and its suckers for nutrients and light as well as by pest and animal damage (Sato et al. 1979; Kakuda et al. 2005).

In sago palm cultivation in the mineral soil area of Batu Pahat, Malaysia, the practice of thinning has been adopted to get one sucker from each parent palm every 2 years (except one) (Flach 1977). This sucker management practice enables harvesting sago palms from each clump every 2 years after harvesting the original sago palm (Figure 6-16). Hence, where the sago palm needs 8 years from planting to harvesting, each clump needs to have 4 sago palms growing at various stages. By comparison, the sago palms grown in the peat soils of Riau, Indonesia, require longer to reach maturity, and therefore each clump is managed so that 6–8 sago palms are always developing at different stages (Jong 2001, 2006). When the population density is maintained at this level, the potential starch production of properly managed sago palms is estimated at 25–40 t/ha/yr, which is higher than the potential yields of other cereal crops (Yamamoto 2006a). However, there has been no long-term study of the relationship between the starch yield and sago palm population density maintained in sucker management. There is a need for further information in this regard.

In sucker management, as well as the population density, the positioning of retained suckers after thinning also needs to be considered. The light distribution within a clump may affect the starch accumulation process within the trunk of an individual sago palm (Flach 1977). Sucker positioning is also important for cultivation management, harvesting and transportation. In wild stands, many sago palms are left to wither due to difficulties in harvesting and transporting (Sato et al. 1979). The proper positioning of the retained suckers after thinning is hence an important element of sucker management but it is not easy to achieve since considerable numbers of suckers die prematurely (Shimoda and Power 1992b).

## *6.4.2 Fertilizer application*

### 6.4.2.a Sago palm cultivation in peat soils

Sago palms thrive in mineral soils and thin layers of peat soil (Sato et al. 1979). Sago palms growing in thick peat layers require a longer time to reach maturity, and hence, their starch yield per unit area per year is lower than those growing in thin peat layers (Yamaguchi et al. 1997; Jong et al. 2006). For example, a study of 9-year-old sago palms at a large-scale cultivation site in Sarawak, Malaysia, found that they formed healthy trunks in areas of peat soil depth up to 1.5 m whereas they exhibited growth inhibition and mass withering where peat soil depths were 1.5–3.0 m (Sim et al. 2005). It was suspected that nutrient deficiency was the cause of delayed trunk formation as the thick peat soil layers were found to be low in nitrogen (N), phosphate (P) and potassium (K) contents. Kakuda et al. (2000) surmised that the reduced nitrogen mineralization per unit volume supplied by peat soil is the reason for slower growth in peat soil than in mineral soil. Yamaguchi et al. (1994) found that 8-year-old sago palms growing in thick peat soil layers (> 1.5 m) showed a lower copper (Cu) content in the leaf and the trunk than their mineral soil-growing counterparts and suggested that this was one of the reasons for inhibited growth in thick peat layers. It was shown that the application of zinc (Zn) was needed for ongoing sago palm production because the reclaimed thick peat soil layers accumulated 9.37 g Cu /ha/yr while they lost 146.5 g Zn /ha/yr (Yamaguchi et al. 1994).

Physiological impairments are sometimes observed in wild palms. Nutrition deficiencies, especially combined deficiency, potassium deficiency (Figure 6-17), and deficiency in microelements such as zinc and copper, are widely observed both in the wild and under cultivation, in both peat soil and sandy soil. Obvious boron deficiencies commonly seen in oil palms are yet to be reported in sago palms.

Figure 6-17 Potassium deficiency

### 6.4.2.b Effect of fertilizer application on sago palm

Most of the reports on the effects of fertilizer application on the sago palm have been inconclusive (Kueh 1995). One study of growing young sago palms in culture solution, however, indicated the importance of N, P and K for the growth of young sago palms (Jong et al. 2007).

While appropriate methods of fertilizer application to compensate for nutrient deficiencies in peat soils are yet to be established, some fertilizer application tests are under way in Riau, Indonesia, and Sarawak, Malaysia.

Ando et al. (2007) applied N (urea), P (phosphate rock), K (potassium chloride), calcium (phosphate rock, dolomite) and magnesium (magnesium chloride, dolomite) to sago palms every 4 months from the time of planting in the thick peat soil layer (> 3m) of Riau, Indonesia. Dolomite was spread on the ground surface on the circumference of a circle with a radius of 1m from each of the target sago palms. The other fertilizers were buried in holes 15–20 cm deep that had been dug at 4 spots on that same circle around each sago palm. The chemicals were applied in the following quantities: 0.47 kg N/palm/yr, 0.57–0.73 kg $P_2O_5$ /palm/yr, and 2.4 kg $K_2O$ /palm/yr. An analysis of the sago palms (5 years old) 5 years after the commencement of the fertilizer application found no effect on either palm height or the number of leaves. In a separate plot, a mixed fertilizer containing boron, copper, iron, manganese and molybdenum was applied in addition to N, P, K, calcium and magnesium. Again, no effect was found on either the palm height or the number of leaves 5 years after commencement of the application.

Kueh (1995) tested fertilizer application from the time of planting sago palms in a thick peat soil layer (> 1.5m) in Sarawak, Malaysia. The chemicals were applied in the following quantities: 0.42 kg N/palm/yr of N (ammonium sulfate), 0.36 kg $P_2O_5$ /palm/yr of P (phosphate rock), and 1.2 kg $K_2O$ /palm/yr of K (potassium chloride). The effect on sago palms (1–12 years old) was studied annually for 12 years after the commencement of application. No difference was found between them and unfertilized sago palms in the number of leaves from the second to sixth year of the commencement of application, the trunk diameter (at 1 m from the ground surface) from the seventh to eighth year, and the trunk height from the seventh to tenth year. While N application had no effect on the N content in the leaf from the first to twelfth year, the application of P and K had some effect as there were differences between the fertilized and unfertilized groups in the P and K contents in some years.

Purwanto et al. (2002) applied N fertilizer on 4-year-old sago palms in the peat soils of Riau, Indonesia, and Sarawak, Malaysia, and studied their growth. Holes were dug to a depth of 15–20 cm at 4 spots on the circumference of a circle with a diameter of 1 m from each of the target sago palms and 500 g N/palm/application of LP-100 or urea were buried in them every 6 months. While no difference was found between the fertilized and unfertilized groups in palm height, the number of leaves and leaf N content in the 5th and 17th months of the commencement of application, the leaf N content was found to be significantly higher in the 23rd month.

Kakuda et al. (2005) applied approximately seven times the usual amount of nutrients (N (urea), P (phosphate rock), K (potassium chloride), calcium (Ca) (phosphate rock, dolomite), magnesium (Mg) (dolomite), copper (Cu) (copper sulfate), zinc (Zn) (zinc sulfate), iron (Fe) (iron sulfate) and boron (B) (borate)) twice on 5-year-old sago palms growing in a thick peat soil layer (> 3m) of Riau, Indonesia. The application method used was as described in Ando et al. (2007) above. Their growth was studied 16 months after the commencement of application. The sago palms in the fertilized plot were found to have higher dry matter weights of the petiole and the sucker than those in the unfertilized plot but other organs (leaf blade, rachis and trunk) showed no change of dry matter weight. The ratio of the parent palm dry matter weight to the total above-ground dry matter weight declined in a linear manner as the sucker dry matter weight increased. It declined further upon fertilizer application. This demonstrated that fertilizer application increased the dry matter weight of the suckers rather than the parent palm.

The effects of Cu and Zn application were studied by Nitta et al. (2000b, 2003). Fertilizer application was tested on sago palms growing in the peat soil of Riau, Indonesia, that were suspected of having Cu and Zn deficiencies. It was found that the application of Cu (copper sulfate) and Zn (zinc sulfate) fertilizers increased the leaf Cu and Zn contents. This effect lasted for 6 to 30 months after application.

These test results indicate that the effects of fertilizer application are not reflected in the palm height, the number of leaves or the amount of trunk growth but they may improve the nutrient content in the leaf. It is possible that applied fertilizers do not improve the growth of the parent palm because they promote the growth of suckers instead, as demonstrated by Kakuda et al. (2005). Sucker growth therefore needs to be strictly controlled in future fertilizer application tests.

#### 6.4.2.c Determining the fertilizer application rate

Before the fertilizer application rate can be determined, the amount of natural nutrient supply, the amount of nutrient demand to achieve the target yield, and the recovery rate of applied nutrients must be ascertained. The N balance in sago palm culture was studied by Okazaki and Yamaguchi. (2002). While nutrient demand is derived from a difference in nutrient absorption between fertilized and unfertilized plots, it is expected to vary widely depending on meteorological, soil and cultivation management conditions. Therefore information needs to be gathered at multiple spots. As the low N adsorption capacity of peat soil leads to N fertilizer losses (Kakuda et al. 2000), it may be useful to ascertain the recovery rate of applied N by using other media such as pot culture or solution culture.

### *6.4.3 Water management*

The sago palm is a wet endurance plant that can grow on a site of high groundwater level but experiments have found decreased growth under permanently submerged conditions (Flach et al. 1977). It has also been reported that sago palms growing in peat soil with a water table depth of 40–50 cm did not form trunks (Sim et al. 2005). These findings suggest that the groundwater level has an impact on the growth rate of the sago palm. The groundwater level is also a very important factor for sustained sago palm culture in tropical peat soils. Peat decomposition at lower groundwater levels not only promotes the production of carbon dioxide, one of the greenhouse gases, but also results in the loss of the foundation of sago palm growth.

No consistent relationship was found between the growth of sago palms at the rosette stage and the groundwater level. However, a positive correlation was found between the groundwater level in the range of 52–89 cm below ground and the number of leaves, chest-height diameter, trunk height increase rate (Figure 6-18) and trunk volume increase rate (Figure 6-19), which are growth parameters for sago palms at the trunk formation stage (Sasaki et al., unpublished). This suggests that sago palms grow better at higher groundwater levels. Approximately 50% of the sago palm roots are found to be growing less than 30 cm below ground level (Chinen et al. 2003). It appears that a drop in the groundwater level below the sago palm root growth zone causes moisture stress and inhibits sago palm growth.

Generally speaking, the chest-height diameters of the sago palms growing in tropical peat soil are smaller than those growing in mineral soil (Yamamoto et al. 2003c). While the rates of trunk height increase 1.34m/yr and trunk volume increase 0.166m$^3$/yr in sago palms growing in mineral soil according to Yamamoto et al. (2003c), Figures 6-18 and 6-19 show higher maximum rates at 2.2m/yr and 0.29m$^3$/yr respectively (Sasaki et al. 2007). The growth rate of the sago palm in peat soil when the groundwater

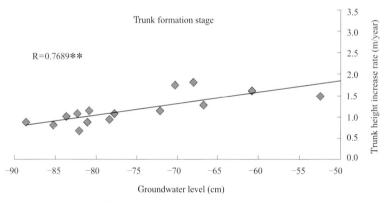

Figure 6-18 Groundwater level and trunk height increase rate

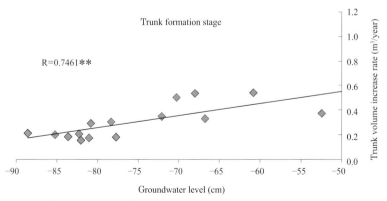

Figure 6-19 Groundwater level and trunk volume increase rate
Source: Sasaki et al. (2007)

level is high appears to exceed that of the mineral soil-growing sago palm. As the groundwater level in the reported study was set to a limited range, further research needs to be carried out with a wider range of groundwater levels. Moreover, different roles that the groundwater level might play in the transplantation, rosette and trunk formation stages deserve further investigation. Another question that requires further study is the relationship between the optimal groundwater level for sago palm growth and the optimum groundwater level for peat sustainability.

### 6.4.4 Weeding

Although the undergrowth is inhibited when it is shaded by the leaves of developed sago palms, suckers must compete with weeds and shrubs until canopies form (Yamamoto 1998a). Weeding must therefore be carried out 2 to 3 times a year for several years after sucker planting (Jong 2001, 2006).

Yanbuaban et al. (2007) reported that among 30 odd plant species found in the peat soil of Narathiwat, southern Thailand, grass weed (*Leersia hexandra*) and sedge weed (*Fimbristylis umbellaris*) were the main weeds that affected sago palm growth. *L. hexandra* severely inhibits sago palm growth by shading the sago palm leaves and preferentially absorbing phosphate and zinc, which tend to be deficient in peat soils. Nitrogen absorption by the sago palm is inhibited by the presence of *L. hexandra*. However, both the sago palm and the weed absorb nitrogen fixed by symbiotic microbes, hence the N problem is possibly not due to competition for soil nutrients unlike in the case of P or Zn.

Figure 6-20 Termites: a pest affecting sago palms

## 6.4.5 Pest and disease management

There are about 40 insect species that are known to be injurious to the sago palm. Many of them infest other palm species as well. The larvae of *Graphosoma rubrolineatum* and *Oryctes rhinoceros* damage the sago palm's trunk by feeding (Kimura 1979). Because they gain entry through the cut surfaces of the gathered rhizomes or suckers, these cut surfaces need to be treated with chemicals, mud or tar for protection (Jong and Flach 1995). The larvae of *Monema flavescens* and the *Psychidae* (locusts and crickets) are known to feed on sago palm leaves (Kimura 1979; Jong 2006).

In intensive sago palm culture in Sarawak or Indonesia, termites (Figure 6-20) and hispid beetle larvae are problematic. The widely reported sago beetles also damage certain parts of the sago palm, as well as laying eggs there. Sago beetles often infest the stump of a cut sago palm (Figure 6-21) and sago logs. Monkeys sometimes damage the petiole by chewing on it (Figure 6-22). Additionally, there is potential for damage by other pests in intensive, large-scale sago palm monocultures. Severely damaging pest infestations of sago palms have not been seen so far, though. Even in intensive sago palm plantations, the undergrowth continues to grow, providing a diversity of both flora and the entire biota which helps maintain a diverse ecology (Mitsuhashi and Kawai 1999).

Figure 6-21 Sago palm stump infested by sago beetles and fungi

No fatal sago palm disease has been reported. Fungal diseases such as white root disease are found in young palms but they do not inflict serious damage.

Figure 6-22 Petiole damaged by monkeys

## 6.5 Harvesting and transportation

Sago palms are harvested by humans using hand-held mechanical tools. Organized transportation systems are yet to be found on sago palm farms. In order to transport sago palms from the harvesting site to a canal with minimal labor, sago trunks are cut to a length of about 1 meter (called a log) and rolled to the destination on temporary tracks made of sago petioles or nearby trees. The average sago palm farmer uses a chainsaw to harvest sago palms (Figure 6-23). He first clears the area around the sago palm with a parang (machete) and then removes any remnants of leaf sheaths and petioles at the position he intends to cut with the chainsaw. He proceeds to cut down the palm with the chainsaw, which usually takes less than 10 minutes depending on site conditions. He uses either the parang or the chainsaw to remove leaves and sheaths from the felled sago palm. He makes holes in the trimmed trunk at regular intervals, then cuts logs with a chainsaw. He then passes ropes through the holes to assemble a raft (Figure 6-24). Log lengths vary from farm to farm, but it is typically around

1 m. The mean diameter of a log is about 40 cm on Tebing Tinggi Island, Riau (Figure 6-25). It took 2 workers 20–25 minutes to complete the process from harvesting to log-making on Tebing Tinggi Island. In Sarawak, a single worker can harvest about 20 sago palms in a day and cut them into logs using a chainsaw (Yamamoto 1998a). The time required for sago palm harvesting appears to be similar regardless of the region.

Logs are transported to the starch processing factory. Logs are rolled from sago palm plantations either directly to the sea or to canals or streams depending on access. At sago palm farms developed on peat soils, sago leaves and sheaths are laid on the ground for ease of transportation (Figure 6-26). Hooks are attached to the cut ends of the logs (Figure 6-27) to push and roll them. Where canals (streams) are accessible, the logs are assembled into rafts and floated downstream to the sea (Figure 6-28). In Riau, as many as 2,000 logs are assembled together into a raft. The logs are gathered in the sea (Figure 6-29) and tugged to the starch processing factory by a diesel-powered boat (Figure 6-30). In Riau, logs are collected from sago palm farms within a 30 km radius of the starch processing factory (according to interviews at the factory). On Tebing Tinggi, the log rafts are stored in the sea adjacent to the starch processing factory.

Figure 6-23 Sago palm logging using the chainsaw

Figure 6-24 The sago palm trunk cut to length for transportation

Figure 6-25 Cut log

Figure 6-26 Sago logs are rolled away on a track of leaf sheaths laid on the ground

Figure 6-27 Hooks are attached to cut ends for ease of transport

Figure 6-28 Sago logs assembled into a raft in the canal

Figure 6-29 Logs gathered in the sea

Figure 6-30 Logs tugged by boats in the sea

## 6.6 Cultivation management and environment

Proper management of water and soil nutrient conditions through controlling groundwater levels and applying fertilizer is necessary for sustainable sago palm cultivation in tropical wetlands. However, if such practices impose a heavy burden on the environment and the local ecosystem, sago palm cultivation cannot be considered to be sustainable agriculture. It is important to understand how sago palm culture impacts on the atmosphere, water and soil and to establish cultivation management techniques that can achieve both high productivity and low environmental impact.

Global warming, the greatest environmental challenge of the twentieth first century, is considered to be caused mainly by the increased tropospheric concentrations of gases such as carbon dioxide ($CO_2$) and methane and nitrous oxide which absorb energy released from the earth's surface (IPCC 2007). Wetlands are the biggest source of methane (IPCC 2007). A large part of methane in the atmosphere is biologically generated by microorganisms called *methanogens* (archaebacteria). These microorganisms are active in a reductive environment free of molecular oxygen. They obtain energy in the process of producing methane mainly from $CO_2$ and hydrogen or acetate. Wetlands provide a suitable environment for methanogens as atmospheric oxygen ingress is limited by high ground water tables while the accumulated plant residue and peat material supply them with energy sources. It is common practice to drain the peatlands as well as removing the natural vegetation in order to improve oxygen supply to plant roots before peatlands are used for agricultural production. Oxygen not only facilitates respiration in the crop roots but also significantly reduces the activity of methanogens while activating methane-oxidizing bacteria. As a result, methane emissions to the atmosphere is markedly reduced (Martikainen et al. 1995). The supply of oxygen activates not only methane-oxidizing bacteria but also many other bacteria and fungi, which accelerates organic decomposition and increases the rate of $CO_2$ production (Martikainen et al. 1995). Consequently, water management in sago palm cultivation has the potential to greatly change greenhouse gas emissions from soil.

A majority of nutrient elements other than nitrogen in natural soil are supplied from minerals in the parent material but the nutrient content in peat soil is inherently low as the surface layer and the mineral soil layer are widely separated by a thickly accumulated peat layer. The drainage of peat lands for agricultural use results in ground surface subsidence and accelerates the leaching of soil nutrients (Laiho et al. 1999). However, few studies have attempted to analyze the physiochemical properties of sago palm cultivated soils (Funakawa et al. 1996, Kawahigashi et al. 2003), and there have been no reports concerning the changes in the amounts of various nutrients in soil under sago palm production over time. The residual amounts and distribution of the nutritional elements applied as fertilizer are also unknown. In addition, there is concern that leached soil nutrients or fertilizer-derived nutrients may escape from drainage canals into rivers and the sea, thereby affecting the riverine and marine ecosystems.

This section discusses the effect of sago palm cultivation management in tropical peatland on greenhouse gas emissions and nutrient concentrations in draining canals and soils, mainly based on studies conducted by Watanabe et al. (2008, 2009) on Tebing Tinggi Island, Riau, Indonesia.

## 6.6.1 Effects of cultivation management on greenhouse gas (methane, $CO_2$) emissions

According to Furukawa et al. (2005), the methane emission rate from peat paddy fields in the Indonesian islands of Kalimantan and Sumatra ranged from 0.05 to 8.0 mg of carbon (C) /m²/hr. Furukawa et al. (2005) and Haji et al. (2005) reported that the $CO_2$ flux from peat soils used as upland fields in Malaysia and Indonesia ranged from 100 to 400 mg C /m²/hr. The methane and $CO_2$ fluxes from sago palm growing soils have been reported to range from 0.02–1.4 and 25–340 mg C /m²/hr respectively (Melling et al. 2005a, b).

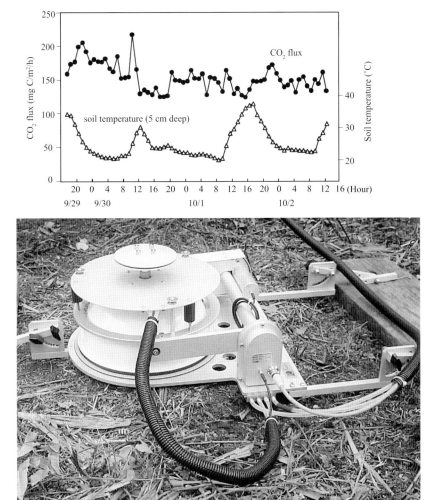

Figure 6-31 The chamber section of the automatic measuring device for $CO_2$ fluxes from soil set up in a sago palm field, Tebing Tinggi (bottom) and diurnal changes in $CO_2$ fluxes and soil temperature (5 cm deep) (29 September–2 October 2006: top)

The gas flux (emission or absorption rate) from the soil surface is frequently measured using a closed chamber method. The open bottom of a cylindrical or other form of chamber is put in the soil or placed over a frame built in the soil (Figure 6-31 bottom). The flux is calculated based on gas concentrations against a time series in the chamber while considering the inner volume and cross-sectional area of the chamber. Chambers come in various sizes and it is possible to measure the gas flux via plant body in the case of short herbaceous plants such as rice and reed. The rate of greenhouse gas emissions in the temperate zone is influenced by soil temperature and exhibits seasonal and diurnal variations. The $CO_2$ flux from trpical peat soils, however, does not show a clear diurnal variation. A similar tendency is found with methane (Watanabe et al. 2008). Soil temperature may not be the main factor determining the variability of gas fluxes in the tropics, probably because it is sufficiently high and fluctuates little throughout the year. The fluxes of methane and $CO_2$ from sago palm growing soils frequently did not differ from those of the nearby secondary forest soils even though soil temperature was consistently higher in the sago palm growing soil.

At this plantation, the annual nutrient loss from the soil was estimated based on the rate of sago palm leaf development and the contents of various elements in the leaf. Fertilizer applications were conducted to compensate for them. This meant that the growth of sago palms was not promoted appreciably under the normal fertilization regime but in a plot where NPK derived from urea, rock phosphate and potassium chloride was applied at 10 times the normal rate (10-fold macronutrient [10NPK] plot), the plant height tended to be larger by 20% or so. A comparison of the methane fluxes from 4 treatment plots shown in Table 6-6 found that the flux from the 10-fold microelement (10MiE) plot in which no increased plant growth was observed, as well as from the 10NPK plot exceeded the fluxes from the control and unfertilized plots while there was no significant difference between the control and unfertilized plots. Little difference was observed between the $CO_2$ fluxes from the 4 treatment plots. In the absence of some events such as a large increase in pH due to a mass application of lime (Murayama 1995), fertilizer application and variations in the amount of applied fertilizer may not be important control factors for $CO_2$ fluxes. In view of the rather inefficient growth enhancement effect on sago palms, increased methane fluxes, and a possibility that N application influences the generation of nitrous oxide, which is another greenhouse gas, it

Table 6-6 Methane and $CO_2$ fluxes from sago palm cultivation soil with different fertilizer treatments

| Treatment plot | Methane (μg C/m$^2$/h) | $CO_2$ (mg C/m$^2$/h) |
|---|---|---|
| No-fertilizer[a] | 43 ± 32[e] | 43 ± 21 |
| Control[b] | 66 ± 75 | 43 ± 25 |
| 10NPK[c] | 90 ± 114 | 49 ± 25 |
| 10MiE[d] | 110 ± 156 | 53 ± 17 |

[a] Dolomite was surface-dressed at a rate of 300 kg/ha/yr in all treatment plots, including no-fertilizer plot.
[b] 467 kg N/ha/yr, 124–160 kg P/ha/yr, 996 kg K/ha/yr, 40 kg Cu/ha/yr, 40 kg Zn/ha/yr, 2.6 kg B/ha/yr.
[c] 10 times more NPK than control; the rest were applied at the same rate.
[d] 10 times more Cu, Zn, B and Fe than control; the rest were applied at the same rate.
[e] Mean ± standard error.
Source: Watanabe et al. (2009)

is reasonable to say that simply increasing the amount of fertilizer application is undesirable.

The methane flux tended to increase during the wet season. Differences in the methane flux between treatment plots with different fertilizer application rates were also observed in the wet season. Increased differences in the methane flux during the wet season is possibly due to changes in the soil redox status due to variations in the groundwater level. In fact, a significant correlation was found between the methane flux and the groundwater level at sites where transplantation was conducted in the same year (Figure 6-32a; Watanabe et al. 2009). While similar relationships are often observed in peat land other than sago palm plantations, it is possible to adequately control the groundwater level at sago palm plantations, unlike a completely natural or a fully drained peat land. For example, Figure 6-32a shows that the methane flux is less than 100 μg C /m²/hr where the groundwater level is about 50 cm or more below ground while no relationship is found between the $CO_2$ flux and the groundwater level (Figure 6-32b). Accordingly, it is considered possible to minimize the sum of the methane and $CO_2$ fluxes by keeping the groundwater level low but not inhibiting sago palm growth.

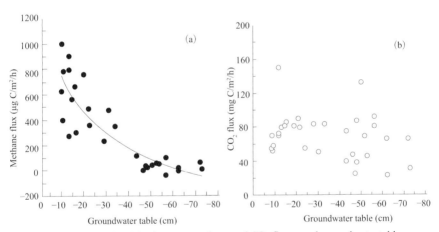

Figure 6-32 Relationships between methane and $CO_2$ fluxes and groundwater table
(a): Methane, (b): $CO_2$. Regression curve: $y = -376x + 1560$ ($r^2 = 0.759$; $P < 0.005$)
Source: Watanabe et al. (2009)

The well-developed lysigenous aerenchyma in the primary roots of sago palm points to the possibility that the plant body of sago palm functions as the route of methane transfer as in the case of paddy rice. To confirm this hypothesis, methane fluxes from soil only and soil with sago palm suckers, which are easier to handle than the parent tree, were compared (the total sucker cross section ranged from 6 to 18% of the measurement area). The methane emission rates were 2 to 50 times greater in the soil with than that without suckers (Watanabe et al. 2009). Further study is required as to whether adult palms produce similar results.

## 6.6.2 Effect of cultivation management on the chemical properties of soil and effluent

The drainage canal water at the sago palm plantation in Tebing Tinggi was tested over 2 years at 7 blocks that were transplanted between 1997 and 2001 (Miyamoto et al. 2009). Data up to 7 years after transplantation were obtained (Figure 6-33). Dissolved organic matter concentration in water is generally represented by dissolved organic carbon (DOC) concentration. Changes in the supply of water-soluble organic matter from soil to the surrounding water environment may have considerable effects on the concentration of polyvalent metal elements such as Fe, Al and Zn which are present in the form of complex organic compounds in natural water to a large extent as well as the concentration of constituent elements of DOM such as C, N, S and P. For instance, an average of 96% of Fe detected in the canal water in Tebing Tinggi was estimated to be present in organically bound form. DOC concentrations in canal water from sago palm growing soil (Figure 6-33a), which were generally higher than those observed in temperate peatlands, and did not increase over the years. Fe, Al and Zn concentrations did not tend to increase either. The absence of temporal changes in DOC concentrations, as in $CO_2$ fluxes, suggests a lack of major changes in the rate of peat decomposition. Concentration of other heavy metals, Cu, Cd, Ni, Pb, Cr and Mo were below 0.05 mg/l throughout the measurement period.

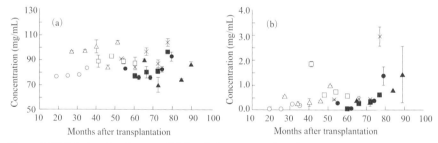

Figure 6-33 Relationships between DOC and Ca concentrations in canal water from sago palm cultivation soil and months after transplantation
(a): DOC, (b): Ca. Different symbols represent data from different blocks.
Source: Miyamoto et al. (2009)

Ca (Figure 6-33b), K and Mg concentrations in canal water showed a positive correlation with the number of months after transplantation. Although the statistical significance of the correlation is not completely reliable, as feed rates from the test blocks were not identical, it is conceivable that part of dolomite and fertilizer components were beginning to leach out. Nevertheless, these 3 elements are unlikely to affect the aquatic ecology of tropical peatlands near the sea as they are already found in sea water in high concentration. No other relationship was found between the number of months after transplantation and the 16 measured elements or pH, but a significant positive correlation was found between P and K concentrations, which appears to warrant continuous monitoring of the effects of fertilizer application.

At the sago palm plantation on Tebing Tinggi Island, fertilizer (granules) is buried in holes (4 or more) dug in the ground to a depth of 10–20 cm at a distance of about 1 m from each sago palm (spot application) rather than surface dressing in order to curtail competition with weeds. Fertilizer components are expected to remain in situ during the dry season, gradually seep downward when the rain starts and then move from the water table into drainage canals. Since no accumulation of fertilizer components was detected just a few dozen cm away from fertilizer application spots even in 10NPK and 10MiE plots (Table 6-6), horizontal dispersion is probably insignificant. While the degree of fertilizer dependence was not mentioned, no dense concentration of sago palm roots has been observed around or directly below fertilizer application spots so far. In fields up to the fifth year after transplantation, little variation was found in the nutritional element content in the soil at distances of 1, 3 and 5 m from sago palms which were considered to have different root mass distributions (Miyamoto et al. 2009). This finding appears to preclude the possibility of rapid soil depletion due to active nutrient absorption by the developing root system from the soil outside of the area of fertilized spots in the course of sago palm growth. This was also suggested by a lack of significant differences in the nutrient content over the years between the sago palm growing soils and the adjacent secondary forest soils.

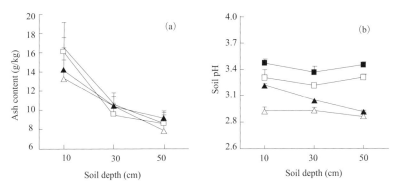

Figure 6-34 Ash contents (a) and pH (b) of sago palm cultivated soils and their adjacent secondary forest soils

▲ and △: Sago palm cultivated soils 2 years after transplantation and their adjacent secondary forest soils

■ and □: Sago palm cultivated soils 5 years after transplantation and their adjacent secondary forest soils

Source: Miyamoto et al. (2009)

An analysis of soils near sago palms at different depths found that the ash content and pH in the surface soil layer (10–15 cm) were higher than those at depths of 30 or 50 cm. It also found that the pH of sago palm growing soils was slightly higher than that of adjacent secondary forest soils, albeit by small amounts (Figure 6-34). The Ca, Mg and Fe contents showed positive correlations with the ash content. The Ca, K and Mg contents in the surface layer soil 2–6 years after transplantation were 0.4–2.5, 0.2–0.6 and 0.9–2.5 g/kg respectively. These are similar to the values reported in a study of sago palm cultivation soil (peat soil) in Mukah, Sarawak, Malaysia, conducted by Kawahigashi et al. (2003). The proportions of Ca, K and Mg in the form of exchangeable cation, which is easily available to plants and prone to leaching, was about 30%, 95% and 60% respectively. These were not influenced by sago palm cultivation. The Fe, Cu and Zn contents were 0.1–0.4 g/kg, < 0.05–2.8 mg/kg, and 1.2–5.0 mg/kg respectively, which were generally lower than the contents in Malaysian peat soils under sago palm cultivation reported by Funakawa et al. (1996). There was little difference between different depths. The contents of these elements are originally very small and they probably avoid leaching by strong adsorption to insoluble organic substances in peat. This is also suggested by the fact that Cu concentrations in canal water stay at very low levels even though it is applied as fertilizer. In fact, when Cu, Fe and Zn were added to a column that was filled with peat soil and intermittently leached with water in a laboratory experiment, the leaching rates of Cu, Fe and Zn were equivalent to only less than 2%, 10% and 25% respectively of the added quantities. As there was no difference in sago palm growth between the aforementioned 10MiE plot and the unfertilized plot or the control plot, and no difference was found in elemental concentrations in the sago palm leaf between treatment plots when sago palms were grown under different application rates and combinations of metal microelements in another field experiment (Ando et al. 2007), it seems unlikely that microelements are adequately absorbed by the sago palm. In future, it is important to develop a new fertilizer application technique which ensures the efficient absorption of fertilizer components by the sago palm while not resulting in their microlocal accumulation in soil.

Authors:
6.1: Foh Shoon Jong
6.2: Hiroshi Ehara
6.3: Yoshinori Yamamoto
6.4: Kenichi Kakuda, Yuka Sasaki, Foh Shoon Jong
6.5: Ho Ando
6.6: Akira Watanabe

# 7
# *Starch Productivity*

Sago palm starch productivity is expressed as starch yield per hectare per year. Starch yield per hectare per year is obtained by multiplying the number of harvested palms per hectare per year by the average starch yield per palm. The number of harvested palms per hectare depends on clump density, population density within a clump, and the age structure of individual palms (years to the harvesting stage). The average starch yield per palm varies according to folk variety (hereinafter called 'variety') as well as the environmental and cultivation conditions such as weather, soil and groundwater.

This chapter first describes aspects of the starch accumulation process that occur in the sago palm trunk pith. Next it discusses variations in the starch yield per palm due to differences in varieties and differences in cultivation and soil conditions. Finally the current sago palm starch productivity is estimated and compared with the productivity of other starch-producing root crops before the yield potential of the sago palm is discussed.

## 7.1 Process of starch accumulation in the trunk pith

### *7.1.1 Measuring starch yield*

In the laboratory, the harvestable portion of the pith was dried and pulverized into fine powder from which soluble sugar was extracted with alcohol. The residue was then degraded with acid or enzyme to derive the glucose content, which was multiplied by 0.9 to obtain the starch content (chemical analysis method). In another method, the pith was cut into small pieces, pulverized in an electric mixer (or juicer) and filtered through a 100–200 mesh. The precipitated starch was dried in the sun and then in a drying oven to obtain dry starch (mixer method). This method could extract about 80% of the amount extracted by the chemical analysis method (Miyazaki et al. 2006).

Local sago palm farmers use various tools to crush the pith into sawdust-like flakes, squeeze them with hands or trample them with their feet as they add water, and use a cloth or coconut fibers as a filter to extract starch containing liquid (see 8.1). The liquid is left standing overnight, the clear upper layer is discarded and the starch that has settled at the bottom is collected. This is the raw starch called 'wet sago' or 'raw sago' and contains 35–45% moisture. Wet sago is sometimes dried in the sun to a moisture content of 12–15%. This starch extraction method employed can reportedly achieve an extraction efficiency of around 60–80%, according to Shimoda and Power (1986), although others have reported that it is only about 50% efficient (Schuiling 2006; Yamamoto et al. 2007a).

It is therefore necessary to take note of the extraction method when assessing starch content measurements.

### 7.1.2 Expressions of starch content

The starch content obtained via the methods described above is expressed in the following ways.

1. Dry starch content per pith fresh weight (%, gDS/gFW) = Sample pith dry starch weight (g) / sample pith fresh weight (g) × 100
2. Dry starch content per pith dry weight (%, gDS/gDW) = Sample pith dry starch weight (g) / sample pith dry weight (g) × 100
3. Dry starch content per fresh pith volume (g/cm$^3$, gDS/cm$^3$) = Sample pith dry starch weight (g) / sample fresh pith volume (cm$^3$)

Sometimes the dry starch weight cited in the formulae above is replaced with wet sago weight or sun-dried starch weight instead. In practice, as sago palm trunks are sold and bought in the form of unpeeled logs, not the pith alone, the weight or volume of the unpeeled trunk is sometimes used instead of the pith weight or volume in the calculation (Jong 1995a). Accordingly, it is necessary to take note of how the starch content has been calculated.

### 7.1.3 Starch accumulation process

The starch accumulation process in the sago palm pith was studied on sago palms growing in shallow peat soil (about 30 cm thick) of Dalat (Jong 1995a), and in the mineral soil and shallow peat soil (30–60 cm) of Mukah (Yamamoto et al. 2003b), both in the State of Sarawak, Malaysia. Differences in the starch accumulation process between mineral soil, shallow peat soil (20–120 cm) and deep peat soil (300–450 cm) in Mukah and Dalat was examined in Yamamoto et al. (2003c). The list of sago palm development stages in Sarawak, Malaysia, is shown in Table 4-1.

Tables 7-1 and 7-2 show variations in the starch content (dry starch weight / pith fresh weight × 100, %) and the starch density (dry starch weight / fresh pith volume, g/cm$^3$) at different levels along the sago palm trunk. The trunks of sago palms at different development stages (according to Table 4-1) were cut at the base and at the leaf sheath attachment of the lowest-positioned green leaf. The base and the top of each trunk were labeled sampling position 1 and 5 respectively and the part between them was divided into 4 equal segments (3 marked positions were labeled sampling positions 2, 3 and 4 from the bottom). Trunk samples were taken from these 5 positions in the shape of discs about 2 cm thick (Jong 1995a). A positive correlation was found between the starch content and the starch density (r = 0.970, $p < 0.001$).

Table 7-1 Starch distribution in sago palm pith at five sampling positions along the trunk

| Sampling position[1] | Starch content (%) at different growth stages[2] | | | | | | | | | |
|---|---|---|---|---|---|---|---|---|---|---|
| | 3 | 4 | 5 | 6 | 7 | 8 | 9 | 10 | 11 | 12 |
| 1 | 5.06 | 13.91 | 19.93a | 19.77a | 25.01a | 23.43ab | 19.94 | 11.95bc | 4.50 | 5.69 |
| 2 | 1.88 | 9.12 | 13.82ab | 19.72a | 23.79a | 25.83a | 25.73 | 15.40ab | 5.74 | 7.24 |
| 3 | 0.78 | 8.32 | 9.20bc | 17.96ab | 25.19a | 22.81ab | 23.91 | 16.69ab | 6.95 | 8.50 |
| 4 | 0.49 | 5.29 | 2.93cd | 13.59ab | 21.51a | 22.33ab | 23.02 | 18.73a | 6.02 | 9.06 |
| 5 | 0.83 | 3.26 | 1.24d | 10.99b | 13.53b | 18.68b | 25.48 | 9.47a | 2.13 | 4.10 |
| Mean | 1.85 | 8.00 | 9.42 | 18.35 | 21.81 | 22.62 | 23.62 | 14.45 | 5.07 | 6.92 |
| S.E. (dif) | 3.12 | 5.03 | 3.07 | 3.79 | 3.35 | 2.22 | 2.93 | 2.69 | 3.93 | 3.79 |
| C.V. (%) | 211 | 89.2 | 39.96 | 28.35 | 18.8 | 12.04 | 17.60 | 26.4 | 101 | 77.5 |

1) The lowest position on the trunk is 1 and the highest position is 5. The segment between them is divided into 4 equal lengths and numbered 2, 3 and 4 from the bottom up.
2) See Table 4-1.
Figures in the columns followed by different letters are significantly different at p=0.05 by DMRT
Source: Jong (1995a)

Table 7-2 Starch density of sago palm pith at five sampling positions

| Sampling position[1] | Starch density (g/cm$^3$) at different growth stages[2] | | | | | | | | | |
|---|---|---|---|---|---|---|---|---|---|---|
| | 3 | 4 | 5 | 6 | 7 | 8 | 9 | 10 | 11 | 12 |
| 1 | .032 | .102 | .120a | .137ab | .180a | .167ab | .153 | .088ab | .031 | .038 |
| 2 | .004 | .069 | .100ab | .143a | .193a | .193a | .183 | .108ab | .042 | .051 |
| 3 | .005 | .064 | .060bc | .130ab | .180a | .180ab | .183 | .125a | .046 | .056 |
| 4 | .004 | .039 | .025cd | .097ab | .157a | .153ab | .180 | .128a | .042 | .064 |
| 5 | .005 | .021 | .009d | .080b | .103b | .143b | .180 | .063b | .015 | .017 |
| Mean | .010 | .059 | .063 | .117 | .163 | .167 | .176 | .102 | .035 | .045 |
| S.E. (dif) | .02 | .04 | .02 | .03 | .02 | .06 | .02 | .04 | .03 | .10 |
| C.V. (%) | 227 | 92.1 | 35.1 | 28.9 | 12.9 | 14.3 | 14.4 | 53.3 | 97.9 | 419 |

1) See Table 7-1
2) See Table 4-1
Figures in the columns followed by different letters are significantly different at p=0.05 by DMRT
Source: Jong (1995a)

It is clear that the starch content and the starch density are higher at the lower levels immediately after trunk formation and at the early trunk elongation stage, which means that starch accumulates from bottom up. During the proper harvesting period from the full trunk growth stage (Stage 7) through to the flowering stage (Stage 9), both the starch content (14–26%) and the starch density (0.103–0.193 mg/cm$^3$) reach a maximum level and differences between different heights along the trunk axis are minimized. Like Jong (1995a), Yamamoto et al. (2003b) also found that starch accumulates in the trunk pith from the base to the top. While the starch content and the starch density declined rapidly at all levels during the fruiting stage (Stage 10), the much greater decline in the topmost section indicates that the starch in this section was consumed in fruit development. The starch content and the starch density continued to decline through to the fruit-ripening stage (Stage 11) and differences between different levels diminished. However, Yamamoto et al. (2008b) reported that the translocation of pith starch to the fruit began at the trunk base contrary to the finding reported in Jong (1995a) that the translocation began at the trunk top.

Regarding variations in starch content (or density) at different trunk levels at the harvesting stage, some studies (Shimoda et al. 1994; Yamamoto et al. 2003b) have reported minor variations similar to Jong's (1995a) findings but others have reported that the highest content is in the middle sections (Sim and Ahmed 1978; van Kraalingen 1986; Osozawa 1990). These differences may be attributable to factors such as differences in defining the harvesting stage, different varieties of sago palm, the effect of environmental conditions and variations in the measurement position on the trunk.

Figure 7-1 shows the relationship between the starch content distribution in the trunk pith cross section, i.e. from the pith center to the periphery, palm age and the trunk (log) position (Yamamoto 1998a). Each palm trunk was divided from the base into 90 cm-long logs and samples were taken from the mid position of the lowest, middle and highest logs in the form of a disc about 2 cm thick. The starch content (dry starch weight / pith dry weight x 100, %) was measured at the pith center (C), the periphery (P, about 2 cm inside of the bark), and the midpoint between them (M). The Figure demonstrates that the starch content tended to be slightly higher at the center than at the midpoint or the periphery regardless of the log position in younger palms (up to 6–8 years) but this difference diminished in older palms.

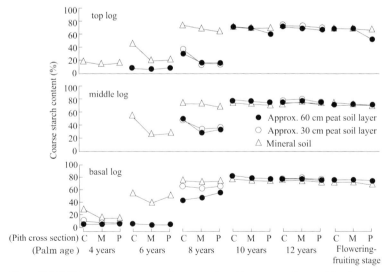

Figure 7-1 Difference in coarse starch content by pith cross section position in sago palms at different palm ages (estimates)
C: center, M: middle, P: periphery
Source: Yamamoto (1998a)

Reports on the starch content in the pith cross section of the harvesting-stage palm also vary significantly. As mentioned above, some have reported that it is uniform (Osozawa 1990; van Kraalingen 1986), while others have assessed that it is higher toward the center (Fujii et al. 1986a; Kelvim et al. 1991), and others report that it is lower toward the center (Yatsugi 1987b). As in the case of the starch content along the trunk axis, the reasons for these diverse findings might be attributable to differences in the definition of the harvesting stage, varietal differences, the effect of environmental conditions and variations in the sampling position.

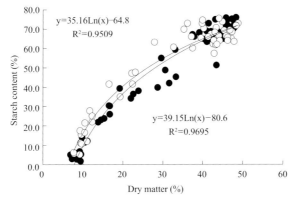

Figure 7-2 Correlation between percentage of dry matter and starch content in pith
○ Mineral soil, ● shallow peat soil
Source Yamamoto et al. (2003b)

The pith starch content at different trunk positions and development stages described above shows a very significant positive correlation with the dry matter percentage (Figure 7-2). It is therefore conceivable that the starch content can be estimated based on the dry matter percentage (Wina et al. 1986; Yamamoto et al. 2003b).

Table 7-3 Variation in mean starch content and mean starch density in the pith and trunk of sago palms at different growth stages

|  |  | Development stage[1] | | | | | | | | | |
|---|---|---|---|---|---|---|---|---|---|---|---|
|  |  | 3 | 4 | 5 | 6 | 7 | 8 | 9 | 10 | 11 | 12 |
| Starch content (%) | Pith | 1.85e | 8.0cd | 9.42c | 18.4b | 21.8a | 22.6a | 23.6a | 14.5b | 5.1e | 6.9c |
|  | Trunk | 1.47e | 6.5cd | 7.6c | 13.7b | 18.6a | 19.0a | 20.1a | 12.1b | 4.2e | 5.8cd |
| Starch density ($g/cm^3$) | Pith | 0.10d | .06bc | .06bc | .12ab | .163a | .167a | .176a | .10b | .04cd | .05c |
|  | Trunk | 0.12d | .05c | .05c | .10b | .15a | .15a | .16a | .09b | .03cd | .04c |

1) See Table 4-1
Figures in the columns followed by different letters are significantly different at p=0.05 by DMRT
Source: Jong (1995a)

Table 7-3 shows the average starch content and the average starch density per pith or trunk based on the starch contents of sago palms at different development stages measured at 5 different positions described above (Jong 1995a). Both the average starch content and the average starch density increased significantly from the trunk formation stage (Stage 3) through to the full trunk growth stage (Stage 7) and reached the maximum level during the harvesting period from the full trunk growth stage to the flowering stage (Stage 9) at 21.8–23.6% and 0.163–0.176 g/cm$^3$ per pith and 18.6–20.1% and 0.15–0.16 g/cm$^3$ per trunk respectively. No significant difference was found at these stages. They declined rapidly from the flowering stage to the young fruit stage (Stage 10) and from the young fruit stage to the fruit ripening stage (Stage 11). As mentioned earlier, this decline might be attributable to the utilization of pith starch for fruit development. This is also evident from the finding that the pith starch content of non-fruit-setting or poor-fruit-setting palms at the supposedly fruit ripening stage was markedly higher than that of normal-fruit-setting palms (Table 7-4) (Jong 1995a). These results indicate that sago palm pith starch reaches its maximum concentration from the full trunk growth stage to the flowering stage.

Table 7-4 Comparison of starch content (%) in mature palms (stage 11) with normal and abnormal fruit development

| | Sampling position[1] | | | | | |
| --- | --- | --- | --- | --- | --- | --- |
| | 1 | 2 | 3 | 4 | 5 | Mean |
| Normal fruit development | 4.50 | 5.74 | 2.45 | 6.02 | 2.12 | 4.16 |
| Abnormal fruit development | 18.97ab | 21.50a | 22.59a | 21.51a | 14.56b | 19.9 |

1) See Table 7-1
Figures in the columns followed by different letters are significantly different at p=0.05 by DMRT
Source: Jong (1995a)

Yamamoto et al. (2003c) reported that the timing of a rapid increase in the trunk pith starch content varies depending on the soil type. They estimated that it happened 3–6 years after trunk formation in mineral soil and shallow peat soil but 4–8 years after trunk formation in deep peat soil. However, they also reported that the rapid increase in the pith starch content in sago palms growing on deep peat soil happened later because they grew at a slower rate than those growing on mineral or shallow peat soils. They found that the rapid increase in the accumulation of trunk pith starch commenced when trunk length and trunk weight reached 3 m and 250–300 kg respectively, regardless of the soil type.

Starch yield per palm is derived by multiplying the average starch content or starch density in the pith or the trunk by pith (trunk) weight or pith (trunk) volume respectively.

Table 7-5 shows starch yield per palm and starch yield per year (after planting). Starch yield per palm increased from the trunk formation stage (Stage 3) to the full trunk growth stage (Stage 7) and reached the maximum yield

(203.4–219.4 kg) at the flowering stage (Stage 9). There was no significant difference in starch yield between Stages 7, 8 and 9. However, starch yield diminished rapidly from the flowering stage through to the fruit ripening stage (Stage 11). Starch yield per year, which took into account the number of years after planting, reached the maximum level at 17.55–18.04 kg/yr from the full trunk growth stage to the flowering stage as with starch yield per palm. Consequently, the period from the full trunk growth stage to the flowering stage constitutes the proper time for harvesting in this example from Sarawak.

Table 7-5 Starch yield of sago palms at different development stages

|  | Development stage[1] | | | | | | | | |
|---|---|---|---|---|---|---|---|---|---|
|  | 3 | 4 | 5 | 6 | 7 | 8 | 9 | 10 | 11 | 12 |
| Estimated age (years) | 7 | 8 | 9 | 10 | 11.5 | 12 | 12.5 | 13 | 14 | 14.5 |
| Starch yield (kg/palm) | 3.6d | 36.9de | 49.2cd | 128.7b | 203.4a | 216.6a | 219.4a | 93.1bc | 24.8de | 41.8de |
| Yield unit/time (kg/palm/year) | 0.52e | 4.62cd | 5.47cd | 12.87b | 17.69a | 18.04a | 17.55a | 7.16c | 1.77de | 2.88de |

1) See Table 4-1
Figures in the column followed by different letters are significantly different at $p=0.05$ by DMRT.
Source: Jong (1995a).

As the starch content or concentration in the pith generally reaches the maximum level at around the full trunk growth stage, an increase in starch yield per palm at the subsequent stages is attributable to an increase in trunk (pith) weight or trunk (pith) volume (Figure 7-3) (Sim and Ahmed 1978; Yamamoto 1998a).

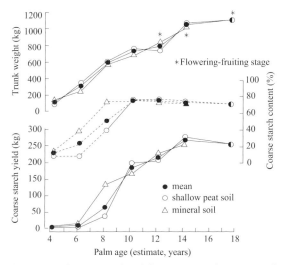

Figure 7-3 Palm age, trunk weight, coarse starch content and yield
Source: Yamamoto (1998a)

## 7.1.4 Sugar types and levels

Sago palm starch is a product of photosynthesis in the leaf, which translocates to the pith in the form of sugar. The movement of sugars and starch in the trunk pith of sago palms has been studied in terms of changes in total sugar content and starch content in the pith at different levels along the trunk length at different ages of growth (Yamamoto et al. 2003b). The study found that the total sugar content in the trunk pith was higher towards the top and that the differences between the levels diminished as the starch content approached its maximum (Figure 7-4). The average pith total sugar content per palm was around 40% in young palms shortly after trunk formation. It declined rapidly as the starch content rose with palm age, being as low as 2–3% by the time the starch content had reached its maximum. This suggests that, in sago palms, starch synthesis proceeds quickly once the trunk reaches a certain stage of development and a high synthesis rate is maintained thereafter. The total sugar content at different trunk positions and development stages described above showed a very significant negative correlation with the starch content (Figure 7-5).

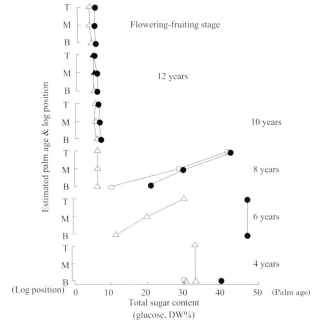

Figure 7-4 Difference in pith total sugar content by log position in sago palms at different palm ages (estimates)
B: base, M: middle, T: top
○: Approx. 60 cm peat soil layer
●: Approx. 30 cm peat soil layer
△: Mineral soil
Source: Yamamoto et al. (1998a)

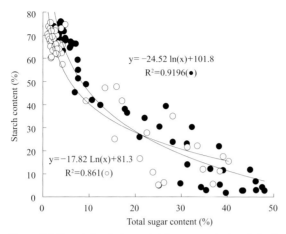

Figure 7-5 Correlation between total sugar content and starch content in pith
○ Mineral soil
● Shallow peat soil
Source: Yamamoto et al. (2003b)

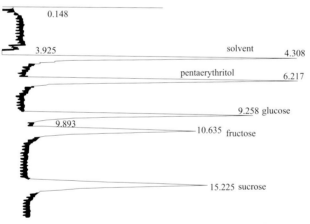

Figure 7-6 Example of sugar measurements by high-performance liquid chromatography
Source: Yamamoto et al. (2010)

There have been few studies on the types of sugar in the sago palm pith or the movement of sugars by palm age. Wina et al. (1986) analyzed sugars and starch in the sago palm pith and found sucrose, glucose and fructose but no maltose. However, they did not study changes in these sugars with palm growth or differences in their concentration distribution across various trunk pith positions. Yamamoto et al. (2010) examined the composition of sugars in the trunk pith and changes within the pith over the course of trunk development in sago palms of different ages sampled from Mukah in Sarawak, Malaysia. The sample included palms 2–10 years after trunk formation. Pith samples were taken from the base, middle and top segments. The total sugar and starch contents were measured and the different types of sugar were identified using high-performance liquid chromatography and the quantity of each type of sugar was measured. The results confirmed Wina et al.'s (1986) findings in that glucose, fructose and sucrose constituted a large part of alcohol-soluble sugar in the sago palm pith throughout the growth period (Table 7-6). A significant negative correlation was found between the starch content and the total sugar content ($r = -0.971$, $p < 0.001$). The starch content was also significantly negatively correlated with glucose, fructose and sucrose separately.

Table 7-6 Sugar and starch contents in sago palm pith

| Palm age[1] (years) | No. of logs | Pith position | Sugar type[2] (High-performance liquid chromatography) | | | | Starch (Somogyi-Nelson) (g/kg) |
|---|---|---|---|---|---|---|---|
| | | | Fructose | Glucose | Sucrose | Total | |
| | | | (g/kg) | | | | |
| 2 | 1 | Middle | 87.9 (23) | 265.5 (71) | 21.1 (6) | 374.5 | 49 |
| 2 | 1 | Middle | 91.9 (21) | 281.8 (65) | 58.6 (14) | 432.3 | 34 |
| 4.5 | 3 | Base | 9.8 (9) | 15.3 (14) | 82.3 (77) | 107.4 | 504 |
| | | Middle | 69.7 (19) | 212.1 (57) | 90.9 (24) | 372.7 | 81 |
| | | Top | 89.1 (31) | 201.6 (69) | 0.4 (0) | 291.1 | 100 |
| 4.5 | 3 | Base | 5.3 (6) | 19.4 (23) | 59.0 (71) | 83.7 | 526 |
| | | Middle | 62.2 (20) | 250.0 (80) | 0.7 (0) | 312.9 | 95 |
| | | Top | 65.6 (18) | 210.5 (58) | 86.9 (24) | 363.0 | 107 |
| 6.5 | 5 | Base | ND | ND | 51.4 (100) | 51.4 | 812 |
| | | Middle | 1.9 (5) | ND | 34.4 (95) | 36.3 | 757 |
| | | Top | 26.4 (11) | 99.3 (42) | 112.5 (47) | 238.2 | 398 |
| 7.5 | 6 | Base | 1.7 (3) | ND | 51.6 (97) | 53.3 | 682 |
| | | Middle | ND | ND | 40.7 (100) | 40.7 | 752 |
| | | Top | 8.3 (15) | 1.8 (3) | 46.1 (82) | 56.2 | 682 |
| 8.5 | 8 | Base | 0.9 (3) | ND | 26.2 (97) | 27.1 | 872 |
| | | Middle | 0.5 (2) | ND | 27.9 (98) | 28.4 | 860 |
| | | Top | 8.9 (12) | 21.5 (30) | 42.6 (58) | 73.0 | 657 |
| 10 | 9 | Base | ND | ND | 3.0 (100) | 3.0 | 686 |
| | | Middle | 1.0 (53) | ND | 0.9 (47) | 1.9 | 841 |
| | | Top | 0.4 (2) | ND | 24.4 (98) | 24.8 | 871 |
| 10 | 9 | Base | ND | ND | 30.2 (100) | 30.2 | 791 |
| | | Middle | ND | ND | 24.0 (100) | 24.0 | 662 |
| | | Top | ND | ND | 6.0 (100) | 6.0 | 773 |
| 10 | 9 | Base | ND | ND | 26.0 (100) | 26.0 | 781 |
| | | Middle | ND | ND | 13.5 (100) | 13.5 | 744 |
| | | Top | 0.7 (23) | ND | 2.3 (77) | 3.0 | 868 |

1) Estimated years after trunk formation
2) The number in parenthesis is the percentage of each sugar in the total sugar content
ND: Below the minimum value of the respective measured sugar contents
Source: Yamamoto et al. (2010)

As indicated in Table 7-6, the pith starch content was low shortly after trunk formation (2 years). The total sugar content at this young age was around 40%, with especially high levels at the middle and top levels of the trunk. The sugar content decreased in all of the trunk levels as the palm grew. Starch began to accumulate at the base during the period of active trunk elongation (5–6 years after trunk formation) and continued to accumulate toward the top over time. Little difference was found in the starch content between the base, middle and top levels 7–8 years after trunk formation. The starch content reached its maximum across all of the trunk levels about 10 years after trunk formation. The relationship between the rising starch content and the rapidly declining total sugar content with palm age was quite similar to that reported by Yamamoto et al. (2003b).

The glucose content was 200–280 g/kg, amounting to 60–80% of total sugar, at the middle and top levels of the trunk with a pith starch content of less than 100 g/kg in relatively young palms (2–5 years after trunk formation) (Table 7-6). Fructose content was 80–90 g/kg, making it the second most prevalent sugar, while sucrose had the lowest content in palms 2 years after trunk formation. In palms 4–5 years after trunk formation, glucose and fructose content were higher in absolute terms, but the sucrose content was relatively much higher in certain levels. In palms 4–5 years after trunk formation, the glucose content was high in the middle and top levels of the trunk. The total sugar content was lower at the base but sucrose comprised as much as 70% of it. The starch content remained quite low during this period and it is assumed that a large part of the synthesized sugars was used for palm trunk growth. In palms 6–7 years after trunk formation, the total sugar content decreased, the relative proportion of sucrose increased and the proportions of glucose and fructose decreased in all levels. The occurrence of the rapid starch accumulation period in the pith varied according to soil type and cultivation conditions but this period consistently entailed a rapid decline in the amount of sugar and an increase in starch accumulation, with starch content exceeding 600 g/kg in all levels. In palms 10 years after trunk formation, the total sugar content was 30 g/kg or less, a large part of which was in the form of sucrose in all segments. Starch was stored in high concentrations and near its maximum level throughout the trunk.

As we have seen, the sugar content of the sago palm pith was very high at around 40% shortly after trunk formation but diminished as starch began to accumulate in the pith, comprising a mere 2–3% by the harvesting stage. Sugars found in the sago palm pith included sucrose, glucose and fructose for the duration of palm development but their composition varied from one growth stage to another. Glucose and fructose were found in high percentages shortly after trunk formation whereas the percentage of sucrose increased with starch accumulation.

## 7.1.5 Starch granule formation in parenchyma

### 7.1.5.a Amyloplast size

Starch accumulates in a cell organelle called amyloplast in the ground parenchyma cells at the center of the stem (or trunk). The type of amyloplast starch granule found in the sago palm is called simple starch grain as a single starch granule is contained in each amyloplast. Accordingly, what is commonly called 'starch granule' in the sago palm is an amyloplast in the botanical sense.

Ogita et al. (1996) studied amyloplasts in the ground parenchyma cells of the stem center in palms aged 3, 5, 8 and 13 years. They found that the major axis of the amyloplast was 5–20 μm in palms aged 3 years, 20–30 μm in palms aged 5 years and exceeded 30 μm in many of the palms aged 8 or 13 years. Jong (1995a) measured the major axis of amyloplasts in parenchyma cells at the stem center in sago palms shortly after trunk formation, at the full trunk growth stage and the fruit ripening stage (see Table 4-1) and reported that it was longer at later stages of development. He also reported that it was smaller at the top trunk portion in young palms but the difference in the long diameter between different portions on the trunk axis decreased as palms aged. Fujii et al. (1986a) reported that young palms had smaller starch granules, indicating slower rates of sedimentation after extraction and hence lower starch recovery efficiencies.

Scanning electron microscopic observations of 'wet sago' and refined sago starch commonly sold at markets in Indonesia and Malaysia find that many amyloplasts are either oval-shaped or spindle-shaped and about 30–50 μm in major axis. According to Kawasaki (1999) and Kawakami (1975), when sago palm amyloplasts are compared with simple starch grains in other plants, they are larger than those in wheat (secondary starch granules 2–8 μm in diameter; primary starch granules 20–40 μm in diameter) or yam (about 20 μm in diameter) and smaller than those in potato (10–90 μm in diameter) or edible canna (40–100 μm in diameter). In comparison with amyloplasts containing compound starch granules, they are larger than those in rice (2.0–8.0 μm in diameter), sweet potato (8.0–36.0 μm in diameter) or taro (0.13–0.42 μm in diameter). Based on the size of amyloplast in 54 crop species reported in Jane et al. (1994), sago palm amyloplasts are considered to fall on the medium to large range of the simple starch grains.

### 7.1.5.b Formation and proliferation of plastid-amyloplast system

Plastids are a precursor of amyloplasts. Plastids are called amyloplasts when they accumulate starch but as they are often indistinguishable even in scanning electron microscopic observations, plastids and amyloplasts are sometimes called the plastid-amyloplast system collectively (Kawasaki et al. 1999).

# STARCH PRODUCTIVITY

According to Nitta et al. (2000a, 2002b), plastids are formed and grow in the part of the ground parenchyma cells from the growing point to 20 cm toward the base in the stem center of sago palm (Table 7-7). After plastids are formed in that 20-cm zone from the growing point, they multiply by separating

Table 7-7 Feature of plastid-amyloplast system at growing point and in proximal tissue

| Distance from growing point | Plastid form (mean diameter) etc. | Amyloplast form (mean diameter) etc. |
|---|---|---|
| Growing point | Spherical (up to 1 µm). Some are dividing. | Spherical (2.4 µm, minor variation). |
| 10 cm | Spherical (up to 1 µm). Some are dividing. | Spherical (4.0 µm, minor variation). |
| 20 cm | Very few. | Spherical (5.5 µm, minor variation). |
| 40 cm | None. | Spherical (6.3 µm, minor variation). |
| 60 cm | None. | Spherical (7.6 µm, minor variation). Some are oval-shaped or with a protuberance on the surface or separated/divided. |
| Stem base | None. | Oval (13.5 µm, major variation). Some are oval-shaped or with a protuberance on the surface or separated/divided. |

Source: Nitta et al. (2002b)

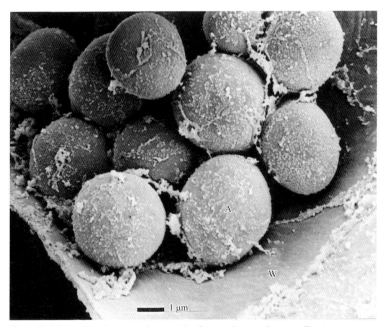

Figure 7-7 Scanning electron micrograph of ground parenchyma cells at stem center in Rotan variety at early growth stage
Amyloplasts are mostly oval-shaped or spindle-shaped with a smooth surface.
A: Amyloplast
W: Cell wall
Source: Nitta et al. (2000a)

Figure 7-8 Scanning electron micrographs of ground parenchyma cells at stem center in Rotan variety at middle growth stage

Many of the amyloplasts have a protuberance, have separated or are about to separate.
A: Amyloplast
▲: Protuberance
*: Amyloplasts appear to be immediately before or after separation.
Source: Nitta et al. (2000a)

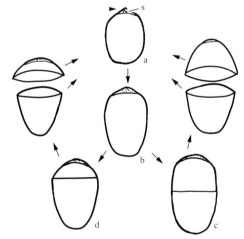

Figure 7-9 Schematic diagram of amyloplast separation/ division process

The amyloplast appears to multiply through the repetition of a process involving: (a) protuberance formation on the surface, (b) protuberance base development and the flattening of the protuberance end, (c) separation at the middle point along the long axis or (d) on the protuberance side.
S: Stroma inside.
▲: Protuberance.
Source: Nitta et al. (2000a)

and dividing in the area of parenchyma cells on the basal side of the level 60 cm below the growing point. In other words, multiplication in the plastid-amyloplast system takes place at two segments in the stem – at the growing point and its adjacent area and the basal side of the line 60 cm below the growing point.

**7.1.5.c Pattern of amyloplast proliferation**
Amyloplasts in sago palms multiply in a similar manner regardless of variety (folk variety). The multiplication pattern is described below using the middle part of the stem of Rotan variety palms sampled in Kendari, Sulawesi Province, Indonesia, as an example (Nitta et al. 2000a).

Amyloplasts in the early stage of their growth are mostly oval-shaped or spindle-shaped with a smooth, round surface (Figure 7-7). Many amyloplasts develop a protuberance in one part of the oval or spindle body at the middle stage of their growth (Figure 7-8). Also, a part of the oval or spindle has separated or is about to separate in a number of amyloplasts (Figure 7-8). The planes of separation are very flat and smooth. Some amyloplasts separate in more than one place. No protuberance is observed on the surface of amyloplasts after separation or immediately before separation in many cases (Figure 7-8).

In short, proliferation of amyloplasts is thought to take place in the following manner (Figure 7-9).
1. Plastids accumulate starch (i.e., plastids become amyloplasts).
2. A protuberance forms on the surface of the amyloplast.
3. The base of the protuberance develops and the protuberance grows out while the curvature of the oval or the spindle on the protuberance side becomes flatter.
4. Separation or division occurs at the middle point along the major axis or on the protuberance side of the midpoint in the developed amyloplast.

The protuberance is believed to hold localized stroma inside. This type of separation or division is observed in amyloplasts at or after the middle growth stage and in large amyloplasts at the early stage of their growth. Amyloplasts are smaller in the parenchyma near vascular bundles and become larger the farther they are from vascular bundles (Figure 7-10). Amyloplasts of various shapes and sizes are found at the late stage of growth with many of them separating or dividing (Figure 7-11).

Thus, variation in the size and shape of amyloplasts in the ground parenchyma cells at the stem center of sago palm is minor at the early growth stage and increases as they develop.

Figure 7-10 Scanning electron micrograph of ground parenchyma cells at stem center in Rotan variety at middle growth stage
Amyloplasts are larger in parenchyma cells farther from the vascular bundle located below the image
A: Amyloplast
Source: Nitta et al. (2000a)

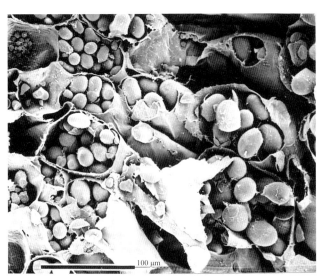

Figure 7-11 Scanning electron micrograph of ground parenchyma cells at stem center in Rotan variety at late growth stage
Many amyloplasts in various shapes and sizes are either after or immediately before separation
Source: Nitta et al. (2000a)

## 7.1.5.d Variation in amyloplast size and number by stem portion and variety

The size and number of amyloplasts vary depending on the portion in the stem or the variety (Mizuma et al. 2007). The following example is based on 6 sago palm varieties growing on the shores of Lake Sentani near Jayapura, Papua Province, Indonesia, close to the place of origin of the sago palm (Figures 7-12, 7-13). The major axis is longer in amyloplasts in the middle and basal portions of the stem than in the top portion in all varieties (Figure 7-12). The major axis is short at the basal stem in Rondo and Para, suggesting that separation and division take place more actively on the basal side of the middle stem. The minor axis is also longer in the middle and basal portions than in the top portions in all varieties (Figure 7-12).

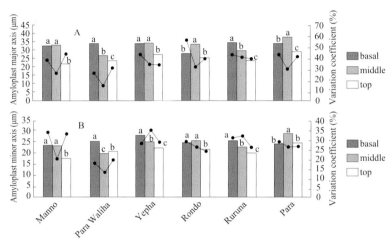

Figure 7-12 Major axis (A) and minor axis (B) of amyloplast by stem portion and variation coefficient

There is no significant difference between samples marked with the same letter within each variety by Fisher's LSD at the 5 % level
Source: Mizuma et al. (2007)

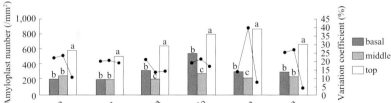

Figure 7-13 Number of amyloplasts per unit cross section area of ground parenchyma cells at stem center by stem portion and variation coefficient

There is no significant difference between samples marked with the same letter within each variety by Fisher's LSD at the 5 % level
Source: Mizuma et al. (2007)

The number of amyloplasts per unit cross section area in the stem center parenchyma cells is greater in the top portion than in the middle or basal portions in all varieties (Figure 7-13).

## 7.2 Starch productivity of sago palm

### 7.2.1 Starch yield per palm

#### 7.2.1.a Variety (folk variety)

A great number of sago palm varieties have been reported to exist on New Guinea Island, which is thought to be the place of origin of the species (see 1.2). As the extent of genetic variation among them is yet to be confirmed, it has been proposed that they should be called folk varieties (Yamamoto 2005). It has been reported that starch yield differs between folk varieties (hereinafter called 'varieties') but careful attention must be paid to the method employed for starch yield measurement as well as whether the measured palms are at the full trunk growth stage or the flowering stage when starch yield reaches a maximum.

Among these varieties, those which are locally called wild varieties show lower starch yield levels than cultivated varieties. Shimoda reported that a wild variety (Wakar) found in the Sepik River region, Papua New Guinea, had lower starch contents, pith volumes and starch yields (0–41 kg) than cultivated varieties (starch yield 109–279 kg) (Japan-PNG Goodwill Society 1985). Yanagidate et al. (2007) reported that two types of a wild variety called Manno ('Manno Besar' [MB: large Manno] and 'Manno Kecil' [MK: small Manno]) among sago palms growing around Lake Sentani near Jayapura, Papua, Indonesia, had lower starch contents (MB, MK), a low trunk weight (MK), and hence lower starch yields than other cultivated varieties, and that MK in particular produced less than 100 kg of starch and has hardly been harvested. These wild varieties are suspected to have less efficient processes of synthesizing starch from sugars, as they have high sugar content in the pith even at the flowering stage when they are supposedly harvestable (Yanagidate et al. 2007).

There have been few studies of variations in starch yield among cultivated varieties. Shimoda and Power (1986) studied the starch yield of 6 cultivated varieties (Ketro, Anum, Makapun, Kangrum, Ambutrum and Awir-Koma) grown in semi-cultivated stands in the Sepik River basin, Papua New Guinea. A certain volume of pith was sampled from each palm and analyzed for starch content using a mixer extraction method (see section 7.1.1). Starch yield per palm was estimated based on the pith volume and the starch density (dry starch weight per 100 ml of pith) of each variety. It was found that the starch yield from the 6 varieties ranged from 106 to 279 kg. The difference is attributable to variations in the pith volume and the starch density as well as to variations in the growth stage of the sample.

Yamamoto et al. (2000, 2005b,c,d, 2006b, 2008b,c) compared starch yields in sago palm varieties in Lake Sentani near Jayapura, Papua, Indonesia (8 varieties), on Seram Island, Maluku (5 varieties), and in Kendari, Southeast Sulawesi (3 varieties), while Yanagidate et al. (2008) did the same in Pontianak, West Kalimantan (2 varieties). These studies selected palms at the harvesting stage (from the flower bud formation stage to the fruit setting stage), collected pith samples using the same method and determined starch yields by chemical analysis. Starch yield in the 8 varieties (Yepha, Para, Ruruna, Osukulu, Folo, Pane, Wani and Rondo) grown around Lake Sentani varied widely from 150 to 975 kg. All of the varieties except for the early maturing Rondo (yield: 150–200 kg) yielded more than 300 kg, although there were marked differences between individual palms of the same varieties (Yamamoto et al. 2005d). The starch yield of Rondo was about the same as that of the wild variety MB. The highest yield (975 kg) was produced by Para (Yamamoto 2006c). As Para has also been reported to yield 858 kg (Yamamoto 2006c) and 835 kg (Saitoh et al. 2004), it is likely to be the highest yielding variety among the sago palm varieties that have been studied so far. No significant difference in the yield (500–600 kg) was found between 4 major cultivated varieties (Ihur, Tuni, Makanaru and Molat) on Seram Island, Maluku (Yamamoto et al. 2008b). An early maturing variety found on the island (Duri Rotan), which was not included in the study, was considered to have a lower yield (Dr. Louhenapessy, private correspondence). In Kendari, starch yield was clearly lower in the early maturing variety of Rotan (142 kg) than in Molat and Tuni (270–365 kg)

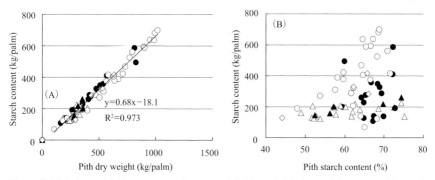

Figure 7-14 Relationship between starch content (yield) and pith dry weight (A) or pith starch content (B)

△: Batu Pahat in Johor, and Mukah and Dalat in Sarawak, Malaysia
◇: Tebing Tinggi Island in Riau, Indonesia
●: Kendari in Southeast Sulawesi, Indonesia
▲: Ambon in Maluku, Indonesia
○: Sentani, near Jayapura in Papua, Indonesia
Source: Yamamoto (2006a)

Table 7-8 Starch yield per palm in Malaysia, Indonesia and Papua New Guinea

| Country | District, state, province | Starch yield[1] (kg/palm) | Sample size (palms) | Authors (year) | Note |
|---|---|---|---|---|---|
| (Traditional or mixer method of extraction) | | | | | |
| Malaysia | Mukah, Sarawak | 83–179 | 5 | Sim and Ahmed (1978) | Peat soil |
| | Mukah, Sarawak | 123–189 | 4 | Sim and Ahmed (1978) | Mineral soil |
| | Mukah, Sarawak | 166 | - | Tie et al. (1987) | Deep peat soil |
| | Batu Pahat, Johor | 185 | - | Flach and Schuiling (1989) | |
| | Mukah, Sarawak | 209–227 (218)* | 2 | Maeda et al. (1992) | Shallow peat soil |
| | Mukah, Sarawak | 184* | 1 | Maeda et al. (1992) | Mineral soil |
| | Dalat, Sarawak | 203–219 (214)* | 10 | Jong (1995a) | Shallow peat soil |
| Indonesia | Seram Island, Maluku# | 272 | - | Wallace (1885) | |
| | Irian Jaya# | 113–158 | - | Barrau (1959) | |
| | Seram Island, Maluku ♭ | 165 | - | Ellen (1979) | |
| | Luwu, South Sulawesi# | 95-445 (203) | 31 | Osozawa (1990) | |
| | Bogor, West Java | 55 | - | Haska (2001) | |
| | Seram Island, Maluku ♭ | 18–188 (68) | 41 | Sasaoka (2006) | |
| Papua New Guinea | East Sepik# | 219–240 (229) | 2 | Lea (1964) | |
| | East Sepik# | 13–99 (48) | 18 | Dornstreich (1973) | |
| | Sepik River upper stream basin, East Sepik# | 28–205 (87) | 7 | Townsend (1974) | |
| | Fly River basin, Western Province# | 62–303 (221) | 10 | Rhoads (1980) | |
| | Purari Delta, Gulf# | 38–359 (134) | 18 | Ulijaszek (1981) | |
| | Oriomo, Western Province# | 29–104 (66) | 8 | Ohtsuka (1983) | |
| | Sepik River lower stream basin, East Sepik# | 106–278 (198)* | 8 | Shimoda and Power (1986) | Cultivated varieties only |
| | Siuhamason, Western Province ♭ | 28–265 (125) | 19 | Suda (1995) | |
| (Chemical analysis method) | | | | | |
| Malaysia | Mukah, Sarawak | 277 | - | Kueh et al. (1991) | Deep peat soil |
| | Sarawak | 219 | - | Kueh et al. (1991) | Mineral soil |
| | Batu Pahat, Johor | 124–200 | 4 | Yamamoto (1998b) | |
| | Mukah, Sarawak | 126–226 | 13 | Yamamoto et al. (2003c) | |
| Indonesia | Halmahera, North Maluku | 201–608 (384) | 6 | Ehara et al. (1995b) | |
| | Kendari, Manado etc., Southeast/North Sulawesi | 28–512 (231) | 5 | Ehara et al. (1995b) | |
| | Tebing Tinggi Island, Riau | 129–416 | 4 | Yamamoto (1998b) | |
| | Palembang, South Sumatra | 231–305 (268) | 2 | Ehara and Mizota (1999) | |
| | Padang, West Sumatra | 182–189 (186) | 2 | Ehara and Mizota (1999), Naito et al. (2000) | |
| | Kendari, Southeast Sulawesi | 109–587 | 17 | Yamamoto et al. (2000) | |
| | Siberut, West Sumatra | 151–245 (195) | 3 | Naito et al. (2000) | |
| | Tebing Tinggi Island, Riau | 225 | 1 | Saitoh et al. (2004) | |
| | Pontianak, West Kalimantan | 100 | 1 | Saitoh et al. (2004) | |
| | Near Jayapura, Papua | 835 | 1 | Saitoh et al. (2004) | |
| | Near Jayapura, Papua | 34–975 | 37 | Yamamoto (2006c) | |
| | Pontianak, West Kalimantan | 250–380 (283) | 6 | Yanagidate et al. (2008) | |
| | Seram Island, Maluku | 339–747 (515) | 11 | Yamamoto et al. (2008b) | |

1) Dry starch. Mean yield in brackets
#: Cited in Osozawa (1990)
♭: Cited in Sasaoka (2006)
*: By mixer extraction method

(Yamamoto et al. 2000). There was little difference in starch yield (280–330 kg) between two varieties (Bembang and Bental) in Pontianak. The starch yield in these two varieties was more or less the same as in Mukah and Dalat, Sarawak, Malaysia (Yanagidate et al. 2008).

Ehara et al. (1995b) also found 57% variance in starch yield (28–608 kg/palm) on the eastern islands of Indonesia.

When these intervarietal starch yield differences are divided into trunk weight (pith weight) and starch content per pith dry weight, it becomes evident that such differences largely stem from differences in trunk weight (pith weight) (Figure 7-14) (Yamamoto 2006a).

Table 7-8 is a list of yields per sago palm that have been reported so far. It should be noted that starch yields vary widely depending on the measurement method employed.

### 7.2.1.b Soil type

Sago palms are able to grow in both mineral soils and in peat soils that are oligotrophic, low in pH, with high groundwater levels all year round and flooding during the wet season. Regardless of soil type, however, sago palms growing under permanently submerged conditions fail to form a trunk or grow poorly after trunk formation and yield very little or no starch (Yamamoto 1998a).

Comparative studies on the starch productivity of sago palms by soil type (namely, mineral and peat soils) was conducted in the Malaysian state of Sarawak. Sim and Ahmed (1978) reported that starch yield reached a maximum immediately after the flowering stage in both soils and that there was no significant difference in starch yield per trunk between mineral soil (189 kg) and peat soil (179 kg). Kueh et al (1991) compared starch yield in sago palms growing in the Anderson Series peat soil (peat layer over 150 cm deep) and mineral soil and found that sago palms growing in peat soil have a low starch content at 56.6% but their trunk length was 24% longer and hence their starch yield was 277 kg, which was higher than the 219 kg yielded by sago palms growing in mineral soil. However, the average number of years to harvest was 12.7 years in the former and 9.8 years in the latter. Hence, there was little difference in starch yield per year between the two groups (21.8 kg and 22.3 kg). Yamamoto et al. (2003c) conducted a comparison of starch yield in sago palms growing in mineral soil, shallow peat soil (20–120 cm), and deep peat soil (350–450 cm). The trunk weight was lower in the deep peat group (899, 830 and 734 kg respectively) but the starch content was higher in the deep peat group (55, 59 and 70% respectively) and hence there was no significant difference in starch yield (164–180 kg) between these soil types. Nevertheless, as the deep peat group takes 4–5 years longer to harvest than the mineral soil and shallow peat groups, starch productivity per year was lowest in the deep peat growing sago palms. This suggested that the starch content in sago palms growing in deep peat soil was highest because this particular sago palm farm, which had been operating for a short time, had a low trunk population density and good light conditions.

### 7.2.1.c Shading in the clump

Sago palm suckers are planted, take root and initiate many suckers at the base as they grow. The initiated suckers develop, form trunks and grow to be harvested after the parent palm is harvested. However, if they are not thinned out, competition within the clump intensifies and results in delays in trunk formation, poor trunk growth and low starch yield. Figure 7-15 shows the effect of shading within the clump (the subject palm is surrounded by trunked palms) on trunk length, trunk diameter and trunk volume in relation to estimated palm age in sago palms over 3 m in trunk length at a semi-cultivated sago palm farm on Tebing Tinggi Island, Riau, Indonesia (Yamamoto 1998b). It was clear that in comparison to unshaded palms, shaded palms had smaller trunk diameters although their trunk lengths were similar. Consequently, trunk volume also tended to be smaller in shaded palms. When starch yield was measured in the palms aged 7 years after trunk formation and at the harvesting stage, the starch content was around 60%, which was no different from the starch content in unshaded palms, but starch yield was only about one-third of that in unshaded palms because of the lower trunk volume and weight.

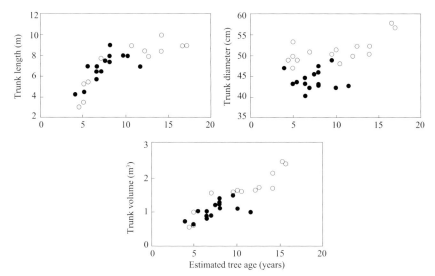

Figure 7-15 Relationship between palm age and trunk length, trunk diameter or trunk volume
○: Unshaded
●: Shaded
Source: Yamamoto (1998b)

The above findings clearly indicate the importance of sucker control at sago palm farms in improving starch productivity.

### 7.2.1.d Distance from the sea

Sago palms show a considerable degree of salt tolerance and are able to grow in brackish waters. Nevertheless, they grow poorly and fail to form trunks in areas of frequent seawater invasions. Yamamoto et al. (2003a) studied the relationship between the distance from the sea (2, 15, 20 and over 100 m), growth and starch yield in sago palms growing on Tebing Tinggi Island, Riau Province, Indonesia. The island is flat with marked coastline erosion. Sago palm colonies are found on the seashore and are subjected to seawater invasions at spring tide. The sago palms in the study were all at the harvestable stage, ranging from the flower stalk formation stage to the flowering stage. Those growing within 20 m of the sea line had a trunk length of 5–6 m and a trunk diameter of 42 cm. They were much smaller than those found more than 100 m from the sea line (inland sago palms, which were 9–11 m high and 46–52 cm in diameter). Accordingly the trunk weight of the ones on the shoreline was 470–570 kg, significantly less than 1,140–1,710 kg of the inland sago palms. At the same time, however, the pith starch content of the coastal sago palms was 65–84%, which was generally higher than the 66% of the inland sago palms. Consequently, the starch yield in the sago palms growing within 20 m of the sea line was 86–150 kg, which was about 20–50% of the 284–416 kg of the inland sago palms, primarily due to difference in trunk (pith) weight. Analysis of the mineral content in sago palm leaflets found that those growing within 20 m of the sea line exhibited some effects of seawater: they had higher soluble sodium (Na) and magnesium (Mg) levels (Yamamoto et al. 2003a).

These findings suggest that sago palms growing near the sea and subject to seawater invasions suffer poor trunk growth due to the effects of seawater and therefore produce lower starch yields.

### 7.2.1.e Altitude

Sago palms are reportedly found at altitudes ranging from 0 to 700 m (Flach 1997; Rasyad and Wasito 1986). There are few studies, though, on variation in starch productivity by altitude. Sasaoka (2006, 2007) surveyed starch productivity in sago palms growing in Manusela (730 m above sea level), one of the villages scattered across Manusela Valley sandwiched by Mount Kobipoto (1,577 m) and Mount Binaiya (3,027 m) in the central mountain range on Seram Island, Maluku Province, Indonesia. This village is considered to be near the upper altitude limit for sago palm growth. The average starch yield per palm based on the traditional extraction technique during the harvesting stage (flower stalk formation–fruit setting) was 68 kg (n = 41), which was markedly lower than the average yield of 165 kg in the coastal area on the island (Ellen 1979). The average trunk length and trunk diameter of the sample were 7.4 m and 51 cm; slightly shorter and slightly thicker than their lowland counterparts but their trunk volume was comparable. Accordingly, the low yield level in Manusela Village at a high altitude is likely to be attributable to a low pith starch content but this question requires further analysis.

### 7.2.1.f Planting density

There has been no study of the effect of planting density on starch yield in sago palms. Planting density is thought to influence starch yield through the number of harvested palms per area and starch density in individual trunks. Jong (1995a) examined the effect of planting density on growth in square planting with a spacing ranging from 4.5 m to 13.5 m (Table 7-9). It was found that the number of trunk forming palms per area and the pith volume reached their maximum in the 10.5 $m^2$ planting plot (91 palms/ha). While there has been no study on the relationship between the starch content and planting density, starch yield is expected to peak at around the planting density that provides a maximum pith volume assuming that the starch content at the harvesting stage does not fluctuate widely except at extreme planting densities.

Table 7-9 Effect of spacing treatments on trunk production and pith volume of sago palms at 9 years after planting

| Spacing (m × m) square | Number of palms planted (/ha) | Trunk formation rate (%) | Number of leader trunks (/ha) | Number of follower trunks (/ha) | Mean pith volume of leader palm ($m^3$/palm) | Pith volume ($m^3$/ha) | | |
|---|---|---|---|---|---|---|---|---|
| | | | | | | Leader palm | Follower Palm | Total |
| 4.5 | 494 | 35.2 | 174 | 5 | 0.2 | 35.7 | 1.0 | 36.7 |
| 7.5 | 178 | 80.6 | 144 | 45 | 0.3 | 43.5 | 9.9 | 53.5 |
| 10.5 | 91 | 94.4 | 86 | 137 | 0.4 | 33.0 | 35.9 | 68.9 |
| 13.5 | 55 | 100.0 | 55 | 110 | 0.4 | 20.8 | 29.2 | 50.0 |

Source: Jong (1995a)

### *7.2.2 Starch yield per area*

A majority of sago palms currently in use are either wild or semi-cultivated. There is only a small number of cultivated sago palm forests. Moreover, growing conditions in wild and semi-cultivated sago palm stands vary greatly from those of pure or almost pure forests to mixed forests of a wide variety of trees. Zwollo (1950) and Wttewall (1954) reported that other tree species accounted for 30–50% of the area in natural (wild) sago palm stands in West Irian, Indonesia (Flach 1980).

#### 7.2.2.a Sago palm clumps and population by development stage in natural (wild), semi-cultivated and cultivated sago palm stands

There have been very few studies on the number of sago palm clumps or the sago palm population by development stage within a sago palm stand. Table 7-10 shows the number of clumps per hectare and population by growth stage in natural (wild), semi-cultivated and cultivated sago palm stands in Indonesia, Malaysia and Papua New Guinea. The number of clumps per hectare ranged from 100 to 417 clumps, a more than four-fold difference. The differences tended to be greatest in natural stands but the overall differences between semi-cultivated and cultivated stands was unclear. Variations in the number

## STARCH PRODUCTIVITY

Table 7-10 Number of clumps and palm population by development stage (per hectare) in natural (wild), semi-cultivated and cultivated sago palm stands

| Province, country | Study site | Type of stand | Clump number | Population before trunk formation | | | Population at trunk formation stage (a) | Adult palm (b) | Harvestable palm (c) | (a) + (b) + (c) | Total population | Authors |
|---|---|---|---|---|---|---|---|---|---|---|---|---|
| | | | | Early R[1)] | Late R | Total | | | | | | |
| West Irian, I[2)] | West Irian | Natural | 217 | - | - | - | - | - | 24 | - | - | Wtewaal (1954) |
| Papua, I | Kaure | Natural | 380 | 1,605 | 135 | 1,740 | 75 | 245 | 60 | 380 (1.0)[4)] | 2,120 | Matanubun and Maturbongs (2006) |
| Papua, I | Inanwatan | Natural | 417 | - | - | 4,898 | 387 | 326 | 75 | 788 (1.9) | 5,686 | Luhulima and Maturbongs (2006) |
| East Sepik, PNG | Sepik River lower basin | Semi-c | 236 | - | - | - | 46 | - | 14-17 | - | - | Japan-PNG Goodwill Society (1984) |
| East Sepik, PNG | Sepik River lower basin | Semi-cultivated | 136 | - | - | - | 212 | - | - | - | - | Japan-PNG Goodwill Society (1984) |
| South Sulawesi, I | Luwu | Semi-cultivated | 100 | 878 | - | 1,004 | - | - | 14-34 (21.5)[3)] | 187 (1.9) | 1,191 | Osozawa (1990) |
| Sarawak, M | Sarawak | Semi-cultivated | 239 | - | - | 1,887 | 203 | 77 | 42 | 322 (1.3) | 2,209 | Tie and Kalvim (1991) |
| Papua, I | Waropen | Semi-cultivated | 117 | - | - | 468 | 173 | 91 | 125 | 389 (3.3) | 856 | Istalaksana et al. (2006) |
| Johor, M | Batu Pahat | Cultivated | 120 | 2,984 | 623 | 3,607 | - | - | - | 297 (2.5) | 3,904 | Watanabe (1984) |
| Southeast Sulawesi, I | Kendari | Cultivated | 169 | - | - | 2,256 | - | - | 11.0-34.3 (22.0) | 228 (1.3) | 2,484 | Yanagidate et al. (2009) |

1) Rosette stage
2) I: Indonesia, PNG: Papua New Guinea, M: Malaysia
3) Mean
4) Mean number of palms per clump

Table 7-11 Number of clumps and ratio of constituent varieties in natural and semi-cultivated sago palm stands

| Type of sago palm stand | Sago palm variety | Number of clumps | |
|---|---|---|---|
| | | (clumps/ha) | (%) |
| Natural stand[1] | Bosairo | 5 | 1.2 |
| | Mola/Igo | 21 | 5.0 |
| | Edidau | 9 | 2.2 |
| | Bibewo | 382 | 91.7 |
| | Total | 417 | 100.0 |
| Semi-cultivated[2] | May | 64 | 54.7 |
| | Ndosa | 48 | 41.0 |
| | Umbei | 5 | 4.3 |
| | Total | 117 | 100.0 |

1) Luhulima et al. (2006)
2) Istalaksana et al. (2006)

of clumps in natural and semi-cultivated stands are likely to be attributable to intervarietal differences in palm size and the proportion of other tree species in the stand. Natural sago palm stands in Inanwatan, Papua, Indonesia, had 4 varieties (Luhulima et al. 2006) and semi-cultivated stands in Waropen in the same province had 3 varieties (Istalaksana et al. 2006) and their composition ratios were heavily skewed (Table 7-11). In particular, Bibewo comprised over 90% in the natural stands in Inanwatan. May was the most prevalent in the semi-cultivated stands in Waropen, but it was rarely harvested; the variety most commonly harvested was Ndosa, which produced white starch.

The population of untrunked sago palms at the rosette stage ranged from about 500 to 5,000 palms/ha but no consistent pattern was observed on the basis of stand type (Table 7-10). When the rosette stage was divided into two phases, the population dropped rapidly from the early phase to the late phase of the rosette stage regardless of stand type (Table 7-10). The sudden decline in population between the early and late rosette stages is likely to be attributable to death from competition, pest damage and collateral damage from the harvesting of mature trunks.

The population of trunked sago palms (after the so-called trunk formation stage) varied greatly from 187 to 788 palms/ha. It tended to be higher in natural stands than in cultivated stands. Due to variation in the number of clumps per ha, the average population per clump after the trunk formation stage was 1.0–3.3 palms. There was no significant difference between the three stand types. Shimoda and Power (1992a) highlighted the importance of management practices that improve light conditions within the clump, including sucker thinning and removing dead leaves and other trees for a smooth transition from the rosette stage to the trunk formation stage. The absence of clear differences in the trunked population per ha between different sago palm stand types in Table 7-10 serves as clear evidence that cultivation management is lacking even in cultivated stands.

The population at the harvesting stage showed an almost 10-fold difference from 14 to 125 palms/ha but there was no clear difference in the number of harvested palms between the three sago palm stand types (Table 7-10). The wide variety of factors that may be involved in explaining the differences in the harvestable sago palm population, including harvesting frequency, intervarietal differences in the harvestable palm size, planting density and the level of cultivation management, need to be studied in detail in order to secure planned harvest quantities for sago palm plantations in the future.

### 7.2.2.b Starch yield in natural (wild), semi-cultivated and cultivated sago palm stands

Table 7-12 lists annual starch yields reported for sago palms from various provinces in Indonesia, Papua New Guinea and Malaysia. All of the figures are dry starch yields using traditional local extraction methods. The annual starch yield is 0.8–15.5 t/ha in natural (wild) stands, 1.5–37.0 t/ha in semi-cultivated stands, and 2.8–6.6 t/ha in cultivated stands. The starch yield of 0.8–1.9 t/ha/yr in natural and semi-cultivated stands in the Sepik River basin, Papua New Guinea, is markedly lower than in other areas. This is primarily due to the low number of harvested palms (10–17 palms/ha). In comparison, it is markedly high in semi-cultivated stands in Waropen, Papua, at 37 t/ha. When these outliers are excluded, variation in starch yield between stand types is small and there is no apparent tendency for cultivated stands to yield higher than natural stands. The fact that starch productivity in cultivated and semi-cultivated stands is no higher than in natural stands suggests that adequate cultivation management is not being practiced.

Flach (1980) argued that the starch yield of 2.9 t/ha/yr in natural sago palm stands could be increased to 4.4 t by improved extraction techniques and to 10.2–25.2 t through an increase in the number of harvested palms by thorough cultivation management (Table 7-13). A starch yield of 25.5 t is said to be possible in cultivated stands through the proper management practices. This is estimated on the basis of a planting density of 6 m × 6 m (278 clumps/ha), alternate year harvesting from each clump, harvesting 138 palms/ha and an average starch yield of 185 kg/palm. Others have noted, however, that it would be difficult to achieve alternate year harvesting of one trunk per clump at such a high planting density as well as such a high starch yield per trunk (Sato 1986).

Table 7-12 Annual sago palm starch production (per ha)

| Province, country | Study site | Stand type | Clump number (palms/ha) | Harvestable palm number (palms/clump) | Harvestable palm number (palms/ha) | Starch yield[1] (kg/palm) | Starch yield[1] (t/ha/year) | Authors |
|---|---|---|---|---|---|---|---|---|
| West Irian, I[2] | West Irian | Natural | 217 | 0.11 | 24 | 120 | 2.9 | Wttewaal (1954) |
| East Sepik, PNG | Sepik River down stream basin | Natural | - | - | 14–17 | 60 | 0.8–1.0 | Japan-PNG Goodwill Society (1984) |
| Papua, I | Inanwatan | Natural | 417 | 0.18 | 75 | 130–207 | 9.8–15.5 | Luhulima et al. (2006) |
| Papua, I | Kaure | Natural | 380 | 0.16 | 60 | 130–259 | 7.8–15.5 | Matanubun and Maturbongs (2006) |
| East Sepik, PNG | Sepik River down stream basin | Semi-cultivated | - | - | 10 | 154–192.5 | 1.5–1.9 | Japan-PNG Goodwill Society (1984) |
| Sarawak, M | Sarawak | Semi-cultivated | 202 | 0.50 | 102 | 166 | 17 | Kueh et al. (1991) |
| South Sulawesi, I | Luwu | Semi-cultivated | 100 | 0.14–0.34 (0.22)[3] | 14–34.2 (21.5)[3] | 200 | 2.8–6.8 (4.3)[3] | Osozawa (1990) |
| Papua, I | Waropen | Semi-cultivated | 117 | 1.07 | 125 | 56–296 | 7.0–37.0 | Istalaksana et al. (2006) |
| Riau, I | Tebing Tinggi | Cultivated | 100 | 0.24–0.45 | 24.1–44.7 | 118–197 | 2.8–6.6 | Yamamoto et al. (2008a) |

1) Dry starch yield produced by the local traditional extraction method
2) I: Indonesia, PNG: Papua New Guinea, M: Malaysia
3) Mean

Table 7-13 Production capacity of sago palm

| Situation | Clump interval (m) | Mature trunks number (No./ha) | Recoverable starch (kg/trunk) | Possible production of dry starch (t/ha/year) |
|---|---|---|---|---|
| Natural stand[1] | 7 × 7 | 24 | 120 | 2.9 |
| At: Improved starch extraction method | 7 × 7 | 24 | 185 | 4.4 |
| In: Improved natural stand | 7 × 7 | 55 | 185 | 10.2 |
| In Fully: natural stand | 7 × 7 | 136 | 185 | 25.2 |
| At full cultivation | 6 × 6 | 138 | 185 | 25.5 |

1) Zwollo (1950), Wttwaal (1954)
Source: Flach (1980)

## 7.2.2.c Estimated long-term starch productivity of sago palm stands

The starch yields reported above were estimated on the basis of the number of harvestable sago palms and the average starch yields which were current at the time of the studies. For sago starch to be used as industrial material in the future, annual starch production per area needs to be estimated for a period of several years. The long-term estimate of annual starch yield per area can be obtained based on:
1. trunked population per area per year
2. trunk growth rate per year
3. trunk length at the harvesting stage
4. average starch yield per palm.

The trunked population per area per year is determined by (a) number of emerged suckers × (b) survival rate × (c) trunk formation rate. There have been few studies of these factors but according to Shimoda and Power (1986), a survey in the Sepik River basin, East Sepik, Papua New Guinea, found that variation in the rate of sucker emergence was attributable more to the palm vigor of an individual clump than to intervarietal differences and that a substantial percentage of emerged suckers withered and died. The trunk formation rate was reported to be low where planting density was high; i.e., where excessive sucker growth or dead leaves blocked sunlight from reaching the forest bed in sufficient amounts.

For the population at and after the trunk formation stage in existing sago palm stands, it is possible to estimate the number of harvested palms per area per year based on the trunk lengths of individual palms growing within an area, the average trunk growth rate per year, and the average trunk length at the harvesting stage. Osozawa (1990) estimated the number of harvestable palms over time by applying this method to semi-cultivated sago palm stands in Luwu, South Sulawesi, Indonesia. First, the trunk lengths of trunked palms within a small area (0.16–0.24 ha) were measured at 4 places within a sago palm stand to obtain frequency distributions at every 50 cm of the trunk length per hectare as shown in Figure 7-16. The frequency distribution by

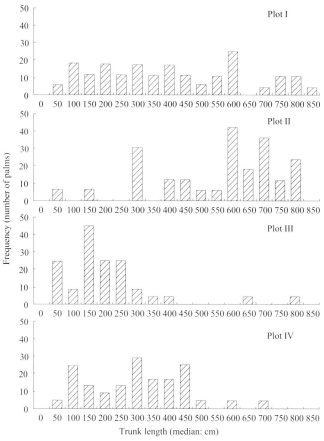

Figure 7-16 Frequency distribution of trunk lengths by study plot (palms/ha)
Source: Osozawa (1990)

trunk length varies from one plot to another: Plot I shows a relatively uniform distribution whereas Plots II–IV show uneven distributions. When these 4 plots were combined, the average trunk length distribution was smoothed out considerably (Figure 7-17). The average trunk growth rate was 1 m/yr and the mean trunk length at the harvesting stage was about 8 m at the study field in Luwu District. The trunk length distribution at every 1 m is shown in Figure 7-18. Palm numbers in this diagram indicate the number of harvestable palms/yr over the next 9 years which ranges from 14 to 34.2 palms/ha (21.5 palms/ha on average). When these figures are multiplied by the local mean starch yield of 200 kg, the annual starch yields become 2.8–6.8 t/ha (4.3 t/ha on average). The more than two-fold difference in annual starch production poses a problem from the viewpoint of starch supply stability.

Yanagidate et al. (2009) conducted a similar study in cultivated sago palm stands in Kendari, Southeast Sulawesi, Indonesia, and estimated that the average annual starch production from 2006 to 2015 would be 9.0 t/ha (starch yield in this study was higher as it was based on chemical analysis) but starch production tended to decrease from 13.5 t/ha to 4.3 t/ha over time. They argued that these variations in annual starch production stemmed from variations in the number of harvested palms and pointed out the importance of proper management of trunk numbers by age through sucker control for the ongoing stability of annual starch production.

A similar study was carried out at 3 sago palm farms with different soil types (shallow and deep peat soils) and years of operation (7–25 years) on Tebing Tinggi Island, Riau, Indonesia (Yamamoto et al. 2008a). Starch production per hectare every two years for a period of six years was estimated according

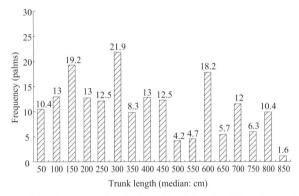

Figure 7-17 Frequency distribution of trunk lengths (50 cm intervals)
The average of 4 study plots (palms/ha)
Source: Osozawa (1990)

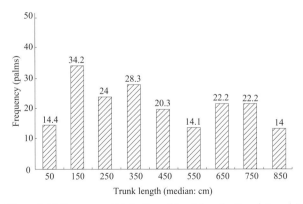

Figure 7-18 Frequency distribution of trunk lengths (1 m intervals)
The average of 4 study plots (palms/ha)
Source: Osozawa (1990)

to local harvesting practice. The study found that annual starch productivity was determined primarily by the number of harvestable palms and that the sago palm farms showed three main patterns of change: yearly increase (a 25-year-old farm on shallow peat), yearly decrease (an 18-year-old farm on deep peat), and yearly increase followed by decrease (a 7-year-old farm on shallow peat). It was suggested that management practices such as sucker thinning were important for sago palm farms in order to secure annual harvest numbers, which would determine starch productivity over time.

Shimoda proposed the following formula to calculate annual starch production per hectare (Y) for the average duration (years) from sucker planting (or emergence) to harvesting (A) at sago palm stands (Japan-PNG Goodwill Society 1984).

$Y = (C \times S) y / A$
(where C: clumps per hectare; S: trunked palms per clump; y: starch yield per palm)

He argued that the well-planned S was the most important factor in improving yield and that sucker control was essential for that purpose.

## 7.3 Productivity comparison with other starch crops

Since sago palm is a starch resource crop, its starch productivity has been studied in comparison with other starch crops such as cereal crops, tuber and root crops.

Nagato and Shimoda (1979) compared starch productivity on a caloric basis between sago palm and the world's major starch crops (corn, potato, sweet potato, cassava, rice and wheat) assuming sago starch production of 7 t/ha from wild stands, 11 t/ha from semi-cultivated stands and 17.6 t/ha from cultivated farms (based on Flach 1977) (25.5 t/ha in Flach adjusted for years from planting to harvest). They found that when the average productivity of the world's starch crops was set to 100, sago palm's productivity index was 685–1,679 for cultivated farms, 428–1,050 for semi-cultivated stands, and 272–668 for natural stands. It is evident that sago palm's caloric productivity is markedly higher than the world's other major starch crops. Flach (1977) stated that difference in productivity between sago palm and rice was attributable to superior distribution of converted carbohydrates to the utilized part.

Flach (1980) also conducted a caloric-based comparative study of productivity between sago palms and other tropical tuber and root crops, including banana, on the basis of sago palm's starch productivity of 25 t/ha/yr (Table 7-14). Against the recorded high yields from tropical tuber and root crops, sago palm's productivity ($275 \times 10^3$ Kcal/ha/day) was lower than that of cassava, sweet potato and taro and higher than that of yam and banana. Sago palm was markedly superior to these tuber and root crops in productivity in terms of the average world yield. Assuming that starch production from natural and semi-cultivated sago palm stands was 7 t/ha/yr and 11 t/ha/yr respectively (Flach 1977), caloric production per day was $77 \times 10^3$ Kcal and $121 \times 10^3$ Kcal respectively; sago palm's productivity not only from semi-cultivated stands but also natural stands was considerably higher than that of tuber and root crops.

Yamamoto (2009) compared sago palm's starch productivity with that of major tuber and root crops, including cassava, sweet potato and potato (Table 7-15). Table 7-15 lists the average world yields and the highest yielding countries based on FAO statistics for 2002 as well as recorded high yields to date for these three types of tuber and root crops. Sago palm's yields are based

Table 7-14 Comparison of productivity of the main moisture-rich starchy staples

| Crop (Reference) | Reported high yield | | High productivity of vegetation ($10^3$ kcal/ha/day) | Average world yield (1996) | | Average productivity of vegetation ($10^3$ kcal/ha/day) |
|---|---|---|---|---|---|---|
| | (t/ha) | Growth duration (days) | | (t/ha)[1] | Growing period (days) | |
| Cassava (CIAT 1969) | 100.0 | 305 | 416 | 9.2 | 330 | 35 |
| Sweet potato (IITA 1976) | 43.1 | 122 | 354 | 9.2 | 135 | 68 |
| Taro (Pucknett et al. 1971) | 128.7 | 365 | 339 | 5.4 | 120 | 43 |
| Sago palm (Flach 1977) | 25.0[2] | 365 | 275 | - | - | - |
| Yam (Rehum et al. 1976) | 60.0 | 275 | 193 | 9.8 | 280 | 31 |
| Banana (Purseglove 1972) | 75.0 | 365 | 155 | 12.5 | 365 | 26 |

1) FAO Production Year Book (1976)
2) Dry starch
Source: Flach (1980)

Table 7-15 Comparison of starch productivity between sago palm and major tuber and root crops

| Crop | Yield[1] (fresh weight, t/ha) | | Recorded high yield (fresh weight, t/ha) | Water content (%) | Starch content[2] (%) | Starch yield (dry, t/ha/year) | | |
|---|---|---|---|---|---|---|---|---|
| | Average | Highest | | | | Average | Highest | Recorded high |
| Cassava | 10.7 | 25.6 (India) | 100 (CIAT)[2] | 70.3[5] | 30–33[7] | 3.2–3.5 | 7.7–8.4 | 30–33 |
| Sweet potato | 13.9 | 26.6 (Egypt) | 50 (PNG)[3] | 66.1[6] | 15–30[8] | 2.1–4.2 | 4.0–8.0 | 7.5–15 |
| Potato | 16.1 | 45.9 (Netherland) | 126 (USA)[4] | 75.8[5] | 10–30[9] | 1.6–4.8 | 4.6–13.8 | 12.6–37.8 |
| Sago palm A[10] | 127.6[11] | 251.2 | 350 | 56.6 | 22.1 | 28.2 | 55.6 | 77.4 |
| Sago palm B | 63.8 | 126.9 | 175 | 56.6 | 22.1 | 14.1 | 28.0 | 38.7 |
| Sago palm C | 42.5 | 83.9 | 117 | 56.6 | 22.1 | 9.4 | 18.5 | 25.9 |

1) FAO statistics (2002)
2) Flach, M. (1980)
3) Evans, L. T. (1996). Dry weight with 66.1% water content was converted to fresh weight based
4) Evans, L. T. (1996)
5) Yatsugi (1987a)
6) Kagawa (2001)
7) Maeda (1998)
8) Sakai (1999a)
9) Umemura (1984)
10) A, B and C assume harvesting of mature palms every year from all, 1/2 and 1/3 of the total number of clumps at a planting density of 10 m × 10 m (100 palms/ha)
11) Trunk fresh weight
Source: Yamamoto (2009)

on planting density of 10 m × 10 m (100 clumps/ha) at sago palm farms at the harvesting stage, harvesting one palm per year from all clumps (Case A), one-half of clumps (Case B) and one-third of clumps (Case C) and an average trunk weight of 1,276 kg and the highest trunk weight of 2,517 kg based on prior studies from major sago palm growing areas in Indonesia and Malaysia. The recorded high yield figure of 3,500 kg was observed in Para variety around Lake Sentani near Jayapura in Papua, Indonesia (Yamamoto 2006c). Dry starch content in each tuber and root crop is per fresh weight while sago palm's starch content is dry starch content per trunk (bark + pith) fresh weight. Dry starch yields from the world average, the highest and recorded high yields were calculated based on the fresh weight yields and starch contents of the tuber and root crops and sago palm. The average starch yield from sago palm was 9.4–28.2 t/ha. Even the lowest yield of 9.4 t/ha from Case C with the lowest estimate of harvested palms was two or more times higher than the average starch yield of tuber and root crops of 3.5–4.8 t/ha. Sago palm's highest starch yield of 18.5–55.6 t/ha was also markedly higher than the 8.0–13.8 t/ha of tuber and root crops. In contrast, the starch yield based on recorded high yield was 15–38 t/ha in tuber and root crops. Cassava and potato in particular showed high figures at 33 t/ha and 38 t/ha respectively. Starch yield based on recorded high yield in sago palm was 26–77 t/ha. Case A yielded markedly higher than tuber and root crops, Case B yielded an almost comparable amount to recorded high yields in cassava and potato while Case C yielded clearly higher than sweet potato but lower than cassava and potato.

Based on these findings, sago palm tends to yield higher levels of starch than tuber and root crops but as demonstrated by recorded high yields, the yield potential of sago palm will need to be determined through cultivation experiments using high-yielding sago palm varieties.

## 7.4 Yield potential of sago palm

The theoretical high-yielding ability of a crop (yield potential) is important as an indicator in improving real yield standards through breeding of high-yielding cultivars (varieties) and improvements in cultivation conditions. Table 7-16 shows starch production per hectare where the average starch yields per palm of reported varieties are classified by every 100 kg between 100 and 800 kg (dry starch) and the number of harvested palms per year from each clump is divided into three levels (one palm every year from all clumps, one-half of all clumps and one-third of all clumps) under practically possible sago palm planting densities (10 m × 10 m, 8 m × 8 m, 6 m× 6 m) (Yamamoto 2006a). According to the table, 3.3 t/ha of starch can be harvested from one-third of the clumps every year at a planting density of 10 m × 10 m where the average starch yield per palm is 100 kg while 222.4 t/ha of starch can be harvested from all clumps every year at a planting density of 6 m× 6 m where the average

Table 7-16 Yield level and yield potential of sago palm

| Level | Starch yield (kg/palm) | Planting density | | | | | | | |
|---|---|---|---|---|---|---|---|---|---|
| | | 10 × 10 m (100 clumps/ha) | | | 8 × 8 m (156 clumps/ha) | | | 6 × 6 m (278 clumps/ha) | | |
| | | Harvestable clumps (clumps/ha/year) | | | Harvestable clumps (clumps/ha/year) | | | Harvestable clumps (clumps/ha/year) | | |
| | | 100 | 50 | 33 | 156 | 78 | 52 | 278 | 139 | 93 |
| 1 | 100 | 10 | 5 | 3.3 | 15.6 | 7.8 | 5.2 | 27.8 | 13.9 | 9.3 |
| 2 | 200 | 20 | 10 | 6.6 | 31.2 | 15.6 | 10.4 | 55.6 | 27.8 | 18.6 |
| 3 | 300 | 30 | 15 | 9.9 | 46.8 | 23.4 | 15.6 | 83.4 | 41.7 | 27.9 |
| 4 | 500 | 50 | 25 | 16.5 | 78.0 | 39.0 | 26.0 | 139.0 | 69.5 | 46.5 |
| 5 | 600 | 60 | 30 | 19.8 | 93.6 | 46.8 | 31.2 | 166.8 | 83.4 | 55.8 |
| 6 | 800 | 80 | 40 | 26.8 | 124.8 | 62.4 | 41.6 | 222.4 | 111.2 | 74.4 |

Source: Yamamoto (2006c)

starch yield per palm is 800 kg. The difference between them is about 67-fold. This table is useful in establishing yield targets for the future development of sago palm plantations. At higher planting densities, however, competition within each clump and between clumps intensifies, making it difficult, especially for high starch yielding varieties, to achieve good growth, which is closely correlated to starch yield. Hence it is likely to become increasingly difficult to harvest from each clump every year. To what extent the selection of varieties for planting and the cultivation management practices and techniques can improve the number of harvested palms and the average starch yield per palm per year at the planting densities listed in Table 7-16 is considered to be one of the most important subjects for sago palm research.

Authors:
7.1.1–7.1.3, 7.2–7.4: Yoshinori Yamamoto
7.1.4: Tetsushi Yoshida
7.1.5: Youji Nitta

# 8
# Starch Extraction and Production

## 8.1 Traditional Extraction Methods

### 8.1.1 Basic form of traditional starch extraction

Methods of extracting starch stored in the trunk (stem) of the sago palm are simple yet quite diverse. Locally distinct forms of work have developed, perhaps due to differences in starch use, local resource use and economic factors among different ethnic groups. This section will consider the factors behind local variations based on an analysis of traditional extraction methods in various locations.

The basic process of traditional extraction methods involves palm felling, cutting the trunk into smaller segments (logs), log splitting, pith crushing, starch filtering (washing away crushed particles), starch extraction, straining and packaging. Traditional resources and indigenous tools are used in these tasks. Extraction techniques used in some of the locations are explained below.

### 8.1.2 Different forms of extraction

The starch content in sago palm reaches its maximum shortly before flowering. People in each locality use their own criteria to determine the proper time for harvesting. In Southeast Sulawesi in Indonesia, for example, it is when the apical leaves turn white (bloom) or when the midribs of young leaves turn black, or they make a small hole in the trunk to check the starch accumulation and bark thickness (Nishimura 1995). Extraction either occurs on-site where sago palms are logged or else the crushed pith fibers are taken to an extraction workshop. An extractor is set up on the side of a river or a swamp where water is readily available and the work is usually performed by a group of several workers.

### 8.1.3 Pith crushing

In Kendari City, Southeast Sulawesi Province, Indonesia, the starch extraction process begins with crushing either trunk slices or the pith of logs into fine pieces. The crushing tool is a handmade chipping axe (hammer) made of wood or bamboo. Its handle length and shape vary slightly from location to location, but it invariably has an iron ring attached to the point of impact. Natural materials such as a single gnarled tree or bent bamboo are used in making this tool which is also used in the Philippines. The pith is manually crushed from one end with the axe. This form of crushing is widely practiced in New Guinea, eastern Indonesia and the Philippines. In Luwu Regency in South Sulawesi Province, Indonesia, Sumatra Island and regions to the west, Kalimantan, and Sarawak in eastern Malaysia, the pith is ground or shaved by grating or rasping (Yamamoto 2007a). Spinning drum-shaped graters (called raspers) are used for pith shaving. Motor-powered versions of these raspers have appeared in recent years to make the task faster and easier. The pith grating methods are classified in two ways: fixed pith or moving pith.

## 8.1.4 Starch filtering

To extract starch, crushed pith fibers are washed in water. The cloudy water containing starch is stored in a trough and the layer of starch that settles at the bottom is collected. After the removal of its pith, the remaining hull-shaped sago palm trunk is often used as the trough. There are local variations in starch extraction equipment and procedures. In Indonesia and Malaysia, the dark milky cloudy water is placed into a vessel that is lined with a strainer or a net which is large enough for a worker to step into. The worker stomps pith while pouring water over it to extract starch. Alternatively, the bottom of a shallow square box is covered with a net on which raw material is placed and kneaded with feet while water is poured over it. Natural materials such as young coconut bark fibers (similar to palm stem particles) were traditionally used as the filter, but nylon netting is widely used these days. While sago particles are typically washed with feet as described above in Southeast Asian locales such as Sulawesi Island and Malaysia, in New Guinea they're more typically washed by hand. The hand-washing extraction technique employs a rectangular or gutter-shaped trough that is set at waist height and slightly tilted. A net is attached to its lower end. Bark fibers are sometimes placed in front of the net as in the case of the feet technique. Another container (tub) is placed below the netted end to receive starchy water. Pith fibers are placed in the trough in front of the net and kneaded with hands while water is poured over them to release starch. A simpler technique involves a longitudinally-halved cylinder with netted ends in which pith is washed by hand. A hand-squeezing technique is also employed in Mindanao Island, the Philippines, where a net filled with pith is hung up and squeezed by hand or the pith is placed on a piece of mesh cloth and hand-squeezed while water is poured over it.

## 8.1.5 Packaging

Sago palm starch extracted from pith settles at the bottom of the water receptor. The receptor may be a hollowed out sago palm stem or a rectangular box. An interesting case is the utilization of a boat-shaped container, which is a common utensil used by the local people in Butuan, Mindanao Island. The settled layer of starch is taken out of the container and packaged in the form of drained and semi-dried starch. Starch is packed in containers made from locally available materials such as the woven epidermis of the sago palm trunk or banana leaves which are tied with string. Starch may be processed further by drying again and grinding into flour.

## 8.1.6 Local variation in extraction method and classification

There are local variations at various stages in this extraction system (Nishimura and Laufa 2002). The sago palm is believed to have originated in New Guinea Island, gradually spreading to the neighboring Southeast Asian regions and Pacific islands (Ehara et al. 2003a). Considering starch extraction technologies and systems from this perspective, two aspects stand out: first, pith crushing tools and techniques and second, extraction equipment and methods. It is possible to classify pith crushing and washing into four types – two forms plus two intermediate/mixed forms. The two basic forms include the New Guinea technique and the Malay technique. The intermediate/mixed forms entail a mixture of these techniques. These forms are summarized in Table 8-1 and described in detail below.

Table 8-1 Patterns of sago pith crushing and starch extraction work

| Location | Sago pith crushing | Starch washing | Type |
|---|---|---|---|
| New Guinea Island | Chipping axe | Hands | New Guinea |
| Malaysia (Kalimantan) | Grater | Feet | Malay |
| Sulawesi Island | Chipping axe | Feet | Intermediate |
| Philippines (Mindanao) | Chipping axe | Hands | Mixed |

Source: Nishimura and Laufa (2002)

### 8.1.6.a New Guinea form

The sago palm pith is crushed with a handheld chipping axe. Pith fibers are placed in an extraction device for washing and squeezed by hand to release starch (fixed pith type). This task is generally performed in a sitting position. Starch is extracted by manually squeezing pith fibers in the washing equipment. This is a horizontal water-pour pattern (Figures 8-1, 8-2 and 8-3).

Figure 8-1 Sago pith crushing work using a chipping axe
Papua New Guinea
(Photo: Laufa M.)

Figure 8-2 Extraction work: Sago pith washing by hands
Papua New Guinea
(Photo: Laufa M.)

Figure 8-3 Extraction work: Washing by hands and an extractor
Sorong, New Guinea Island, Indonesia
(Photo: Yoshihiko Nishimura)

### 8.1.6.b Malay form

The bark is removed from the sago palm trunk and the pith is pulverized by a grater/rasp-type tool. This technique has been modified to mechanize the process by using a rasper (a motor-powered clawed drum) for pulverization (moving pith type). This is a vertical collection pattern by which the pith is placed in a basket or a box and stomped with feet while water is poured over it vertically (Figures 8-4, 8-5 and 8-6).

Figure 8-4 Grater-type pith crushing work
Riau Province, Indonesia
(Photo: Yoshinori Yamamoto)

Figure 8-5 Crushing work using a rasper machine
Southeast Sulawesi Province, Indonesia
(Photo: Yoshihiko Nishimura)

Figure 8-6 Extraction work: Washing by feet using pumping water
Southeast Sulawesi Province, Indonesia
(Photo: Yoshihiko Nishimura)

### 8.1.6.c Intermediate form (Sulawesi type)

This is considered to be an intermediate form as the pith is crushed using an axe while the pith is fixed. Washing is then performed by feet in a vertical water-pour pattern (Figures 8-7 and 8-8).

Figure 8-7 Crushing work using a chipping axe
Southeast Sulawesi Province, Indonesia
(Photo: Yoshihiko Nishimura)

Figure 8-8 Extraction work: Pith washing by feet
Southeast Sulawesi, Indonesia
(Photo: Yoshihiko Nishimura)

### 8.1.6.d Mixed form (Philippine type)

The pith is fixed and fibers are crushed using an axe (standing pattern). Washing involves hand squeezing either in a vertical or horizontal water-pour pattern (Figures 8-9 and 8-10).

Figure 8-9 Pith crushing, Mindanao Island
(Photo: Yoshihiko Nishimura)

Figure 8-10 Starch extraction, Mindanao Island
(Photo: Yoshihiko Nishimura)

Possible reasons for the local variations are discussed next.

## 8.1.7 Background to technological and methodological variation

The study of variations in extraction method found that hands are used to release starch in the New Guinea technique whereas feet are used in the Malay technique. The geographical distribution of the hand technique and the feet technique can be delineated with Wallace's Line and Weber's Line, which represent ecological boundaries (Figure 8-11). For crushing the sago palm pith into fine pieces, the chipping axe used on New Guinea Island is thought to be the original technique. Grating the pith with crushing claws or rasping teeth is a technique that is thought to have evolved into the drum crushing technology rasper to improve the efficiency of the crushing work, mostly on Sulawesi Island and areas further west. These two techniques suggest that the method used at the home of the sago palm in New Guinea was modified to improve sago starch extraction as it crossed Wallace's and Weber's Lines.

The most pertinent question here is how sago starch is used. In New Guinea, it is a staple food produced in a small scale for home consumption. The Malay technique can be regarded as a more commercial form in pursuit of efficiency. From a cultural perspective, the commercial proficiency of the Bugis people might have been important for the development of the Malay technique. A point of contact between the two techniques seems to be situated around Southeast Sulawesi. The Philippine technique is considered to be a mixed form. It is similar to the New Guinea technique in that sago starch is

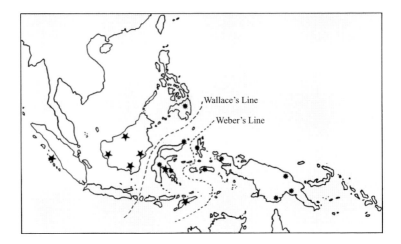

Figure 8-11 Distribution of different sago starch extraction methods and boundaries
● Extraction by hands. ★ Extraction by feet.

generally regarded as an emergency crop or as an ingredient for confectionary which is produced in small amounts and in small-scale operations, and hence efficiency is less important. Yet it adopts the Malay-style mechanism (washing by feet) for extraction.

Thus, it seems likely that the New Guinean method was developed into more commercial technologies and techniques as it disseminated westward from its place of origin. This change appears to have occurred near Wallace's and Weber's Lines. The choice of technique also depends on whether the extraction operation is for home consumption or commercial use. It appears that the New Guinea technique is for small scale production for home consumption while the Malay technique has evolved as a more commercial form.

The course of development of the extraction technique can also be explained by the diffusion path of *M. sagu* varieties.

Ehara et al. propose the diffusion path shown in Figure 8-12 based on a genetic analysis of *M. sagu* varieties (Ehara et al. 2003a). According to this hypothesis, two genetic groups in the New Guinean place of origin (B1, B2) diverged. One of them spread to Malay Peninsula, Kalimantan Island and Sumatra Island to produce A1 varieties and the other to Sulawesi Island and the Philippines to give rise to A2 varieties. This section has examined various sago starch extraction methods that are employed by local farmers and categorized them in an attempt to determine the technological development process. The varietal diffusion path appears to resemble the development path of the starch extraction methods. The development path of starch extraction methods and their categorization are described below.

There is a clear regional distinction between the New Guinea type and Malay methods of starch extraction, which are considered to be basic types. The two methods coexist in Indonesia and Sulawesi Island, which are situated between the two regions. The locally employed methods are considered to be of an intermediate type in the process of development. The coexistence of the two types is also found on Mindanao Island. However, the local methods here are regarded as a mixed type as they are distinct from both the New Guinea and Malay types in that hands and vertical water-pour are used in starch extraction (Figure 8-13).

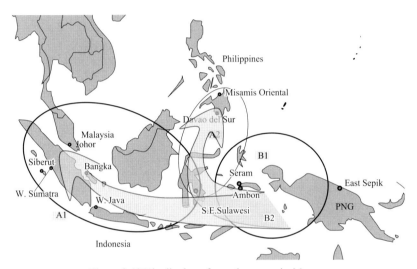

Figure 8-12 Distribution of genetic groups in *M. sagu*

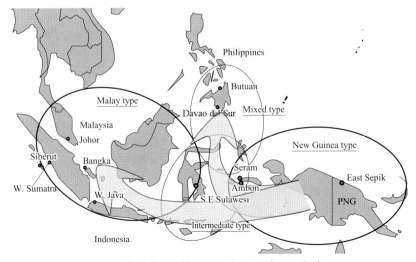

Figure 8-13 Type of Sago starch extraction methods

It is likely that the local variations of the traditional sago starch extraction method with different combinations of techniques and equipment were developed according to the importance of sago starch in the diet, the scale of operation, and the extent to which the operations have commercial (cash earning) purposes in each locality.

## 8.2 Extraction methods and production processes in starch factories

The sago palm pith contains fiber, soluble sugars and polyphenols as well as abundant starch, which needs to be isolated from the non-starch components.

The Malaysian state of Sarawak is one of the areas that is actively pursuing intensive sago starch production. Its factories have been growing in recent years, and have reached a production capacity of 300–500 t/month. In Indonesia, starch refineries are found in Irian Jaya, Halmahera and Sumatra. No large-scale starch factories are yet to be found in Papua New Guinea or southern Thailand.

### 8.2.1 Starch refining process

The industrial starch refining process is described in detail in Ohno (2003). This section shall explain the process by separating it into several stages: transportation, debarking, milling, extraction/separation, and refining.

#### 8.2.1.a Transportation stage

The mature sago palm is felled and cut into sections about 1 m in length which are called sago logs (Figure 8-14). The sago logs are loaded on a truck or assembled into a raft and transported to a starch factory.

Figure 8-14 Sago logs
(Photo: Mei Ohno)

The sago palm pith contains sap of soluble sugars such as glucose and sucrose in addition to starch. If logs are left unprocessed for too long after cutting, the starch quality deteriorates due to oxidation, which produces polyphenols that cause discoloration of starch granules, or microbial decomposition, which causes a foul odor or a decline in pH.

## 8.2.1.b Debarking stage

The bark is removed from the transported sago logs. The hard bark is about 15–30 mm in thickness and contains little starch. While debarking is performed manually using an axe and other tools at small starch factories, mechanization through the use of screw mills and rotary cutters is increasingly common. Where a screw mill is used, each log is split lengthwise into 4–6 pieces with the bark left on and fed into the mill for crushing and separation. This operation requires several workers. The rotary cutter (Figure 8-15), in contrast, scrapes the bark away with the cutter as it rotates the logs. With the use of a conveyor system, a single operator can continuously handle the entire task. Hydro mills, rotary slicers and cylinder-driven extruding debarkers have been trialed in some factories but they do not seem to have been very successful.

Figure 8-15 Rotary cutter
(Photo: Mei Ohno)

Increasing mechanization has reduced the processing time at the debarking stage and improved the starch yield and quality.

When the pith comes in contact with iron (in cutting devices and other machinery), it takes on a purple tint possibly due to iron oxidation (Fujii et al. 1986b).

## 8.2.1.c Milling stage

The sago starch that is stored in the pith is trapped in fibrous tissue. To release starch, the pith is fed into a milling machine called a rasper (a drum fitted with long narrow teeth; Figure 8-16) and milled into a mash in the rotating drum as water is added. This milling separates some of the starch from the fibers. The fibers are then fed through a hammer mill (a rotating drum equipped with hammers; Figure 8-17) in order to release the remaining starch. This process improves the starch yield, but the finely crushed fibers must be carefully removed at the refining stage.

Figure 8-16 Rasper
(Photo: Mei Ohno)

Figure 8-17 Hammer mill
(Photo: Mei Ohno)

### 8.2.1.d Extraction/separation stage

The milled pith contains starch and fibers which must be separated. There are coarse, medium and fine fibers. Coarse and medium fibers are passed several times through a sieve bend (Figure 8-18) and then a rotary screen. More advanced factories use a super decanter system to separate starch milk (a suspension of starch in water) and fibers.

Figure 8-18 Sieve bend
(Photo: Mei Ohno)

As the fibers still contain residual starch after separation, they are retrieved using a rotary screen (a horizontal rotating separator: Figure 8-19) and dewatered in a screw press (continuous pressing by rotating screws in a cylindrical strainer; Figure 8-20) before being discarded. A considerable amount of starch still remains in the fiber waste at the time of disposal.

Figure 8-19 Rotary screen    Figure 8-20 Screw press
(Photo: Mei Ohno)            (Photo: Mei Ohno)

### 8.2.1.e Refining stage

The starch milk obtained via the process described above still contains minute dietary fibers. The refined starch milk is processed further with a multistage hydrocyclone (starch milk is fed to a cylindrical machine which spins on its axis at a high speed to separate and condense starch grains by sedimentation; Figure 8-21), a DeLaval centrifugal separator (continuous centrifugal condensing of starch milk) and a super decanter (Figure 8-22).

Figure 8-21 Multistage hydrocyclone
(Photo: Mei Ohno)

Figure 8-22 Super decanter
(Photo: Mei Ohno)

After refining, starch milk is dried with a blast of hot air from a flash dryer. The dry starch is then collected in a cyclone, sifted and packaged. This process is summarized in Figure 8-23 (Ohno 2004 [adapted from Wagatsuma 1994]).

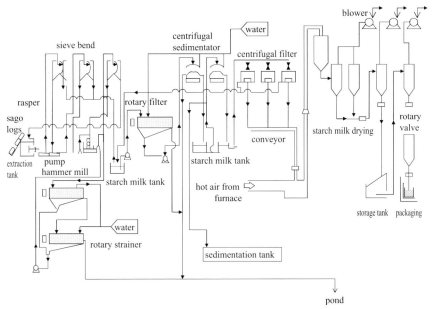

Figure 8-23 Manufacturing process in a sago starch factory
Source: Ohno 2004 (adapted from Wagatsuma 1994)

## 8.2.2 Unresolved issues

### 8.2.2.a Raw material control
Sago logs, the raw material in the manufacturing process, are often left moored in rivers adjacent to sago starch refineries. The sago pith has a high polyphenol content and the polyphenol oxidase reacts with oxygen, causing browning. Prolonged storage promotes browning, which is detrimental to starch quality. Further storage may see putrefaction and insect damage by sago grubs, rendering the raw material unusable. Accordingly, it is desirable to use sago logs within a week after cutting.

### 8.2.2.b Water management
Clean water is preferred for starch production. River water near sago starch factories in Sarawak, Malaysia, is turbid. Some factories use processed water treated with precipitants such as alum but in view of the possible polyphenol-induced discoloration, it is considered beneficial to use sulfur dioxide with bleaching properties to improve starch quality. However, careful attention must then be given to avoiding water pollution. Bleached starch is treated as a food additive in the United States and as a food product in Japan and the European Union as at 2008.

Wastewater from refining also needs to be managed. Wastewater contains finely milled fibers, unextracted starch and polyphenols and is colored and prone to putrefaction. Existing treatment methods such as the use of sedimentation ponds have had detrimental effects on neighboring areas, such as a foul odor. More positive wastewater treatment operations are hoped for. On the flip side, wastewater contains abundant carbohydrates. The application of this material as a biomass resource might lead to new products instead of waste treatment.

## 8.3 Sago palm production in the world

### 8.3.1 Sago palm growing/cultivation area
The sago palm is one of the oldest crops to be used by humans along with taro, banana and breadfruit (Takamura 1990) but it has never been treated as a major crop in its growing areas across Southeast Asia and Melanesia. The agricultural statistics from sago-producing countries still do not include sufficient information about the sago growing and cultivation areas or production quantities. According to Flach (1997) and the International Plant Genetic Resources Institute (IPGRI), there is a total of about 2.5 million ha of sago palm growing areas in the world (Table 8-2), of which 2.25 million ha are natural stands and 224,000 ha are cultivated (a high percentage of which are semi-cultivated) stands. A breakdown by country shows about 1.4 million ha in Indonesia, 1.02 million ha in Papua New Guinea (PNG), and about 45,000 ha in Malaysia. About 40,000 ha are found in East Malaysia and 5,000 ha are

situated in West Malaysia on the peninsula. Many of the sago palm stands in these areas are natural stands. Only 10 percent or so are cultivated or semi-cultivated.

Table 8-2 Estimates of sago palm growing areas

| Country/region | Wild stands (ha) | Cultivated stands* (ha) |
|---|---|---|
| Papua New Guinea | 1,000,000 | 20,000 |
| East Sepik Province | 500,000 | 5,000 |
| Gulf Province | 400,000 | 5,000 |
| Other provinces | 100,000 | 10,000 |
| Indonesia | 1,250,000 | 148,000 |
| Papua Province | 1,200,000 | 14,000 |
| Maluku Province | 50,000 | 10,000 |
| Sulawesi Island | | 30,000 |
| Kalimantan | | 20,000 |
| Sumatra Island | | 30,000 |
| Riau Islands | | 20,000 |
| Mentawei Islands | | 10,000 |
| Malaysia | | 45,000 |
| Sabah State | | 10,000 |
| Sarawak State | | 30,000 |
| West Malaysia | | 5,000 |
| Thailand | | 3,000 |
| Philippines | | 3,000 |
| Other countries | | 5,000 |
| Total | 2,250,000 | 224,000 |

Source: Flach (1997)
*: Including semi-cultivated stands.

According to the Land Custody and Development Authority (LCDA), the existing area of commercial sago palm plantations in Malaysia is approximately 10,000 ha. Small-scale sago palm farms owned by farming families make up 60,000 ha; a total area of about 70,000 ha (Sahamat, private correspondence). The sago palm growing area in West Malaysia in 1974 was 6,158 ha, 4,362 ha of which were in Johor State (JICA 1981a). However, the state's sago palm cultivation area in 1994 was 270.7 ha according to the agricultural statistics of Johor State (Jabatan Pertanian Negri Johor 1994), which means that the sago palm growing area decreased to 0.02% of the total agricultural land. Malaysia's sago palm growing area of about 70,000 ha as at 2008, as mentioned earlier, was more than the estimate made in the aforementioned IPGRI publication, suggesting an increasing trend over the previous decade.

The sago palm cultivation area (interpreted as the actively utilized sago palm growing area) reported in statistics published by Indonesia's agricultural ministry (Secretariat of Directorate General of Estates 2006) in 2006 was 99,445 ha across the country, all of it owned by small farmers (Table 8-3). This indicates a decrease of just under 49,000 ha from the area of cultivated (semi-cultivated) stands reported in the IPGRI publication but the exact details remain unclear as the sources of information may have used different classification systems. In South Kalimantan for example, the provincial agricultural authority breaks down the total sago palm growing area of 5,572 ha into 2,397 ha of wild stands and 3,175 ha of cultivated stands according to the 1981 report of JICA. As shown in Table 8-2, the IPGRI places all sago palms in Kalimantan in the cultivated stand category while the statistics from the Indonesia's agricultural ministry in Table 8-3 shows 5,847 ha in South Kalimantan as productive sago palm stands. Despite these differences, it is perhaps not too far off the mark to consider that a total sago palm growing area of 100,000–150,000 ha are in relatively active use across Indonesia. According to Flach (1997), the global sago palm growing area of 2.25 million ha, which are counted as natural stands, are predominantly sago palm stands in what

Table 8-3 Sago palm cultivation areas and starch production in Indonesia (2006)

| Province | Cultivation area (ha) | Production (t/year) |
|---|---|---|
| Aceh | 10,372 | 2,585 |
| North Sumatra | 0 | 0 |
| West Sumatra | 0 | 0 |
| Riau (incl. Riau Islands) | 59,174 | 9,409 |
| Jambi | 4 | 1 |
| South Sumatra | 0 | 0 |
| Bengkulu | 0 | 0 |
| Lampung | 0 | 0 |
| Total Sumatra Region | 69,550 | 11,995 |
| Jakarta | 0 | 0 |
| West Java | 0 | 0 |
| Banten | 0 | 0 |
| Central Java | 0 | 0 |
| Yogyakarta | 0 | 0 |
| East Java | 0 | 0 |
| Total Java Region | 0 | 0 |
| Bali | 0 | 0 |
| West Nusa Tenggara | 0 | 0 |
| East Nusa Tenggara | 0 | 0 |
| Total Nusa Tenggara Region | 0 | 0 |
| West Kalimantan | 4,980 | 922 |
| Central Kalimantan | 0 | 0 |
| South Kalimantan | 5,847 | 810 |
| East Kalimantan | 15 | 5 |
| Total Kalimantan Region | 10,842 | 1,737 |
| North Sulawesi | 3,692 | 498 |
| Gorontalo | 62 | 6 |
| Central Sulawesi | 7,467 | 898 |
| South Sulawesi | 3,987 | 1,001 |
| West Sulawesi | 2,534 | 288 |
| Southeast Sulawesi | 480 | 18 |
| Total Sulawesi Region | 18,222 | 2,709 |
| Maluku | 22 | 5 |
| North Maluku | 294 | 10 |
| Papua | 515 | 132 |
| West Papua | 0 | 0 |
| Total Maluku & Papua Region | 831 | 147 |
| Total Indonesia | 99,445 | 16,588 |

Source: Secretariat of Directorate General of Estates, *Tree Crop Estate Statistics of Indonesia 2004-2006 SAGO.* The productive sago palm growing area is counted as the cultivation area.

is called good condition. The total sago palm growing area, including other types of stands, is estimated to be as much as 6 million ha. Haryanto (1987) estimates that the total growing area is about 740,000 ha (the total area listed in Table 8-3 is 449,215 ha).

Table 8-4 Sago palm growing areas in central and eastern Indonesia

| Province | Area (ha) |
|---|---|
| Papua (Irian Jaya) | 270,300 |
| Maluku | 50,000 |
| Riau | 31,605 |
| North Sulawesi | 19,890 |
| Central Sulawesi | 75,000 |
| West Kalimantan | 2,420 |

Source: Haryanto (1987). Including data from 1979-1985.

Bintoro (2008) cites a wide range of estimates of sago palm growing areas in Indonesia, including 716,000 ha (Soedewo and Haryanto 1983), 850,000 ha (Soekarto and Wiyandi 1983), 418,000 ha (Manan and Supangkat 1984) and 6 million ha from the provincial forestry authority of Papua (former Irian Jaya Province) (Table 8-5). In Papua Province, 15,000 ha of sago palm stands are found in Manokwari Regency and 100,000 ha in Sorong Regency and Merauke according to Haryanto and Suharijito (1996). It was reported in *The Jakarta Post* on 17 October 2006 that a plan was under way to develop 31,360 ha of sago palm plantation in Maluku Province where 6,000 ha were already being cultivated and managed.

Table 8-5 Estimated sago palm growing aresa in Indonesia

| Island(s) | Area (ha) | Source |
|---|---|---|
| Papua (Irian Jaya) | 4,183,300 | Darmoyuwono (1984) |
|  | 800,441 | Henanto (1992) |
|  | 1,471,232 | Kertopermono (1996) |
|  | 4,371,590 | Haryanto and Pangloli (1994) |
| Maluku | 30,108 | Darmoyuwono (1984) |
|  | 47,600 | Universitas Pattimura (1992) |
|  | 41,949 | Kertopermono (1996) |
|  | 30,048 | BPPT* (1982) |
| Sulawesi | 45,540 | Kertopermono (1996) |
|  | 49,700 | Haryanto and Pangloli (1994) |
| Sumatra | 31,872 | Kertopermono (1996) |
|  | 71,900 | Haryanto and Pangloli (1994) |
| Kalimantan | 2,795 | Kertopermono (1996) |
|  | 2,000-50,000 | Haryanto and Pangloli (1994) |
| Java | 262 | BPPT* (1982) |

Source: Bintoro (1999)
*BPPT: Badan Pengkajian dan Penerapan Teknologi (Agency for the Assessment and Application of Technology).

## 8.3.2 Starch production

According to an IPGRI publication, 11,000 t of starch was produced in West Malaysia in the early 1990s (Othman 1991). The center of production has now shifted to East Malaysia. The state of Sarawak exports about 50,000 t of dry starch according to Jong (1995a). A research facility was set up in Mukah, Sarawak in 1982 to facilitate the efficient utilization of peaty wetlands amounting to 1.5 million ha. The world's first commercial plantation was developed over a 7,700 ha site in the basin of the Mukah River. In 1993, the second plantation was opened on a 1,600 ha site in the basin of the Oya River. In Sabah State, a factory with a small engine-powered rotary rasping machine produces 200–500 kg/day of dry starch. The whole sago palm industry in Malaysia produces 102,600 t/year of starch, including starch for various industrial uses as well as dietary use (Bujang and Ahmad 2000a).

According to an IPGRI publication, P. T. Sagindo Sari Lestari, a private enterprise in Bintuni, Papua (Irian Jaya) Province, Indonesia, operates a floating plant with a production capacity of 36,000 t/year. According to Japan Food Industry Association (1991), factories in Papua (Irian Jaya) owned by Djajanti Group have a production capacity of 60,000 t/year. INHUTANI I, a semi-governmental corporation, operates a factory with a dry starch production capacity of 30 t/day or 6,000 t/year in Kao on Halmahera Island, North Maluku Province, but the factory has been producing below full capacity since 1994 as the company has trouble maintaining its equipment. Japan Food Industry Association (1991) reported the following production capacities: 9,000 t/year for P. T. INHUTANI in North Maluku Province, 36,000 t/year for Stage 1 (500,000 t/year is planned for Stage 2) for P. T. Sagindo Sari Lestari in Papua (Irian Jaya) Province, 72,000 t/year for P. T. Sari Alam Guna Utama, and 40,000 t/year for P. T. Bumi Sempurna Tani. At that time, the Indonesian government had a policy to increase sago starch production to more than 100,000 t/year with a second-stage target of about 600,000 t. If achieved, this would have surpassed Malaysia's exports of sago starch several fold. According to Indonesia's agriculture ministry, however, the current national starch production is 16,588 t as shown in the official statistics in Table 8-3 (Secretariat of Directorate General of Estates 2006). However, this output figure for the whole of Indonesia appears to be too small. By region, Table 8-3 shows 9,409 t in Riau, 2,585 t in Aceh, followed by 1,001 t in South Sulawesi. It has been reported that South Sulawesi has 25,000 ha of sago palm stands around Bone, Parepare and Luwu (Maamun and Sarasutha 1987) and produces 8.4 t/ha of dry starch (Flach 1997). In short, different sources report different production outputs. Barie (2001) estimates Indonesia's potential production at 500,000 t and actual production at 200,000 t. Considering that Sarawak State in East Malaysia alone produces more than 50,000 t, a production output of 200,000 t from the whole of Indonesia appears to be a reasonable estimate.

Looking at information supplied by local or regional authorities, a publication from the industry authority of Bengkalis Regency, Riau Province (Ehara 1996), reported that the sago palm growing/cultivation area (including cultivated, semi-cultivated and natural stands where sago palms are used as a crop) was 12,576 ha of which 9,460 ha were harvested and 22.8 t/ha of wet sago were produced in 1992 while 9,494 ha were harvested and 23.4 t/ha were produced in 1993. For the entire province of Riau, the growing and cultivation area was 21,794 ha and the average wet sago production was 20.2 t/ha (newspaper report, *Riau Pos* dated 10 December 1996). The yield/ha/year is about 3 t higher in Bengkalis than in Riau. It appears that the supply of raw material is relatively stable in Bengkalis Regency as its sago palm growing sites and starch refineries are situated in close proximity. This may have some bearing on the variation in yield (Ehara 1997). It was reported that there were 59 businesses in Bengkalis' sago starch refining industry with a total starch refining capacity of 40,388 t/year in 1996. The industry reportedly operated at almost full capacity in 1995 to produce 39,426 t. In 1995, Bengkalis' four noodle (vermicelli-like dry noodle called '*sohun*') factories and four confectionery factories were established, which altogether produced 110,952 kg (Ehara 1997) of sago starch products. All of them were cottage industry factories.

According to a study conducted by Indragiri Hilir Regency of Riau Province and the University of Riau, the sago palm growing area in the regency, including Tembilahan, was 12,366 ha and the wet sago yield was low at 1.125 t/ha with a production output of 13,920 t in 1995 (BPPD TKII INHL–UNRIFAPETA 1996). On Tebing Tinggi Island in the regency, National Timber and Forest Product, a subsidiary of Kea Holdings of Singapore, has been operating a plantation business since 1996 with a plan to cultivate sago palms on 20,000 ha of land.

According to the Australian National University, sago starch production in PNG was 82,962 t in 2000 (Bourke and Vlassak 2004). East Sepik Province had the highest production output at 23,484 t/year followed by 16,711 t in Sandaun Province and 5,288 t in Madang Province; production tended to be higher in the provinces on the northern side of New Guinea Island along the Pacific coast (Table 8-6).

Table 8-6 Sago palm production by province in PNG

| Province | Production (t/year) |
|---|---|
| Western | 12,940 |
| Gulf | 10,369 |
| Central | 588 |
| Milne Bay | 1,676 |
| Oro | 1,624 |
| Southern Highlands | 2,405 |
| Enga | 104 |
| Western Highlands | 7 |
| Chimbu (Simbu) | 166 |
| Eastern Highlands | 3 |
| Morobe | 572 |
| Madang | 5,288 |
| East Sepik | 23,484 |
| Sandaun | 16,711 |
| Manus | 4,575 |
| New Ireland | 1,797 |
| East New Britain | 0 |
| West New Britain | 222 |
| Bougainville | 431 |
| Total | 82,962 |

Source: Bourke and Vlassak (2004)

Authors:
8.1: Yoshihiko Nishimura
8.2: Takashi Mishima
8.3: Hiroshi Ehara

# 9
# *Starch Properties and Uses*

## 9.1 Properties of sago starch

Sago palm (*M. sagu*) accumulates starch in its trunk. Approximately 200 kg of dry starch can be harvested per trunk. While sago palm has been a staple food for people in the tropics for a long time (Sato 1967), it is now being recognized as an underutilized source of starch. It grows in tropical low lands without competing with other cultivated crops and offers higher productivity than paddy rice or cassava.

The physiochemical and structural characteristics of sago starch have been studied from numerous perspectives, including microscopic observation, X-ray diffraction, amylose content and amylopectin chain length distribution, measurement of the degree of gelatinization by the β-amylase-pullulanase method, swelling power and solubility, and viscosity measurement by the Rapid Visco Analyzer (RVA). Its properties have been compared with starches derived from potato, corn, mung bean, kudzu, bracken, sweet potato and wheat. Its gelatinization and retrogradation properties have been examined based on physical changes of the starch sol and gel. Studies have shown that sago starch has a very interesting set of characteristics which are shared by both root-tuber starches and cereal starches. For example, sago starch's viscosity characteristics are similar to those of potato starch but its gel properties are similar to those of corn starch. As sago starch exhibits transparency, viscosity and gel hardness similar to those of Japanese kudzu and bracken starches, it is seen as a possible substitute for these starches. In fact, sago starch has been used in the preparation of Japanese sweets such as *warabimochi*, *kuzukiri* and *kuzuzakura* and proven to produce delicious products that are transparent, highly viscoelastic and easy to make at home (Takahashi and Kainuma 2006). Similarly, its suitability for producing blancmange, pie fillings, noodles and puffed food is promising as it improves the eating quality of these products. This chapter will report on these and other interesting findings such as the extent of variation in starch characteristics between different starch accumulating palms and sago palm types (landraces, hereinafter called 'varieties').

## 9.1.1 Physiochemical characteristics of sago starch

### 9.1.1.a Starch granules

The scanning electron micrograph of the parenchyma cells of the sago palm pith collected in Sarawak by Kainuma (Figure 9-1) shows that sago starch granules are elliptical, or bell-shaped with part of the ellipse missing. The relatively large average granule size of 35μm is close to sweet potato and potato starches. Starch is distributed unevenly inside the sago palm trunk, with a majority of it found a short distance away from the vessels. Many of the starch granules inside the trunk or removed from the trunk are found to be damaged (Figure 9-28) possibly due to microbial growth and enzymatic reactions during the storage in water of harvested sago palm logs for prolonged periods (Takahashi et al. 1981).

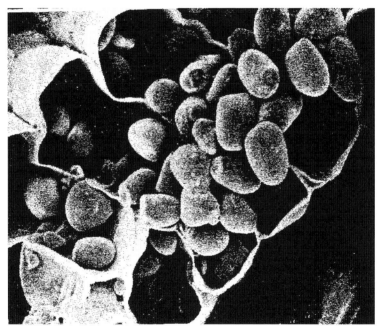

Figure 9-1 Starch granules in cells of the sago palm pith (SEM) ×200
Source: Takahashi et al.(1981) (Photo: Keiji Kainuma).

### 9.1.1.b Gelatinization behavior observed by photopastegraphy

Figure 9-2 shows that the temperature at which transmittance of starch suspension drops at 56 °C for potato starch, 58 °C for sago starch, 64 °C for corn starch and 65 °C for mung bean starch determined by photopastegraphy. Kainuma et al. (1968) contend that photopastegraphic observations reflect not only changes in transmittance due simply to starch granule swelling at the time of heat gelatinization but also minute structural changes inside the starch granules that occur prior to the swelling based on the view that the birefringence disappears due to disturbed molecular orientation inside the starch granules. According to his theory, the transmittance change of sago starch begins to drop at 58 °C because that is the temperature at which micelle orientation is disturbed by heat and starch granules begin to swell. This increase in starch granule volume reduces transmittance. Sago starch exhibits the greatest reduction under these conditions. The birefringence begins to disappear gradually from this temperature. At the next inflection point of 72 °C in the photopastegram almost all starch granules seem to lose polarization.

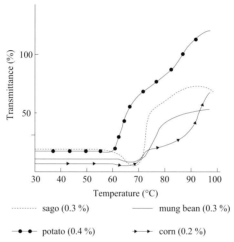

Figure 9-2 Photopastegram of sago, mung bean, potato and corn starches
Source: Takahashi et al. (1981)

### 9.1.1.c Swelling power and solubility

Swelling power refers to how many grams of water is absorbed by 1g of dry starch. Solubility refers to what percentage of starch dissolves in hot water at 60, 70, 80 and 90 °C. Potato starch has a high swelling power of 100 as well as a high solubility of 100% at 90 °C. Corn starch, with a swelling power of 22 and 26% solubility, is slow to swell and dissolve even at 90 °C. The swelling power and solubility of sago starch are noteworthy properties. At 40 and 53% respectively they are valued between tuber-root starches and cereal starches and second only to potato starch.

### 9.1.1.d Amylose content and amylopectin chain length distribution

The amylose content of sago starch is 26%, which is close to that of corn starch. The long-chain fraction Fr. II in the amylopectin chain length distribution obtained by the gel filtration method, or Fr. III/Fr. II, which are considered to be very important physical properties, are close to those of tapioca starch (Takahashi and Hirao 1994).

### 9.1.1.e Degree of gelatinization determined by β-amylase-pullulanase method

Kainuma et al. (1981) determined the degree of gelatinization using the β-amylase-pullulanase method for various pastes that had been prepared with a RVA or gelatinized under various conditions. They found that sago starch gelatinized at the second lowest temperature after potato starch. The degree of gelatinization increased sharply at around 70 °C. However, sago starch subsequently showed a slow gelatinization process similar to that of corn and mung bean starches and proved to have a wide temperature range for gelatinization (Takahashi and Watanable 1983).

### 9.1.1.f Viscosity measured by Rapid Visco Analyzer

Figure 9-3 shows the viscosity change of various starches during heating. It was found that commercial sago (lab-refined), spineless sago (*M. sagu*) and spiny sago (*M. rumphii*) had high viscosity; second only to potato starch and close to kudzu starch. By comparison, the viscosity curve of sugar palm (represented by *Arenga pinnata*) starch, another tropical palm-derived starch, was the lowest among the samples and close to that of wheat starch.

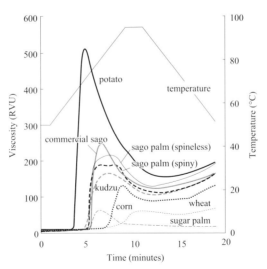

Figure 9-3 Rapid Visco Analyzer (RVA) curves of various starches
Source: Hamanishi (2002)

## 9.1.1.g Physical properties of sago starch gel

The hardness of sago starch gel is similar to that of sugar palm, potato and kudzu starch gels. Its adhesiveness property is close to that of sugar palm, corn, wheat and kudzu starch gels as shown in Figure 9-4 (Miyazaki 1999). While mung bean and corn starch gels expel a lot of water in low-temperature storage, sago starch gel shows the second lowest syneresis level after potato starch gel which is beneficial for cooking and processing applications (Takahashi et al. 1981).

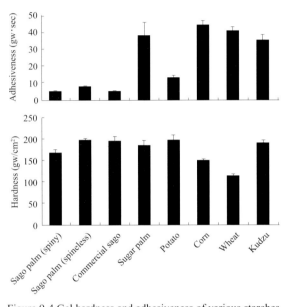

Figure 9-4 Gel hardness and adhesiveness of various starches measured by Tensipresser
Source: Miyazaki (1999)

Static measurement by a creepmeter has found that sago starch gel is softer and more viscous than sweet potato starch gel and less runny than potato starch gel.

## 9.1.1.h Dynamic viscoelasticity of sago starch gel

According to dynamic viscoelasticity measurements using Rheograph Gel (Table 9-1), sago starch gel is softer than potato, sweet potato and corn starch gels with low E' and E", corresponding to hardness and viscosity respectively. It has a strong internal viscosity element with a large tanδ (E'/E"), as opposed to the small tanδ of corn starch, pointing to an elastic body (Takahashi and Hirao 1994).

Table 9-1 Dynamic viscoelasticity of various starch gels

| Starch gel | Storage elastic modulus $E'\times10^3$(dyn/cm$^2$) | Loss elastic modulus $E''\times10^3$(dyn/cm$^2$) | Loss tangent (tanδ) $E''/E'$ |
|---|---|---|---|
| Sago | 2 | 1.4 | 0.70 |
| Potato | 10 | 4.2 | 0.42 |
| Sweet potato | 7 | 3.5 | 0.50 |
| Corn | 45 | 0.9 | 0.02 |

Source: Takahashi and Hirao (1994)

## 9.1.2 Difference between genera and varieties and starch characteristics

Within the genus *Metroxylon*, the starches of the spineless and spiny sago palms resemble one another in terms of granular shape, amylose content, thermal characteristics and gel physical properties. In contrast, the starch of the sugar palm of the genus Arenga, is notably different from those of sago starches in terms of viscosity properties. Sugar palm starch expresses a very low viscosity during heat gelatinization (Figure 9-3) but its paste hardness is similar to that of spineless sago starch (Figure 9-4) and it requires a very short time for gelling. As the transparent gel turns a turbid white color similar to corn starch over time during low-temperature storage (Figure 9-5), it can be used in cooking in methods similar to corn starch. It also shows promise as a gelling agent due to its rapid gelation time. Spineless sago starch gel, in contrast, has transparency similar to that of potato starch gel and shows little whiteness change during low-temperature storage. It is clear that the viscosity, gel properties and transparency of the palm trunk-accumulated starch vary greatly between different genera (Hamanishi 2002c). Ehara (1998) reports that 'sugar palm starch is said to be of good quality, non-browning and tastes similar to wheat flour' and is sold at higher prices than sago palm starch in areas where sugar palm starch is eaten.

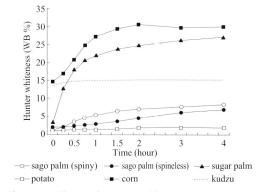

Figure 9-5 Changes in Hunter whiteness of various starch gels during storage at room temperature

Source: Hamanishi (2002)

### 9.1.3 Changes in physicochemical characteristics of starch by sago palm growth stage and site

Jong (1995a) reported that the sago starch yield increases up to the flowering stage and that the site of starch accumulation shifts from the base to the apex of the trunk during the life of the sago palm. The maximum starch yield is reached at the age of 11.5–12.5 years – between the maximum trunk growth stage and the flowering stage – when starch comprises 18–20% of the total trunk weight. At the final growth stage, most of the starch transfers to the apex for fruit production and the amount of starch left in the trunk drops rapidly to 4–9% of total trunk weight. Hamanishi et al. (1999, 2000) studied 35 types of sago palm starch supplied by Jong and reported clearly observable differences between growth stages and sites as follows.

Starches taken from the trunk base are faster to gelatinize, more viscous when heated and formed a highly transparent paste and soft gel compared with starches from the upper part of the trunk; the most distinctive characteristics were exhibited by starch taken from the upper part of a palm aged 14.5 years at the final growth stage for it had a low degree of transparency and formed the hardest gel. This suggested that the starch in the upper part remained in the trunk as it was small granule, highly crystalline and slow to metabolize. These findings provide some clues to the hitherto unknown sago starch accumulation mechanism.

Thus, sago palm starch is found to have excellent physiochemical characteristics and cooking properties and offers promising prospects for application to a broad range of food products, including gel-like food, blancmange, pie fillings, noodles and puffed food (Takahashi et al. 1995). When the ongoing expansion of sago palm plantation (Ohno 2003) improves the availability of high-purity sago palm starch and promotes the wider utilization of its characteristics, sago palm starch may be seen as a more useful starch material for cooking and processing.

## 9.2 Utilization of sago palm starch

### 9.2.1 Current uses of sago palm starch

The oldest description of sago palm is perhaps a passage in *Travels of Marco Polo* which says: 'They have a kind of tree that produces flour, and excellent flour it is for food. These tree are very tall and thick, but have a very thin bark, and inside the bark they are crammed with flour' (Kainuma 1981). Sago palm starch has been eaten in various forms for many centuries in sago palm growing areas. As shown in Table 9-2, sago palm starch is still cooked and eaten in a variety of foods such as *lempeng* and *papeda* in East Indonesia, the western coast of Halmahera Island and Seram Island in Maluku Province (Takahashi and Hirao 1992). It is also made into various cakes and baked confectionery

(Yamamoto 2006b). However, the area in which sago palm starch is consumed as a staple food is shrinking as it is being replaced by other cereals such as rice (Takahashi and Hirao 1992). (See section 2.2.)

Table 9-2 Food uses of sago starch

| | Name | Place | Cooking method | Current use |
|---|---|---|---|---|
| Sago starch | Lempeng | East Indonesia, west coast of Halmahera Island | Raw sago starch is baked in earthenware tray for long storage. | ○ |
| | Papeda | Seram Island of Maluku Province | Raw sago starch is mixed with equal amount of boiling water to make a paste which is added to soup. | ○ |
| | Lempeng | Sumatra, Riau Province | Raw sago starch is dissolved in water, seasoned with coconut and salt, and steam-roasted in a wok. | × |
| | Kurupun | | (1) Added to soup in the same way as papeda. (2) Mixed into soup to make congee. | × |
| Secondary processed sago products | Sagurendang | | Dried sago in pearls 2 mm in diameter. Added to boiled coconut milk sweetened with brown palm sugar. | ○○ |
| | Sagun | | Resembles Japanese rice crackers. | |
| | Sago noodles | | Sago starch in noodle form. | |

○○ High usage

Source: Takahashi and Hirao (1992), Masuda (1991)

### 9.2.1.a Use for cooking in Japan

In Japan, the term *sagobei* found its way into dictionaries as early as the eighteenth century. There is a record of sago grains being eaten as porridge after the country closed its doors to foreign trade (Ichige and Ishikawa 1984). Sago grains or sago pearls, together with tapioca pearls, have been used in jellies and puddings and as soup garnishes. They have been recorded in cook books, such as *perles du Japon* (Akiyama 1966), and they have been appreciated for both their visual appeal and unique texture (Takahashi and Kainuma 1989). Sago pearls are used in dishes such as *consommé au perles, consommé au sagou, pudding de sagou l'Anglaise, sagou au vin rouge*, sago custard pudding, veal and sago soup, and *xian nai shi mi ruan gao* (Takahashi and Kainuma 1989). Note, however, that all of these are examples of the use of sago pearls; no record of sago palm starch use in Japan has been found.

### 9.2.1.b Sago starch uses in Japan

Japan currently imports sago starch from Indonesia and Malaysia. According to the trade statistics of Japan for 2012 published by the Ministry of Finance, 17,283 t (3,578 t from Indonesia and 13,705 t from Malaysia) were imported per year. Sago starch for use in the manufacture of starch sugar, dextrin, dextrin glue, soluble starch, roasted starch or starch glue is not subject to tariffs. A majority of imported sago starch undergoes some form of processing such as oxidation prior to its use. Compared with raw starch, oxidized starch begins to gelatinize at lower temperatures, exhibits lower viscosity and retrogradation rate, and its bleaching effect improves whiteness. Modified sago starch is largely used as dusting flour for noodles such as *udon*, *ramen* and *soba* or dumpling skins such as *gyoza* and *shumai* in Japan. Oxidized sago starch only dissolves a little in boiling water and hence moderates the turbidity and viscosity of cooking water so that it does not need to be replaced often. It also appears to be marketed as a wheat substitute for allergy-free food products although no detailed study is available. In addition to food, it is used for industrial and animal feed purposes (see 9.2.3 and 9.2.4). Sago starch has attracted interest in recent years and efforts are being made for its practical application as a raw material for bioethanol production due to its high starch productivity.

As mentioned, sago starch is not actively used as food in Japan. The reasons for this include quality issues, competition with other starches and import tariffs on raw starch. Once the availability of high-quality sago starch improves and its cooking properties become widely known, its application is bound to expand greatly.

### 9.2.1.c Starches from sago palm varieties

It used to be conventional to classify sago palms broadly into the spineless (true sago palm) and spiny groups but they are now treated as one species, whether they have thorns or not. Many folk varieties have been reported in the region from Indonesia's Maluku Islands to New Guinea Island (see 7.2). Research on varietal differences in sago starch is important for the future expansion of sago starch use as it provides fundamental data concerning variety selection for sago palm plantations, such as the characteristics and yield of starch to be produced. Towards these ends, Hamanishi et al. (2006, 2007) and Hirao et al. (2006) recently reported on varietal differences in sago starch. They studied and compared the physiochemical characteristics and physical properties of starches from 10 sago palm varieties collected in Jayapura and surrounding areas of Indonesia's Papua Province where a particularly diverse range of folk varieties has been identified. Attempts were made to classify these varietal starches and their cooking properties using cluster analysis (nearest neighbor method) based on their amylose contents, viscosity properties and gel properties (Figure 9-6). The distance shown on the horizontal axis of the dendrogram represents similarity between samples. The cluster analysis was able to classify the varietal starches into four groups (Table 9-3).

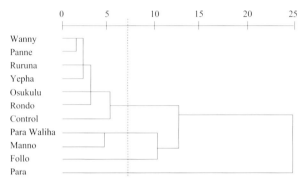

Figure 9-6 Dendrogram by cluster analysis of physiochemical properties of starch (basal)
Source: Hamanishi et al. (2007)

Table 9-3 Starch characteristics of folk varieties by group

| Group | Sample varieties | Amylose content (%) | Viscosity properties and starch gel physical properties | |
|---|---|---|---|---|
| 1 | Wanny Panne Ruruna Yepha Osukulu Rondo | 23.4-25.2 | - Medium-level viscosity and gel physical properties.<br>- Similar characteristics to the control. | |
| 2 | Para Waliha Manno | 27.1<br>22.5 | - High viscosity.<br>- High gel cohesiveness.<br>- slow retrogradation. | → Starchy sauce |
| 3 | Follo | 26.7 | - Hard gel.<br>- Fastest retrogradation. | → Gel-like food |
| 4 | Para | 24.8 | - Low viscosity.<br>- Soft gel.<br>- Slowest retrogradation. | → Sol-like food |

Source: Hamanishi et al. (2007)

1. Wanny, Panne, Ruruna, Yepha, Osukulu and Rondo: Amylose content 23.4–25.2%. Their viscosity and physical properties were at the intermediate level among 10 samples.
2. Para Waliha and Manno: While they were found to be of the same variety in subsequent studies, their amylose contents differed significantly (Para Waliha 27.1%, Manno 22.5%). Their viscosity and physical properties were similar: high viscosity, high gel cohesiveness, and slow retrogradation. They could be used in products requiring high viscosity such as thickened sauce.

3. Follo: High amylose content at 26.7%, hard starch gel, and fast retrogradation. Considered suitable for gel-like foods such as blancmange.
4. Para: Amylose content 24.8%, low viscosity, soft starch gel, and slow retrogradation. It could be used in sol-like products such as sauces and thickeners.

It is evident that sago starches taken from different varieties have different characteristics. This knowledge is expected to provide valuable data for the future utilization of sago starch.

### 9.2.1.d Effects of additives on sago starch characteristics

When starch is used in cooking, it combines with proteins, lipids and sugars to form a complex system. Research on their interactions is important for developing an understanding of the characteristics of cooked and processed food products using starch. Takahashi et al. (1983, 1985a), Takahashi and Watanabe (1983), and Hirao et al. (1998, 2002, 2003, 2004a) have shed light on the interactions between these components. The following is a brief account of their findings.

### 9.2.1.d.1 Added protein

When soy protein was added to different kinds of starches, potato starch exhibited a marked drop in viscosity while its gel hardness and rupture force increased. Corn starch exhibited only a minor change in viscosity but its gel texture decreased noticeably. Sago starch, however, appeared to be less susceptible to the effect of soy protein than potato and corn starches; its viscosity and gel texture changed little although its swelling power and solubility tended to be inhibited (Takahashi and Watanabe 1983). When the degree of gelatinization was measured by the β-amylase-pullulanase (BAP) method, all starches showed slower gelatinization after the addition of soy protein. This was observed at the early stage of heating in sago and potato starches. In contrast, although corn and mung bean starches exhibited slow gelatinization at high temperatures on their own, they showed rather high degrees of gelatinization at the early stage of heating after the addition of soy protein. Thus it is clear that the effect of soy protein additive on gelatinization varies between different types of starch (Takahashi et al. 1983). The addition of silk fibroin had the same effect on sago starch as the addition of soy protein (Hirao et al. 2004a).

### 9.2.1.d.2 Added protein, oil and sucrose

In studying the effect of adding oil, sucrose and protein, measurement of the viscosity, gel texture, shape retention and syneresis was undertaken along with sensory evaluation at various compounding ratios of sago starch, soy protein isolate and soybean oil through the heating process according to Scheffé's simplex lattice design method. It was found that the maximum viscosity was high, gel hardness and springiness increased and syneresis decreased at high starch levels. There was greater shape retention and adhesiveness at higher soy protein isolate levels but decreased hardness, springiness and shape retention. Syneresis increased at higher soybean oil levels. Gel hardness, springiness and shape retention decreased and syneresis increased at higher soybean oil levels (Hirao et al. 1998). The sensory evaluation of blancmanges prepared with various ratios of sago starch, soy protein isolate and soybean oil together with sucrose found that a blancmange made of 9% starch, 3% soy protein isolate and 4% soybean oil was the preferred sample in terms of color, flavor, hardness, springiness, adhesiveness, smoothness and overall evaluation (Hirao et al. 2002).

## 9.2.2 Use as food

While New Guinea Island is considered to be the place of origin of sago palm starch, *Travels of Marco Polo* contains an account of people in Sumatra who ate sago starch as a staple food at the time (circa 1300 CE). The Indonesian Archipelago to the Malay Peninsula and eastern Java Island are rice-producing areas where rice has long been the staple food. Sago palm is a major produce in northern Mindanao Island in the Philippines, northern Borneo Island, northern Sulawesi Island and Maluku Islands and was utilized as one of the most important food resources in these areas until the mid-twentieth century (Hirao 2001). However, in these places, the use of sago palm starch as a highly promising untapped resource has been decreasing as rice consumption has increased. This can be seen, for example, in Indonesia's Southeast Sulawesi Province under the influence of the rice-eating Javanese political ruling class, even in Serum Island, where sago starch has always been eaten, rice is becoming the staple food (Masuda 1991). Despite this trend, sago starch continues to be used for cooking and processing in some areas. The development of new utilization options will support the continuing use of sago starch for dietary purposes.

### 9.2.2.a Sago starch use in places of production
Sago starch is cooked or processed in the following ways in sago producing areas.
1. Added to soup to make it porridge-like.
2. Mixed into hot water to make a paste that is put in fish soup.
3. Baked to make cookies or breads.
4. Wet sago starch is processed into pearls, noodles or crackers as secondary products.

Specific cooking or product names for the above include *kurupun* for 1, *papeda* and *randang* for 2, *lempeng*, *keropo* and *sinoli* for 3 and sago pearl, *kerupuk sagu* (prawn crackers) and *mie sagu* (sago noodles) for 4. Masuda (1991) reports that *lempeng* is used on the western coast of Halmahera Island, East Indonesia, and *papeda* on Seram Island, Maluku Province. *Lempeng* and *kurupun* are no longer used in Riau Province, Sumatra, but secondary products such as sago pearls, *kurupuk sagu* and sago noodles are still used. Yamamoto et al. (2008b) conducted a field survey on Seram Island, Maluku Province, and found that sago starch was still used as a staple food in the form of *papeda*, *lempeng* and *sinoli* there as well as on Ambon Island. Nishimura (2008) surveyed the eastern part of Mindanao Island in the Philippines, an important point of contact between the cultures of New Guinea and Asia, and reported that sago starch was used there for confectionery and emergency provision rather than as a staple food.

In the Malaysian state of Sarawak, the conventional sago starch uses have been supplemented by some newly developed methods of utilization in recent years, including as a raw material for high-fructose syrup, glutamic acid, caramel and bread (Bujang and Ahmad 2000b) and used with cross-linked wheat in manufacturing noodles (Puchongkavarin et al. 2000).

As the sago palm producing areas are exploring new ways to use and process sago starch while preserving their traditional uses, their sago starch cultures will likely be passed on to the next generation.

The uses and processing methods of sago starch employed in Selat Panjang, Riau Province, Indonesia, are described in more detail below (Hirao et al. 2008).

1. *Tepung kue(h):* Cake flour (cookies and dumplings). The butyric acid odor of sago starch is neutralized by adding powdered clove, jasmine and vanilla and is sold as easy to use flour.
2. *Sohun:* Rice vermicelli-type dry noodles (Figure 9-7). A highly viscous paste is prepared in a pot and formed into fine noodles which are dried in the sun (Figure 9-8).They are soaked in water prior to use just like rice vermicelli.
3. *Mie sagu:* Semidried noodles made of sago starch. Starch and water are mixed with a partially gelatinized sago starch paste and formed into noodles (Figure 9-9). The color varies depending on how refined the sago starch is. These are often served with a sauce just like fried noodles.
4. *Kue(h) bangkit:* Cookies for the lunar New Year. Sago starch and tapioca starch are mixed and cooked in a pot, sugar and coconut juice are mixed in, and then the mixture is molded and baked. Cookies are sometimes cut or colored (Figure 9-10).
5. *Kue(h) pisang:* Wrapped banana cake. Sago starch, sugar, coconut and banana are mixed with boiling water into dough which is wrapped in banana leaves and steamed or stone-roasted (Figure 9-11).

Figure 9-7 Manufacturing method of sago starch noodle '*Sohun*'
(Photo: Hidetake Tanaka)

Figure 9-8 *Sohun*

Figure 9-9 *Mie sagu*

Figure 9-10 *Kue(h) bangkit*

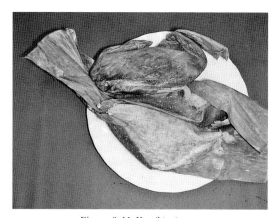

Figure 9-11 *Kue(h) pisang*

Figure 9-12 *Kerupuk sagu*

6. *Kerupuk sagu:* Crackers. Gelatinized sago starch is formed into a thin sheet and cut into pieces. The sago chips puff up when deep-fried in oil. Some are colored (Figure 9-12).

In Selat Panjang, traditional cooking and processing methods are commonly adapted to contemporary lifestyles. These processed food products are sold at food markets.

### 9.2.2.b Exploring sago starch uses

When sago starch is gelatinized, the gel exhibits properties such as transparency, low syneresis, high viscoelasticity and good shape retention. Its potential application to food products has therefore been of great interest to researchers.

*1 Warabimochi*

*Warabimochi* (bracken starch pastry) is a Japanese cake enjoyed in early spring. It is very popular for its smooth texture and its transparent and refreshing appearance. Although the traditional *warabimochi* is made from bracken starch, which is very expensive at around US$ 126/kg, these days the commercially sold bracken starch consists mostly of sweet potato starch. When sago starch was used to prepare '*sagomochi*' instead, the end product had better color, cutting quality, firmness and biting quality than *warabimochi* (Figure 9-13) and had a unique viscoelastic texture. While sweet potato starch has been used as a substitute for bracken starch, sago starch is in fact a superior substitute in cooking that requires a transparent end product. Furthermore, its distinctive pink tinge goes well with the yellowish soybean flour. *Sagomochi* should be promoted widely as a Japanese-style confection that can be readily prepared at home. The key point of making *sagomochi* from sago starch is to keep stirring at high temperatures to promote gelatinization so that the end product has preferred characteristics such as high viscoelasticity with good eating quality and slow retrogradation (Takahashi and Hirao 1994; Hamanishi et al. 2002a).

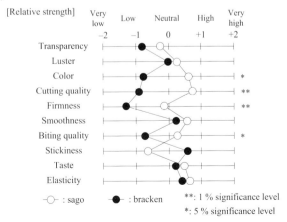

Figure 9-13 Sensory evaluation of warabimochi made of sago starch
Source: Takahashi and Hirao (1994)

## 2 Kuzuzakura

*Kuzuzakura* (kudzu starch pastry) is commonly made with kudzu starch. It requires characteristics such as easy handling of the starch pastry, good formability around the adzuki bean paste filling, good balance in the firmness of the pastry and the filling, and transparency. When *kuzuzakura* made of sago starch was compared with *kuzuzakura* made of a highly workable mix of three parts kudzu starch and one part potato starch (Teramoto and Matsumoto 1966) based on ease of production and sensory evaluation, the sago starch *kuzuzakura* was found to be superior in terms of ease of production, formability and shape retention, smoother and more springy than the control, and highly preferred in terms of transparency, color and cutting quality (Figure 9-14). The low fluidity and syneresis plus the high viscoelasticity and transparency of sago starch give it the malleability, formability and shape retention that are required for *kuzuzakura* production. Sago starch forms a soft gel which achieves the right balance of firmness between the pastry and the filling (Takahashi and Hirao 1994).

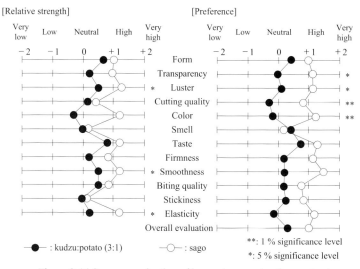

Figure 9-14 Sensory evaluation of kuzuzakura made of sago starch
Source: Takahashi and Hirao (1994)

## 3 Kuzukiri and fen pi

The mung bean starch sheet called *fen pi* (known as *kuzukiri* in Japan) is used in Chinese cooking and sold either fresh or dried (Takahashi et al. 1995). It is made of kudzu starch and/or potato starch as it requires characteristics such as chewiness and easy blending with a dressing or sauce. The firmness of *fen pi* made of sago starch is close to that of potato starch *fen pi*, harder than kudzu starch *fen pi* and approximately half of that of mung bean *fen pi*. Drying gives the sago starch *fen pi* firmness, chewiness and good cutting quality (Ohya et al. 1990).

*4 Gomadofu*
Gomadofu (sesame curd) is made of kudzu starch and ground sesame seeds which are heated and gelatinized. It is an essential dish for a vegetarian diet. When samples were prepared with sago starch or kudzu starch and 30% soybean flour instead of ground sesame seeds, the sesame curd made of sago starch had slightly less luster but exhibited good cutting quality and desirable shape. Sago starch was considered to be a good substitute for kudzu starch in *gomadofu* making (Takahashi and Hirao 1994).

*5 Kuzu mushiyokan*
Kuzu mushiyokan (steamed adzuki bean jelly) is a mixture of kudzu starch or wheat flour and adzuki bean paste and is enjoyed for its thick texture. The *mushiyokan* made of sago starch was softer and less sticky than the one made of kudzu starch and rated on a par with *kuzu mushiyokan* in terms of color, sweetness, taste, springiness and overall assessment (Hamanishi et al. 2002b).

*6 Blancmange*
The English-style blancmange is a pudding made of a mixture of corn starch, sugar and milk which is heated and gelatinized. When sago starch was substituted for corn starch, it gave the English-style blancmange less syneresis, better shape retention, less stickiness and cleaner mouth-feel (Hirao et al. 2002). As it is little affected by the addition of cocoa or green tea powder and offers improved shape retention, taste and texture, sago starch can be used as a gelling agent in a manner similar to corn starch (Hirao et al. 2003).

*7 Pie filling*
Pie filling requires good shape retention and clean cut surface. While corn starch or wheat flour is conventionally used, they have drawbacks such as susceptibility to retrogradation and syneresis during low-temperature storage. Sago starch could be utilized to make pie filling as it has low syneresis and good shape retention, and rated better than corn starch filling in terms of firmness, springiness and overall evaluation (Hirao et al. 2005).

*8 Bread and muffins*
Sago starch can be used in a wide range of puffed food products as it offered better swelling ability, a more uniform texture and more springiness than corn starch or potato starch. When bread was made from 100% sago starch with the addition of activated gluten, the sensory evaluation showed that it was preferable to other starches in terms of appearance, springiness and softness (Ohya et al. 1987).

## 9 Sago biscuits

Biscuits are a baked confection made of flour, butter, sugar and eggs as basic ingredients and enjoyed for their crunchy and crumbly texture. English-style arrowroot biscuits were prepared with 25% or 50% of the wheat flour being replaced by sago starch. The sago biscuits exhibited higher swelling ability, softness and brittleness than potato or corn starch biscuits. These characteristics intensified at the higher ratio of sago starch. Sensory evaluation found that the biscuits with partial sago starch substitution were more crumbly than the wheat-only biscuits and given significantly higher preference in terms of form, taste, hardness, brittleness, mouth-feel and overall evaluation. The 50% substitution in particular was highly preferred (Figure 9-15).

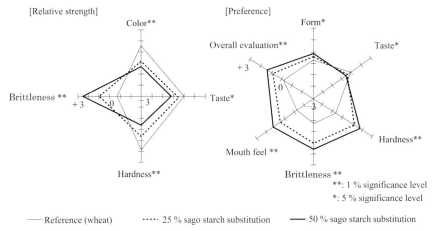

Figure 9-15 Sensory evaluation of biscuits, in which wheat flour was partially substituted by sago starch
Source: Hirao et al. (2004b)

When the amount of butter was increased from 20% to 45%, the sago biscuits improved their swelling power, softness and crumbliness. Preference in sensory evaluation increased significantly for hardness, brittleness, mouth-feel and overall evaluation. As brittleness was achievable with smaller amounts of butter at higher sago starch substitution rates, however, it was possible to reduce the amount of butter by half and hence the energy value (Hirao et al. 2004b).

## 10 Chinese vermicelli

As sago starch was considered to have excellent noodle-making qualities based on its physiochemical properties, the pressurized extrusion of Chinese vermicelli (cellophane noodles) was trialed (Takahashi et al. 1985b, 1986, 1987,

Takahashi 1986). Chinese vermicelli made of sago starch was transparent, firm and non-sticky, which are desirable properties (Takahashi and Hirao 1992). As it tended to rate higher than commercially available extruded vermicelli made in Japan in terms of appearance, texture and overall evaluation, sago starch was considered to be a useful raw material for Chinese vermicelli. The addition of soybean protein isolate had a positive effect as the addition of 5% soybean protein isolate to sago starch was found to inhibit solubility (Takahashi and Hirao 1993) and produce physiochemical properties that resembled those of commercially available Chinese-made vermicelli.

The addition of yolk powder to sago starch produced vermicelli with markedly increased adhesiveness which was considered to be suitable for mixing with mayonnaise or sauce (Takahashi and Hirao 1992).

*11 Cooking sago pearls*
Sago starch is churned into small balls, which are roasted until semi-gelatinized. Sago pearls are a secondary product of sago starch in the form of translucent pearls smaller than tapioca pearls. Starch pearls are used for garnishing in soups and to make jellies and puddings. They are visually pleasing and offer smoothness and good biting quality but they tend to become either mushy or undercooked at the center unless they are cooked properly. As a simple and easy method to cook tapioca pearls to a good texture and biting quality without dissolving the surface starch, it is recommended to use a thermos flask in which 5–6 mm diameter pearls are poured into hot water, stirred and left there for 3–4 hours (or 1.5–2 hours if an electric pot is used) (Hirao et al. 1989). In comparison, sago pearls are about 3 mm in diameter and can be cooked to a springy state in 20 minutes or so by the thermos flask method (Hirao and Takahashi 1996). With sago pearls, the cooking time is expected to be even shorter in an electric pot. The flask method offers many advantages over cooking in a pot as it does not require stirring or additional water during cooking and there is no risk of burning. Rehydrated sago pearls are used to make jellies and puddings or as garnishing for soups.

As we can see, sago starch has characteristics that place it between potato starch and corn starch. It offers many advantages such as high transparency, soft and smooth gel qualities, low susceptibility to the effects of additives and low syneresis. Sago starch and its secondary products have great prospects for application to a wide range of food in the future.

## 9.2.3 Industrial use
Starch is used as an industrial raw material in a very large number (estimated to be more than 2,000) of extremely diverse ways, including small-quantity uses. In the case of sago starch, however, the range of reported applications is limited to processing for consumption as the staple or general food at places of production, processing of wet sago starch into partially gelatinized and dried

sago pearls for domestic consumption and export, and substitution for tapioca starch as a raw material for monosodium glutamate, starch sugar and modified starch (Yatsugi 1987b). The production volume is smaller than other starches (sugar cane, potato, corn, wheat etc.) and production is limited to certain areas. There is at present no reliable global statistics on the production, specific uses and consumption of sago starch. For this reason, this section shall describe the common uses and prospective uses of starch in general as an industrial raw material in non-food industries.

### 9.2.3.a Starch properties and industrial uses
As mentioned earlier, starch has very diverse applications, which can be broadly grouped into the following three categories.

*1 Application of starch polymer characteristics*
Variations in gelatinization temperature, swelling power and viscosity stabilization which are linked to starch's polymer characteristics are exploited in its use as an industrial raw material. These characteristics have a very wide range of applications in food and seafood paste products, adhesives and sizing agents in paper and textile manufacturing as well as for pharmaceuticals, printing, casting and printing ink desiccants. These characteristics differ depending on the origin (type) of the starch.

*2 Application of glucose or maltose derived from starch*
Hydrolysis of starch produces glucose and maltose. These sugars are used as raw materials in food manufacturing in the form of starch syrup and dextrose.

*3 Application of starch as fermentation feedstock*
Either starch itself is used for fermentation as in the case of beer manufacturing or glucose derived in starch hydrolysis is used as fermentation feedstock. They are used as raw materials in food manufacturing in most cases.

### 9.2.3.b Starch uses in Japan
Table 9-4 shows all of Japan's starch demand and supply for starch year 2000 (October 2000–September 2001) (MAFF 2002). The overwhelming majority of Japan's supply (about 84%) was provided by corn starch, followed by potato starch, sugar cane starch and wheat starch. Some sago starch was imported from Malaysia and other places but the quantity was very small. Of the total demand of 3,038,000 t of starch, 2,338,000 t were for food manufacturing, including starch syrup, dextrose, isomerized sugar syrup, seafood paste products, beer, monosodium glutamate, food and others, while 700,000 t or about 23% of the total demand were used as non-food industrial materials, including native starches for textile, paper and cardboard manufacturing and modified starches.

Table 9-4 Total demand and supply for starches for starch year 2000 (thousand tons)

| | Category | Sweet potato | Potato | Corn | Imported starch | Wheat | Total |
|---|---|---|---|---|---|---|---|
| Supply | Balance brought forward | 4 | | 9 | | | 13 |
| | Volume on the market (production) | 64 | 223 | 2,553 | 157 | 29 | 3,026 |
| | Government selling | | | | | | |
| | Total | 67 | 232 | 2,553 | 157 | 29 | 3,038 |
| Demand | Starch syrup, dextrose, isomerized sugar | 61 | 129 | 1,612 | 63 | | 1,865 |
| | Fish paste products | | 19 | 3 | | 12 | 34 |
| | Textile, paper Corrugated cardboard | | | 256 | | 3 | 259 |
| | Modified starches | | 8 | 354 | 79 | | 441 |
| | Beer | | | 153 | | | 153 |
| | Sodium glutamate | | | | 6 | | 6 |
| | Food, other | 6 | 76 | 175 | 9 | 14 | 280 |
| | Total | 67 | 232 | 2,553 | 157 | 29 | 3,038 |
| | Government purchase | | | | | | |
| | Balance carried forward | 0 | 0 | 0 | 0 | 0 | 0 |
| | Total | 67 | 232 | 2,553 | 157 | 29 | 3,038 |

Source: MAFF (2002)

### 9.2.3.c Applications in major industries

*1 Textile industry*

Starches and modified starches, as discussed below, are used as warp sizes, printing pastes and finishing sizes in the textile industry. The warp size is used at the preparatory stage (warp sizing) for weaving in reducing fiber fuzz and improving strength, extensibility, flexibility, smoothness and abrasion resistance (cohesion) for the purpose of improved weaving efficiency and product quality (Takahashi 1984a). As the warp size needs to possess a wide range of properties, it contains a base made of native starches, modified starches, cellulosics and synthetic pastes with additives such as wax, oil and surfactants as well as mildew proofing agents and modifying agents as required. In recent years the use of polyvinyl alcohol has been increasing as a synthetic size but as starches are cheaper than synthetic sizing agents they are blended with polyvinyl alcohol and used for yarn sizing.

Printing is a way to impress a design on fabrics using a printing paste as a carrier for the dyes and pigments. Starch is frequently used as the main component of the printing paste. The necessary properties of the printing paste include high compatibility and stability with dyes and dye-assist agents (acid, alkaline and redox agents) in concentrated solutions, appropriate flowability and permeability to maintain sharpness for sharp and faithful printing of a design on a template, and high transferability and dyeing affinity so that dyes in the starch paste migrate to fabrics readily through hydrothermal treatment.

The characteristics of starches for printing paste are summarized below (Daita 1961).

**Native starches:** Wheat, corn and tapioca starches have been used for a long time but require additives such as tragacanth gum and sodium alginate because on their own these starches are inadequate in terms of permeability, chemical resistance and paste removal rate. Rice flour and rice bran have been used as starch pastes in the *yuzen* silk printing process.

**Dextrin:** British gums made from corn starch have high chemical resistance and are thus useful for dyeing methods such as discharge printing and resist dyeing although its reducing character precludes the use of dyes that are highly reactive to them.

**Starch derivatives:** Native starches are modified in various ways to rectify their shortcomings and improve printing suitability. Hydroxylated starches have excellent coldwater solubility, chemical resistance, coating flexibility and paste removal rate but, as they are slightly low in flowability, they are combined with additives and used as sizing agents for reactive dyes.

Finishing sizes are applied to woven fabrics for proper density and texture after bleaching and dyeing. While sizing may be performed by way of either dipping, applying or spraying depending on the type and intended use of the fabrics, the sizes used in this process are almost the same as warp sizes. Fluorescent dyes or bluing agents are added in some cases to increase whiteness.

Large quantities of starch are used to make laundry starches, which are used as finishing agents as well as to add antifouling properties to fabrics. Carboxy methyl starch and carboxy methyl cellulose are excellent.

*2 Paper manufacturing*

Starches are primarily used for surface sizing and coating in paper manufacturing (Asakura 1984). Surface sizing involves the application of adhesives such as starch, polyvinyl alcohol and carboxy methyl cellulose to the paper surface in order to improve inter-fiber bonding, surface smoothness and printability (anti-bleed). Surface sizes are required to have high solution concentration and low viscosity and therefore oxidized starches or oxygen-denatured starches are commonly used.

Pigment coating involves application of fine pigment particles (clay, sodium carbonate, titanium dioxide etc.) to the paper surface together with binders. The paper manufactured in this way is called art paper or coat paper which has a very smooth surface and high printing suitability. Binders are commonly made of latex mixed with natural materials such as starch, milk casein and soybean protein and synthetic materials such as polyvinyl alcohol. As starches used in binders need to be of high concentration and low viscosity, oxidized starches or oxygen-denatured starches are commonly used.

*3 Corrugated cardboard manufacturing*
Corrugated cardboard is comprised of a fluted corrugated sheet to which one or two linerboards (paperboards on the inside and/or outside of single-facer, double-backer, double-wall or quad-wall corrugated fiberboards) are glued with starch-based adhesives (Ogura 1984a). Corrugated cardboard adhesives require high initial bond strength. As starches alone cannot provide sufficient bond strength, the Stein Hall process is employed. Stein Hall adhesives consist of the carrier portion, which is a starch solution gelatinized in an alkaline medium, and the main starch portion, which is a colloidal starch solution. The carrier portion contributes to reducing the starch gelatinization temperature and accelerating the gelatinization rate at the time of initial bonding while the main portion improves adhesive power by starch gelatinization. They are mixed and applied to the fluted ridges of the core. Applying heat promotes the rapid gelatinization of starch in the main starch portion to manifest a high initial bond strength.

**9.2.3.d Modified starches**
While starch is used widely as an industrial raw material due to its polymer characteristics, new types of starches have been developed by enhancing the existing features or conferring new functions. These starches are collectively called modified starches. Modified starches are grouped into three categories according to the denaturation method shown in Table 9-5. Their modified functions and uses are shown in Table 9-6.

*1 Dextrin*
The primary industrial uses of dextrin are in adhesives and textiles. These applications demand characteristics such as gelatinization and solubility at ambient temperature and appropriate viscosity and adhesiveness for the manufacturing and use of high-concentration solutions. There are three types of dextrin, including white and yellow dextrins that are manufactured by heating starches with inorganic acids and British gums that are processed by high temperature torrefaction (roasting), either uncatalyzed or with alkaline additives. Figure 9-16 shows the process of the inorganic acid-catalyzed dextrinization of starch molecules (Schoch 1967). During the initial low-temperature (110–120 °C) reaction, hydrolysis of amylose and amylopectin produces low-molecular-weight carbohydrates called dextrin. At higher temperatures (over 150 °C), low-molecular-weight starch segments repolymerize and rebranch to a higher degree to produce yellow dextrin. White dextrin has a cold water solubility of about 90% and is highly retrogradable. It is used in textile finishing or sizing, paper surface sizing or clay coating, and adhesive agents. Yellow dextrin demonstrates over 99% cold water solubility and high adhesiveness. It is used in cellulosic adhesive agents, water-soluble films, remoistening adhesives, casting molds, briquette coal, mortar bonds and pharmaceuticals.

Table 9-5 Classification of modified starches by modification method

| Modification method | | Modified starch |
|---|---|---|
| Chemical | Degradation | dextrin, acid-treated starch, oxidized starch |
| | Derivative | cross-linked starch, esterified starch, etherified starch, graft copolymer |
| Physical | | Pregelatinized starch, amylose fraction, moisture-heat treated starch |
| Enzymatic | | dextrin, amylose |

Source: Ogura (1984b)

Table 9-6 Functions and uses of modified starch

| Function | Modified starch | | | | |
|---|---|---|---|---|---|
| | Dextrin | Oxidized starch | Pregelatinized starch | Derivative | Amylose fraction |
| Cold gelatinization, dissolution | A, B, T | | B, T, F | F, T | |
| Heat insolubility (hydrophobized) | | | | F, Ph | F |
| Water retention, viscoelasticity | A, T | | F | F, T | A, F |
| Electric property | | P | | Fl, P | |
| Emulsion protective colloid | | P | | F | |
| Retrograde resistance | A | | | F, P, T, A | |
| Chemical, mechanical shock resistance | T | A | | A, T | A, F |
| Concentrated use | A, B | A, T, P | | | |
| Film property | A | T, P | | T, P | F, T |
| Adhesive, binding property | A, Ph | T, P | B | A, T, P, Ph | T, Ph |
| Thermoplasticity | | | | M | |
| Solvent solubility | | | | C, M | |
| Bioactivity | | | | Ph | |
| Gelation property | | | F | | F |

A: adhesive, B: binding, C: coating, F: food, Fl: flocculating/precipitating agent, P: paper manufacturing, Ph: pharmaceutical, M: molding, T: textile.
Source: Ogura (1984b)

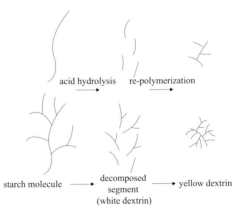

Figure 9-16 Starch dextrinization mechanism
Source: Ogura (1984c)

*2 Oxidized starches*

Oxidized starch is one of the types of modified starch which, together with dextrin, has been put to industrial use for a long time. It continues to be used in many industries due to its ease of production and the diverse applications it provides. Oxidized starches are manufactured through the reaction of raw starches to oxidants. In industrial settings, sodium hypochlorite is used as an oxidant and corn starch is used as a raw starch. A diluted sodium hydroxide solution is added to a starch suspension of about 45% concentration to pH 8–11, then the solution is heated to 40–50 °C and a sodium hypochlorite solution with effective chlorine concentration of 10% is added to initiate oxidation. As the oxidation progresses, carboxyl groups and carbonyl groups are produced within starch molecules (Epstein and Lewin 1962). Oxidation takes place in carbon or hydroxyl groups in the glucose residue constituting starch. The amount of carboxyl groups or carbonyl groups that forms depends considerably on the pH of the reacting system. The rate of carbonyl group formation is high in the neutral range while the rate of carboxyl group formation is high in the alkaline range. As the quantity of these functional groups strongly influences the physical properties of the oxidized starch, the optimization of pH during manufacturing is an important factor of manufacturing.

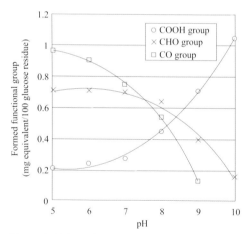

Figure 9-17 Functional group formation by pH of oxidation reaction
Source:Takahashi (1984b)

Oxidized starches are widely used for surface sizing in paper manufacturing. They are also used in cotton yarn sizing and finishing in the textile industry as well as in adhesives and binders for building materials.

## 3 Acetylated starch

Acetylated starch is produced by esterifying starch with reagents such as acetic acid, acetic anhydride and vinyl acetate. Its physical properties change according to the degree of substitution with acetyl groups. A highly acetyl-substituted starch with a mass gain of over 40% becomes insoluble in water and soluble in organic solvents such as glacial acetic acid and halogen hydrocarbon. The acetylated starch with a high degree of substitution is produced using solution techniques. When starch is heated and refluxed in a glacial acetic acid solution at 118 °C, acetylated starch with a mass gain of 42% or so is formed but the degree of polymerization decreases at the same time. When pyridine and acetic anhydride 3.2 times and 3.7 times the amount of starch respectively are added and left to react at 100 °C for 1.5–3 hours, the resultant acetylated starch has a smaller decrease in the degree of polymerization and a degree of substitution close to 3 but the high cost of collecting pyridine and acetic anhydride is a problem.

Acetylated starch with a low degree of substitution (about 0.2 or lower) needs to be heated in order to gelatinize but it gelatinizes at lower temperatures than unprocessed starches. The starch paste does not turn into a gel when cooled but stays transparent and highly resistant to retrogradation. The processes shown in Figures 9-18 and 9-19 are effective for manufacturing acetylated starch with a low degree of substitution. Figure 9-18 describes an acetylated reaction in acetic anhydride. A starch suspension is prepared to pH 7–11 and the reaction takes place at ambient temperature. In Figure 9-19, a starch suspension is prepared to pH 7.5–12.5 with the addition of either sodium hydroxide or sodium carbonate, then vinyl acetate is added, and the reaction takes place at 20–50 °C. Upon completion of an acetylated reaction, it is possible to reduce the pH and utilize acetaldehyde, a by-product, for cross-linking reaction. The low-substitution acetylated starch with a mass increase of 2.5% is used as a thickener or a shape retention agent for food and for textile sizing, paper surface sizing and clay coating for industrial purposes but it is not as competitive as more distinctive etherified starch or cheaper oxidized starch.

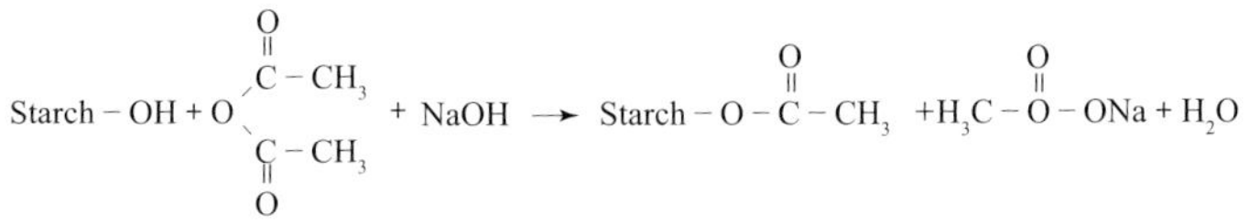

$$\text{Starch} - \text{OH} + \text{O}\begin{matrix}\overset{O}{\underset{\|}{C}} - CH_3 \\ \underset{\|}{\overset{C}{\underset{O}{}}} - CH_3\end{matrix} + \text{NaOH} \longrightarrow \text{Starch} - \text{O} - \overset{O}{\underset{\|}{C}} - CH_3 + H_3C - \overset{O}{\underset{\|}{C}} - O - ONa + H_2O$$

Figure 9-18 Acetylated reaction by acetic anhydride

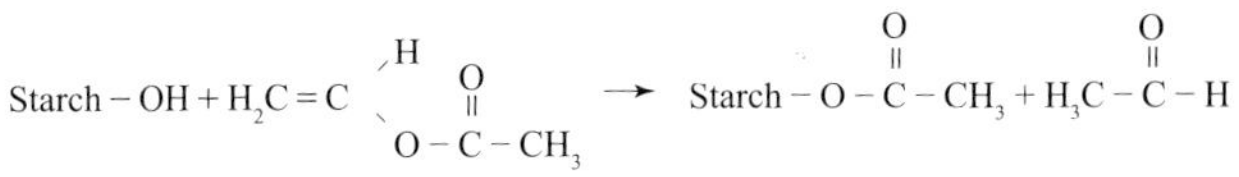

$$\text{Starch} - \text{OH} + H_2C = C\begin{matrix}H \\ O - \overset{O}{\underset{\|}{C}} - CH_3\end{matrix} \longrightarrow \text{Starch} - \text{O} - \overset{O}{\underset{\|}{C}} - CH_3 + H_3C - \overset{O}{\underset{\|}{C}} - H$$

Figure 9-19 Acetylated reaction by vinyl acetate

*4 Carboxymethyl starch*

Carboxymethyl starch is a cold water soluble polyelectrolyte complex. As shown in Figure 9-20, it is prepared by a reaction of starch with monochloroacetic acid in the presence of sodium hydroxide. Even a heat-gelatinizing type with a low degree of substitution has high viscosity, which is a characteristic of polyelectrolyte, and becomes cold water soluble at a substitution degree of about 0.15 or higher. It can be used in food thickeners, textile sizes and binders but it has difficulty competing with carboxymethyl cellulose, which has the same uses and offers superior functionality, including solution stability, viscosity, adhesion and film strength. Since carboxymethyl starch becomes cold water soluble at a lower degree of substitution than carboxymethyl cellulose, however, it is used as a low-cost printing paste in textile manufacturing.

$$\text{Starch} - OH + Cl - CH_2 - \overset{\overset{O}{\|}}{C} - O - H \xrightarrow{NaOH} \text{Starch} - O - CH_2 - \overset{\overset{O}{\|}}{C} - O - Na + NaCl + H_2O$$

Figure 9-20 Carboxymethyl reaction of starch by monochloroacetic acid

*5 Hydroxyethyl starch*

Hydroxyethyl starch is manufactured by a reaction of starch with ethylene oxide as shown in Figure 9-21. Industrial production is done by liquid phase reaction. Hydroxyethyl starch with a low degree of substitution (0.1 or lower) is prepared by adding 2–3% salt in liquid measure as a swelling inhibitor and 0.4–0.5% sodium hydroxide as a catalyst to a starch suspension of 45% concentration, followed by a specified amount of ethylene oxide to trigger a reaction. The introduction of hydroxyethyl groups promotes hydrophilic property and lowers the gelatinization temperature, making molecular dispersion and dissolution possible. As it improves the stability, water-holding ability, transparency and film moldability of the paste, it is used in corrugated cardboard adhesives, internal additives and surface sizes in the paper industry. It is possible to increase flowability or film strength and extensibility by treating the post-reaction starch suspension with acid.

$$\text{Starch} - OH + H_2C \underset{O}{\overset{}{\diagdown \diagup}} CH_2 \xrightarrow{NaOH} \text{Starch} - O - CH_2 - CH_2 - OH$$

Figure 9-21 Hydroxyethyl starch formation reaction

*6 Other modified starches*

Other modified starches, including alpha-starch, dialdehyde starch, starch phosphate, cationic starch, cross-linked starch and grafted starch are used primarily in the textile and paper manufacturing industries.

## 9.2.3.e Prospective industrial applications

*1 Raw material for biodegradable plastics*

Plastic materials have become indispensable for modern living due to its superior moldability, durability and strength characteristics. Figure 9-22 shows a breakdown of Japan's plastics production in 2005. Three major types of plastics – polyethylene, polypropylene and polyvinyl chloride – account for about 60% of the total production. Except for some of the plastics in the

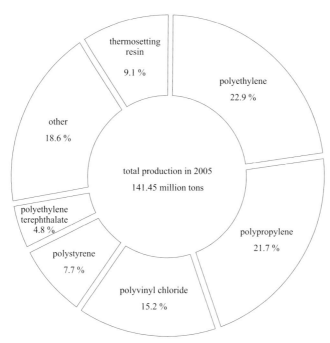

Figure 9-22 Total plastics production in Japan (2005)
Source: Japan Bioplastic Association (2006a)

Table 9-7 Classification of biodegradable plastics

| Type | Material name |
|---|---|
| Microbially produced | poly 3-hydroxybutyric acid (P(3HB)) <br> 3-hydroxybutyrate-co-3-hydroxyvalerate (P(3HB-co-3HV)) <br> 3-hydroxybutyrate-co-3-hydroxyhexanoate (P(3HB-co-3HH)) <br> 3-hydroxybutyrate-co-4-hydroxybutyrate (3HB-co-4HB) etc. |
| Natural material | acetylcellulose, cellulose nitrate etc. |
| Chemosynthetic | polylactic acid (PLA), polycaprolactone (PCL), polybutylene succinate (PBS), polybutyleneadipate (PBA) etc. |

Source: Japan Bioplastic Association (2006a)

'other' category, all plastics produced in Japan are made from petroleum resources. The petroleum-based plastics produced in abundance today consume vast quantities of exhaustible resources that are highly likely to run out in the not-so-distant future. The reserve-to-production ratio of crude oil is estimated to be only another 50–60 years. Plastic's high durability, which is one of its significant characteristics, means that its chemical composition and structure are not easily affected by changes in its environment. It therefore remains in the natural environment after disposal without breaking down. It interferes with the material cycle in the ecosystem and leads to adverse effects on various living organisms as well as shorter operating lives of refuse landfills. Its thermal disposal produces carbon dioxide, which contributes to greenhouse gases. These adverse effects of petroleum-based plastics have prompted the development of plastics that are made from more sustainable raw materials, which are quickly incorporated into the material cycle of the ecosystem after disposal, i.e., biodegradable plastics. Biodegradable plastics which are currently produced on an industrial scale can be divided into three broad types (Table 9-7). Among them, polylactic acid, which falls under the synthetic biodegradable plastic type, is synthesized using starch. One example of its manufacturing process is shown in Figure 9-23 (Japan Bioplastic Association 2006b). Starch is collected from starch crops such as corn and wheat and hydrolyzed with enzyme or acid. The resultant glucose is fermented by lactic acid bacteria to produce lactic acid, which is processed by heat polycondensation to form lactide. Polylactic acid is synthesized by the ring opening polymerization of lactide. The physical properties of polylactic acid are shown in Table 9-8. Compared with petroleum-based plastics such as polystyrene, polyethylene terephthalate and polypropylene, polylactic acid has superior mechanical properties such as high tensile strength, bending strength and bending elasticity but it is a rigid material due to low tensile elongation. It has poor heat resistance due to low softening and heat distortion temperatures which limit its uses. While these shortcomings are being addressed through the development of additives, further improvements to its physical properties are needed. At 400–600 yen (US$ 3.40-5.00) /kg (2007), biodegradable plastics have difficulty competing on price with conventional plastics, which only cost 150 yen (US$ 1.27) /kg. As the Japanese government had a policy target of 200 yen (US$ 1.69) /kg by 2010, further and continuing improvements to the manufacturing process, including better efficiency of raw materials, saccharization and polymerization, will be required. Unfortunately, the target price of biodegradable plastics had not been achieved at the time of writing (in 2014).

*2 Bioethanol materials*
Global warming has become one of the major environmental challenges as the average global temperature continues to rise due to greenhouse gas emissions from various human activities. The Kyoto Protocol to the United Nations

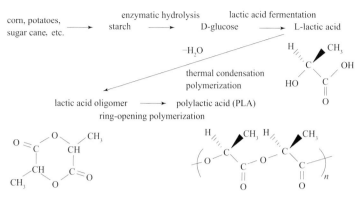

Figure 9-23 Manufacturing of polylactic acid from starch
Source: Japan Bioplastic Association (2006a)

Table 9-8 Physical properties of polylactic acid and general-purpose plastics

| Item | Unit | Test method ASTM | PLA (H-100J) | GPPS | PET | PP (homo) |
|---|---|---|---|---|---|---|
| MFR (190°C) | g/10 min | D-1238 | 11 | - | - | 30 (230°C) |
| Tensile strength | MPa | D-638 | 70 | 45 | 59 | 38 |
| Elongation rate | % | D-638 | 4 | 3 | 300 | 50 |
| Bending strength | MPa | D-790 | 100 | 76 | 90 | 46 |
| Bending elastic modulus | MPa | D-790 | 3,700 | 3,040 | 2,640 | 1,700 |
| Izod impact strength test | J/m | D-256 | 29 | 21 | 60 | 30 |
| Rockwell hardness | (L) | D-785 | 84 | - | - | - |
|  | (R) |  | 115 | 106 | 110 | 100 |
| Vicat softening temperature (9.8N) | °C | D-1525 | 59 | 98 | 79 | 150 |
| Heat distortion temperature (0.45Mpa) | °C | D-648 | 53 | 75 (1.82MPa) | 68 | 120 |

Note: PLA: polylactic acid, GPPS: general-purpose polystyrene, PET: polyethylene terephthalate, PP: polypropylene.

Framework Convention on Climate Change ('Kyoto Protocol') to combat global warming was ratified in 1997. The signatories undertook to achieve agreed greenhouse gas emission reduction targets. The first 5-year commitment period started in April 2008. Among other things, the Kyoto Protocol established the concept of 'carbon neutral' which is the idea that carbon dioxide derived from renewable resources such as plant biomass is cancelled out when plants use it in photosynthesis. It therefore does not contribute to global warming. By positing this concept, the Kyoto Protocol encouraged the use of plant biomass as a substitute for petroleum resources.

Starch has been used as fermentation feedstock since ancient times. For example, rice is prepared with *koji* to saccharize starch, then fermented by yeast to produce sake, i.e., ethanol. When starch-derived ethanol is burned, the emission is not counted in greenhouse gas emissions according to the carbon neutral concept. For this reason, the use of bioethanol as a gasoline substitute has begun worldwide. The bioethanol content of gasoline fuel ranges from 5 to 85% by volume.

The world's total bioethanol production in 2005 was 46.2 million m$^3$, approximately 70% of which were produced in the United States and Brazil. The main raw material in Brazil is sugar cane (low-molecular-weight sugars) while in the United States it is corn (Wright et al. 2006).

The current demand for starches that have conventionally been used in industrial production is relatively stable in terms of both quantity and quality. However, various environmental issues (global warming in particular) have spurred the development of new applications of starch and thus increased demand for the conventional starches (corn, wheat, sugar cane etc.) to be used as industrial resources in addition to the long-standing demand for food resources. Sago starch has been primarily consumed (as food) locally due to its small production and limited growing areas. However, once an adequate framework (stable supply, prices etc.) is established to take advantage of the sago palm's high starch productivity and ability to grow in tropical low lands, the current circumstances present an opportunity to increase its market share not only as a food source but also as an industrial resource. Although sago starch is yet to be utilized as a raw material for industrial production of biodegradable plastics or bioethanol, further development is anticipated in the future. Research is therefore under way that is focused on the characteristics of the sago palm and its possible applications (Ishizaki et al. 2002).

## 9.2.4 Application to animal feed

The commercial use of sago starch in animal feed production in Japan is almost nil. Sago starch poses a range of issues for feed development.

### 9.2.4.a Purpose of starch additive to animal feed

Starch has conventionally been included in animal feed formulas primarily as a source of carbohydrates. Today, however, the majority of carbohydrates are derived from wheat flour with varying degrees of bran content (from whole wheat to white flour) or from tubers and root crops (Morimoto 1979). The direct use of refined starches is usually as a molding compound, filler or spreading agent for special formula feed and, by extension, when the gelatinizing property of starch is required. However, animal feed products containing starches extracted and refined from cereals and root crops are expensive. It is therefore difficult to use them in animal fodder other than in high-value feeds such as milk substitute, artificial milk, pet food and fish food for which the physical properties of feed pellets are important. The special formula feed products include extruded pellet (EP), soft EP, high-absorbent juvenile fish food and moist pellet. Individual feed manufacturers use different starches for their different physical properties, including the water absorption of potato starch, the sponginess of legume, the degradation resistance of tapioca starch, and the hardness of wheat starch. Appropriate types of starch materials are used to complement the physical properties of feed pellets, which vary depending on the number of screws in the extruder (feed extruding and molding machine), screw circumferential velocity, die pressure and processing temperature.

Lipophilic or low-gelatinization-temperature modified starches are used to impregnate feed with oil. Sago starch is being considered for this purpose. Although many researchers and engineers tested sago starch in order to exploit its setting property after gelatinization or insolubility, they failed to find any practical application because of its fluctuating viscosity, price and degree of refinement.

Sago starch is not attractive in terms of price. Starches arranged in order of price are as follows: potato starch > sago starch ≥ sweet potato starch ≥ tapioca starch > wheat starch > corn starch.

### 9.2.4.b Characteristics of sago starch

Sago starch has excellent cryogelling properties. Sago starch rarely goes past the feed designing stage as its characteristics are not considered to be worth the price. The same goes for application to food. Its fluctuating viscosity, high proportion of impurities and higher price than tapioca starch deprives it of any competitive advantage in food product design. The photos in Figure 9-24 show starch filtration residue. It is clear that sago starch contains a high level of insoluble impurities. While the sago starch that is commercially available for food production in Japan is bleached and white in color, it still contains many more impurities than potato starch or tapioca starch. When animal feed grade sago starch is imported and used, the variability in viscosity and high admixture content poses numerous problems. Application to processed food is plagued by similar problems in that sago starch with a high admixture content is particularly low in viscosity and is not white enough. Insufficient cleaning or bleaching of sago starch in feed pellet manufacturing results in a longer heating time after water is added during which raw starch is denatured by enzymes and in some cases the viscosity of the starch paste is changed. This is due to the

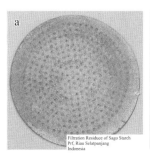
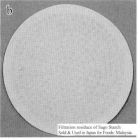
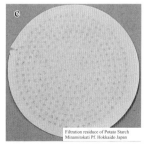

Figure 9-24 Starch filtration residue
a: sago starch made in Indonesia, b: sago starch for food made in Malaysia, c: potato starch.

action of mold-derived enzymes endemic to sago starch. Sago starches of low viscosity are generally found to contain a high level of wood dust or soil-flesh when they are hydrolyzed with liquefying amylase and tested for filtration residue. This has discouraged the direct importation of sago starches.

### 9.2.4.c Characteristics of commercially sold sago starch

A method developed by Kainuma (1986b) has been studied and applied to some of the sago starch products that have been imported to Japan from Malaysia and Indonesia for food in recent years. Immediately after ligneous fibers are separated from the milled sago palm trunk, enzymes are deactivated by acid and pH is adjusted with potassium hydroxide. The products showing high potassium levels in atomic absorption analysis have a high degree of viscosity. This is suggested by the atomic absorption analysis results in Table 9-9. A calcium or sodium content of 100 ppm suggests that bleaching has been carried out. Staining by cationic dyes such as methylene blue can be observed under the microscope if the starch has oxidized at the reducing terminal (Konoo 1998). It is likely that the sago starch harvested in Riau Province (Table 9-9) had extremely low viscosity compared with starches from other areas and a high gelatinization temperature (80 °C) because the development of viscosity, expressed as the gelatinization temperature, was inhibited and slowed by enzymes. On this point, Konoo et al. (1997) report that after starch milk is treated under an acidic condition, viscosity develops once again when the pH level is adjusted to neutral. This type of sago starch from Riau Province was suitable as a base material for the feeding attractants described below due to its ability to gelatinize at high concentrations but it was difficult to obtain products of uniform degradation. For this reason, starch is currently obtained in a raw state (β-starch) and oxidatively-treated before use.

Table 9-9 Sago starch characteristics

| | | Igan, Malaysia ppm | Riau, Indonesia ppm | Commercial product in Japan |
|---|---|---|---|---|
| Cation | K | 47 | 56 | 564 |
| | Na | 105 | 8 | 64 |
| | Ca | 55 | 154 | 162 |
| | Mg | 17 | 30 | 50 |
| | Fe | 5.1 | 4.5 | 2.7 |
| Whiteness | | 82 | 64.5 | 84.2 |
| Ash content % | | 0.14 | 0.16 | 0.14 |
| Crude protein % | | 0.01 | 0.07 | 0.03 |
| Viscosity properties by viscography (6 % concentration) | | | | |
| Viscosity increase initiation temperature (°C) | | 71 | 82 | 70.5 |
| Maximum viscosity (B. U.) | | 560 | 30 | 650 |
| Minimum viscosity (B. U.) | | 300 | 20 | 260 |
| Viscosity at 50 °C (B. U.) | | 500 | 50 | 600 |
| Breakdown (B. U.) | | 46.4 | 33.3 | 60 |

Sago starch is not as viscous as potato starch or as cheap as corn starch. Sago starch is not used for the production of EPs or moist pellets for fish which take advantage of starch's gelatinization properties and use relatively expensive materials. This is perhaps because its gel properties are not as good as those of the more expensive potato starch and its other characteristics are inferior to those of tapioca starch. As a source of carbohydrates, for which starch properties are irrelevant, sago starch is not used even in an enzyme-degraded form with enhanced absorbability (dextrin) because it is more expensive than corn, sweet potato and tapioca starches. Its high soil-flesh content shortens the life of filtering media such as Celite. These days, tapioca, wheat and potato starches are used for their unique physical properties despite their high prices in the manufacturing of those pellets requiring certain physical properties, including expansion pellets, hard pellets and moist pellets among EPs for fish feed and artificial feed for laboratory animals. This tendency is found with fish farming feed and pet food which are marketed at relatively high prices.

### 9.2.4.d Physical properties of EP for fish farming by starch type

There have been many attempts to use sago starch in animal feed. Table 9-10 shows the physical properties of EPs for fish farming prepared with a co-rotating twin screw extruder using 68% brown fish meal, 20% fish oil, 10% starch, 1% dried soy pulp and 1% gluten. While the pellets using sago starch had superior hardness, water absorption time and underwater disintegration time to tapioca, they were inferior to potato starch on these criteria. Although more than one feed manufacturer has reported that raising the degree of acetylated starch substitution to 0.07 can increase the amount of subsequent oil absorption and extend the underwater disintegration time to over two hours, this formulation has not been put to practical use as this level of substitution pushes the material cost to higher than potato starch. Even though sago starch has an interesting viscosity characteristic, in that the starch paste turns into a hard gel when cooled, it has never been utilized in feed making in Japan. This is possibly attributable to a high admixture content, variability in viscosity and unstable quality.

Table 9-10 Physical properties of extruded pellet (EP) for fish culture

| Starch type | Density | Swelling rate | Hardness | Water absorption time | Underwater disintegration time |
|---|---|---|---|---|---|
| | g/cm$^3$ | % | KgF | min | hr |
| Tapioca | 0.25 | 180 | 6.45 | 10 | 0.5 |
| Sago | 0.23 | 160 | 6.45 | 9 | 1.0 |
| Potato | 0.23 | 160 | 7.00 | 8 | 1.0 |
| Bean starch 5: Tapioca 5 | 0.35 | 160 | 7.26 | 7 | 2.0 |

Outside of Japan, however, sago starch is used widely in the preparation of EPs for prawns and fish in Indonesia and Malaysia. Horiuchi (1998), an expert (in aquaculture feed) sent by the Japan International Cooperation Agency (JICA) to the Highlands Aquaculture Development Centre in New Guinea, reported that the use of 5% commercially available wet sago starch in juvenile-stage feed for carp as a source of carbohydrates produced good results. He therefore continues to recommend the use of this formulation.

**9.2.4.e Application to feeding attractant flakes**
Sago starch is currently used commercially in Japan as an additive in cat food and fish food. Japan Flavor Co. Ltd. in Shizuoka Prefecture liquefies fish waste, including tuna guts and dark meat with proteolytic enzymes or by autolysis. After condensing it, starch is added and gelatinized. The mixture is then dehydrated into flakes. As the substance is highly hygroscopic, the normal starch paste cannot increase its concentration sufficiently to form flakes. Dextrin with low viscosity can increase concentration but it becomes difficult to maintain the quality of the final product due to increased hygroscopicity. Sago starch is treated with acid to reduce viscosity and made into a starch paste at a high concentration (around 20%) which is added to the material as a filler and formed into flakes in a disc dryer. The sago starch used here is treated with hypochlorite to reduce viscosity to an oxidation degree of about 10. The material is dried on the heated discs and removed by the scrapers to form sago flakes. When the flakes lose heat after coming off the discs, sago starch's unique setting power enhanced by hypochlorite treatment increases its gelling power and sets the flakes before deliquescing to enable grounding and screening. Mr. Kawashima, Managing Director of Japan Flavor, was generous enough to allow us to photograph his disc dryer. A highly concentrated starch paste is sprayed onto both sides of 10 heat discs and the hose at the top de-aerates the machine to dry the paste under reduced pressure (Figure 9-25a). Scrapers are visible in Figure 9-25b and 20 nozzles are shown in Figure 9-25c. At this degree of oxidation, sago starch on the dryer discs is in a candy-like state but it hardens as soon as it is scraped into a flake form. At any higher degree of oxidation, it remains in a sticky candy-like state which prevents grounding or screening. At any lower degree of oxidation, it forms a film-like substance that absorbs moisture in the air and turns into balls. This is the only

Figure 9-25 Disc dryer

degree of oxidation at which flaky tuna granules can be formed. The product is utilized as a feeding attractant for cat food and some fish farming feed in Japan.

Sago starch with high whiteness and low admixture content can be obtained if a sufficient quantity of clear water with no humic substances is used at the starch extraction stage. If large amounts of high quality sago starch refined with rain water were imported into Japan, the existing shortcomings would be resolved and its distinctive physical properties could be exploited in the production of a wide range of animal feed and human food.

## 9.3 Potential uses of sago starch

### 9.3.1 Research on sago palm and sago starch in Japan

Prior to the first International Sago Symposium entitled 'Sago – The Equatorial Swamp as a Natural Resource', the sago palm was merely a little known tropical plant studied by a small number of tropical agronomists and anthropologists in their respective fields. The symposium attracted about 70 participants from around the world and held a comprehensive discussion on such topics as the cultural and scientific backgrounds of the sago palm for the first time. The author was asked to provide some talking points about the characteristics and utilization properties of sago starch and gave a presentation on the starch's scientific properties and potential uses of starch. The objective of the symposium was to determine the usefulness of biomass production in tropical swamps exposed to powerful solar energy. Sponsored by the Asia Foundation and Malaysia's Sarawak state government and presided over by Professor W.R. Stanton of the University of Malaysia, the symposium also presented broad-based reports on sago as well as recommendations to the Sarawak state government pointing out the future possibilities of sago research and industrialization (Kainuma 1977b).

The proceedings of the symposium were published with the title Sago '76, including two papers from Japan: 'Present status of starch utilization in Japan' (Kainuma 1977) and 'The sago eater's adaptation in the Oriomo Plateau, Papua New Guinea' (Ohtsuka 1977).

Subsequent symposiums were held in various countries and the number of Japanese participants increased. An FAO-sponsored sago expert consultation was held in Jakarta, Indonesia, in 1984 and the sago palm's profile as a tropical crop has been on the rise internationally ever since.

Also in Japan, the sago palm sparked strong interest in the utilization of tropical resources. The Japan International Cooperation Agency launched a research cooperation program for the development of the sago palm. The objective of this research was to investigate the distribution, reserves and utilization status of the sago palm in the field in order to gather basic information required for future development cooperation from private companies. The research team made several field trips under the leadership of Emeritus

Professor Takashi Sato of Kobe University. The author was a member of the first research team to Malaysia and Indonesia and the second research team to Papua New Guinea, Singapore and the Malay Peninsula. The resultant reports constituted valuable resources for the early stage of sago palm studies in Japan (Kainuma 1981; JICA 1981b, 1981c).

The international cooperation fostered at the Sago '76 symposium has continued to date. Two symposiums were organized in Japan: 'Sago '85: Protect mankind from hunger, and the earth from devastation' in Tokyo in 1985 and 'Sago 2001: A new bridge linking south and north' in Tsukuba in 2001. The proceedings of both symposiums were published with the latest information at the time.

## 9.3.2 Characteristics and utilization properties of sago starch

### 9.3.2.a Starch productivity of sago palm

The sago palm is a highly efficient starch producing plant found in the humid tropical zone. While unmanaged naturally-growing sago palms yield 2–6 t/ha/yr of starch, managed sago palms produce starch in much greater quantities. Sato et al. (1979) estimates that a sago palm plantation with planting intervals of 10 m by 10 m and the commencement of harvesting in the 12th year will harvest 100 sago palms/ha each of which will produce 200 kg of starch or a total starch production of 20 t/ha. In the subsequent years 40 sago palms/ha are harvested each year for an annual starch production of 8 t/ha which will continue semi-permanently in well managed sago plantations.

It is clear that the sago palm has a very high starch production potential. A plantation experiment in Batu Pahat on the Malay Peninsula, for example, demonstrated that 100–138 sago palms could be harvested from one hectare of land and that if more productive clumps were selectively planted, it was possible to produce 24 t/ha of starch (Flach 1977).

Another starch producing crop cultivated in the tropics is cassava, which has considerably lower starch productivity than sago palm. The world average harvest of cassava is only 12 t/ha with a starch content of only 20–30%. The productivity of sago starch is higher than corn, the most widely cultivated source of starch in the world, at a unit yield of 10 t/ha in the United States. The sago palm thus has a tremendous potential as a starch producing crop. Its ability to grow in tropical swamps where soil conditions are unsuitable for a majority of crops offers wonderful prospects.

One drawback of cultivating sago palms is that they take 10 years to mature to the harvesting stage. This is perhaps one of the main reasons that industrial-scale sago palm cultivation did not happen for such a long time. Once a sago palm plantation is placed under proper sucker management, however, harvesting can be carried out every 2 years which offers a great advantage (Kainuma 1982).

## 9.3.2.b Structure and utilization properties of sago starch

As the production of sago starch is still in small quantities, the world market is yet to make full use of its unique characteristics. Japan imported sago starch from Malaysia for use as a low-cost raw material for manufacturing modified starches or sugars which were converted into other products such as crystalized glucose, sorbitol and vitamin C. In Europe, sago starch has been used as a home cooking starch for a long time but the supply route was interrupted when the port of Singapore was shut down during WWII. Cassava, potato and corn starches seem to have become the main types of starch used in home cooking in Europe since then.

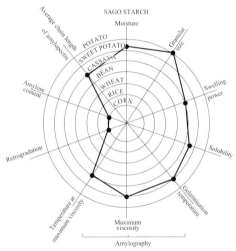

Figure 9-26 Various characteristics of sago starch on starch diagram
Source: Kainuma (1977a)

Figure 9-26 is a diagram summarizing the characteristics of sago starch's physical properties and composition as reported at Sago '76 (Kainuma 1977a). Sago starch exhibits rheological characteristics that are similar to those of sweet potato and cassava starches but it is closer to corn starch in terms of retrogradation properties and amylose content. Observations by the latest instruments such as X-ray photoelectron spectrometer and atomic force microscope have revealed the structure of the outermost surface (Figure 9-27) with myriad small projections that resemble the surface structure of potato starch (Hatta et al. 2002).

The active research conducted by Takahashi Laboratory at the Kyoritsu Women's University shed light on the useful properties of sago starch as a polymer, particularly its cookery-scientific characteristics and possible uses, as discussed in this chapter. One of the characteristics they found is that sago starch is very similar to kudzu and bracken starches, which have become quite expensive due to their scarcity in Japan, as discussed elsewhere (Takahashi 1986; Hamanishi et al. 2002a, 2002c).

Figure 9-27 Atomic force microscope (AFM) image of the structure of the outermost surface of sago starch
Source: Hatta et al. (2002)

### 9.3.2.c Black mold isolated from sago palm – Chalara paradoxa

On one visit to a small Sarawak starch factory in 1976, the author observed that the cut surfaces of sago logs moored in a river were covered by a type of black mold, which was destroying the cut surfaces and proliferating inside the sago logs. When he examined samples from the factory under a scanning electron microscope, he found that many of the starch granules had been broken down by enzymes (Figure 9-28).

He later isolated some microorganisms from samples of sago palm trunk brought back from a survey in Papua New Guinea and discovered a strain of black mold among them, which was identified as *Chalara paradoxa*. This mold released enzymes that exhibited an abnormally high ability to hydrolyze

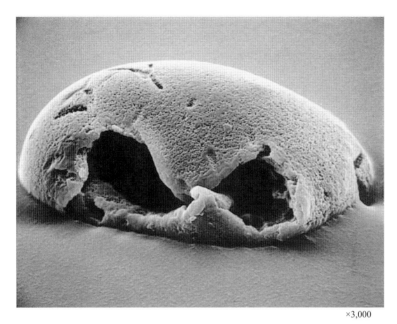

×3,000

Figure 9-28 Scanning electron microscopic image of an enzymatically degraded sago starch granule during starch production
Source: Kainuma (1986b)

raw starch compared with other microbial amylases. The isolated strain was named *C. paradoxa* PNG-80 and its enzymatic properties and possible uses were studied. It was found that *C. paradoxa* PNG-80 released two types of enzymes, α-amylase and glucoamylase, which acted together to digest unheated starch granules efficiently (Kainuma et al. 1985; Kainuma 1986a; Monma et al. 1989). As a result of a basic experiment of intermediate industrial-scale enzyme production in a 10 kl tank, it became clear that the production of these enzymes was strongly induced by the presence of sago starch granules in a culture medium.

A basic study of alcohol fermentation of raw starch using these enzymes (Mikuni et al. 1987) was conducted. The results were offered to an experiment which was underway at that time as part of the Biomass Conversion Project of the Ministry of Agriculture, Forestry and Fisheries of Japan to manufacture sake, rice *shochu* (white distilled spirits) and sweet potato *shochu* from raw rice and raw sweet potato (MAFF 1991). It would be interesting to test these enzymes in the production of bioethanol, which is currently being undertaken as a national project.

## 9.3.3 Future of sago starch

### 9.3.3.a Sago starch in international market
Sago starch still has a low profile in the international market. The following reasons are commonly given:
1. The total volume of production is small.
2. It has lower purity and more variable quality than other industrially produced starches.
3. Sago palms are scattered over large areas and collection requires time and effort.
4. It takes too long to become harvestable.

When these issues are resolved, more specific uses for sago starch will be developed.

According to Ohno (2003), who advised on starch manufacturing in Sarawak, a sago palm plantation project was launched under the direct control of the Sarawak state government in 1988, Around 9,000 ha of the scheduled 25,000 ha site had been planted by 1998 and further planting was taking place. This was all in addition to the state's existing farmer-owned sago palm stands of about 20,000 ha. The project will ultimately plant sago palms in 85,000 ha of land in Sarawak state alone.

In Indonesia, a large project of a 20,000 ha plantation has been planned in Riau Province. Approximately 8,000 ha of this had been planted by 2001 with the planted area expanding at a rate of around 2,000 ha/yr (Jong 2002b).

It is anticipated that these plantation projects will eventually lead to the stable supply and industrial mass production of high quality starch and that sago starch will be recognized as a starch resource from Asian tropical swamps which compares favorably with corn starch, which is predominant in today's international market and widely used in North America and Europe.

### 9.3.3.b Improvements to sago starch factories
Recommendations in JICA's 1981 research report and the author's recommendations at the FAO-sponsored 1984 sago expert consultation meeting in Jakarta for the removal of impediments to the utilization of sago palm as an industrial raw material can be summarized as follows.
1. Selective planting of high-performance palms and plantation management.
2. Provision of canals and transport vessels for smooth transportation of logs.
3. Introduction of efficient starch production equipment, including mobile milling plant.
4. Establishment of integrated production systems such as the construction of a fermentation plant connected to a starch plant by a pipeline.

When these conditions are met, the sago palm can be used as an industrial raw material on a global scale. The sago palm plantation projects that are taking place in Sarawak and Sumatra are moving closer to achieving them. According to Wagatsuma (1994) and Ohno (2003), modern starch factories equipped with sieve bends, centrifugal sedimentators, centrifugal dehydrators and spray dryers have been built in Sarawak, each of which has a monthly production of 300–500 t of starch of increasingly stable quality (Figure 8-23).

### 9.3.3.c Sago palm as a tropical biomass

As mentioned, the sago palm has high starch productivity per unit area in comparison with other crops. While the world is currently focusing on alcohol production from corn, the sago palm can rival it. The sago palm is a perfect raw material for alcohol fermentation due to the large quantities of biomass it produces. As with all biofuel production, though, caution must be exercised to ensure that it does not compete with its use for food for the numerous people who consume sago as their staple food. While it is possible to use coarsely refined starch as a raw material for alcohol fermentation, further refinement poses a problem of high cost in the case of conversion to high-value-added saccharification products such as isomerized glucose syrup and oligosaccharide.

One way to avoid competition with its use for food is to utilize the large quantities of residual starch left in the pulp after starch extraction as a raw material for enzymic hydrolysis. Unlike woody plants, the trunk of sago palm is largely comprised of parenchyma. The enzymatic digestability of its components will be a major research subject in the future as only a small number of studies have been done so far.

According to Sasaki et al. (2002a), the filtration residue after starch extraction has a low lignin content while still comprised of more than 30% starch. The most difficult issue of the hydrolysis and ethanol fermentation of woody biomass lies in the pretreatment stage where lignin is isolated and cellulose and hemicellulose are digested with enzymes. In this regard, the utilization of residual starch can be made more efficient if an alcohol factory is annexed to a sago starch factory.

When the vast areas of sago palm plantation currently under way begin to bear fruit, the promise extolled by the 1976 symposium 'Sago: The Tropical Swamp as a Natural Resource' will finally become reality after four decades. For this to happen, the question of saccharification of the trunk needs to be tackled. The true success of the current plantation projects can be assured only when the treatment process and starch production are operated viably hand in hand.

Authors:
9.1: Setsuko Takahashi
9.2.1: Tomoko Kondo (Hamanishi)
9.2.2: Kazuko Hirao
9.2.3: Masaharu Ohmi
9.2.4: Shigeki Konoo
9.3: Keiji Kainuma

# 10
# *Diversity of Uses*

In areas where sago is the staple food, it is not only used as food: almost all parts of the sago palm are used in some way. This chapter describes the uses of the different parts of the sago palm. Some of the ways that various parts are used in specific areas are described through specific examples (Ellen 1977, 2004, 2006).

## 10.1 Uses of foliage

The peoples who have traditionally eaten starch from the sago palm have also made extensive use of its foliage to make housewares. They exploit a variety of characteristics of different parts of the large sago palm leaf for a wide range of purposes. This section will describe one example of leaf utilization: how the Melanau people in the Malaysian state of Sarawak use the sago palm foliage.

### *10.1.1 Melanau names of leaf parts*

The diet of the Melanau people consists principally of fish and sago palm. They cultivate sago palms and use their leaves in daily living. Different parts of the leaf are used for different purposes and given different names.

The locals often call the sago palm *rumbia*, which is a Malay term. The whole leaf is referred to as *da'an balau*. The word *da'an* means the sago leaf and *balau* means the sago palm plant. *Da'an* includes the leaflet but it mainly refers to the petiole and the rachis. When the locals use English, they call the petiole and rachis 'branch' and the leaflet 'leaf'. It may be considered that *da'an* is equivalent to the 'branch' but sometimes includes the leaf.

The skin part of the leaf sheath of the rachis is widely used. It is referred to as *smat*. It is torn into narrow strips, which are woven like bamboo. The base of the leaf sheath in particular is called *ukap*, or *ukap da'an balau* to be more exact. The skin feels like a thick sheet of wood. Vascular bundles that are arranged at intervals beneath the outer skin surface of the leaf sheath base are called *tajui ukap*.

The leaflet is called *daun* and is primarily used for roofing. The midrib of leaflet is called *tagai*, which is taken out of the leaflet and used like a whittled bamboo stick.

The locals use sago palm starch in a wet state, which is called *sey*. This is formed into granules, roasted and eaten as what is basically called sago.

## 10.1.2 Specific uses of sago palm leaf

### 10.1.2.a *Smat* (leaf sheath skin)

The skin of the sago palm leaf sheath is torn longitudinally into narrow strips, which are used for weaving. The material is hard and looks like bamboo but it is softer than bamboo and produces springy products. It has no nodes. The *smat* is dark on the outside (epidermis side) (turning ocher/brown over time) and pale on the inside (parenchyma side). The difference in color tone is sometimes used to create patterns (Figure 10-1).

Figure 10-1 Articles of everyday use made of sago palm leaves (the under laid mat is *idas*)

**Idas:** A tightly woven *smat* mat, which is used to drain water from precipitated sago starch. The edges are hemmed with tree bark and stitched with the skin of rattan (vine-like palm species mainly of the genus *Calamus*).
**Tapan:** A tightly woven *smat* basket, which is used to make sago starch granules or to screen rice. The edges are hemmed with rattan.
**Paka:** A large loosely woven *smat* basket.
**Kilak:** A tightly woven *smat* basket.
**Gelagang:** A deep woven *smat* basket, which is used to carry fish.
**Babat:** A woven *smat* mat set in a strong bamboo frame. It is used to partition rooms or as a decorative reinforcement for walls. It is also used to partition fish ponds.

### 10.1.2.b *Ukap da'an balau* (skin of leaf sheath base)

The skin at the base of the leaf sheath resembles a thicker version of bamboo shoot skin (leaf sheath) and feels like wood. It is peeled into thin sheets, which are dried and stored. They are sold in this state. They are soaked in water for 1–2 hours before use so that they are soft enough to fold like paper. This material is used to make watertight products.

Figure 10-2 A variety of *upak* made of the skin of the sago leaf sheath base

**Upak:** As drinking water cups and plates (Figure 10-2).
**Tarusong:** A water ladle (Figure 10-1 top right).
**Tapau ukap:** A lampshade-shaped hat made of *ukap*. *Tapau* refers to a flat conical hat worn directly on the head.

Large trapezoidal *ukap* sheets are arranged in alternate directions and connected together to make walls for sheds and huts.

### 10.1.2.c *Tajui ukap* (vascular bundle at the base of leaf sheath)

The vascular bundle at the base of the leaf sheath is hard and straight and is used as a blow dart (*adang sput*). The base of the leaf sheath is cut to a suitable size and a vascular bundle on the back of the epidermis is exposed by shaving off the surrounding tissue. The bamboo stick-like vascular bundle about 2 mm in diameter can be obtained relatively easily (Figure 10-3). The stick is trimmed to 15–20 cm in length, one end is sharpened and a block is attached to the other end to make a blow dart. The block is a piece of parenchyma inside the leaf sheath which is pared down to suit the internal diameter of the blow tube (about 8 mm). It is called *ulow adang*. The parenchyma of a

 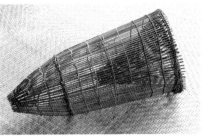

Figure 10-3 The vascular bundle (circled) is removed from the base of leaf sheath to make a blow dart (indicated by the arrow)

Figure 10-4 A fishing tool called *biga* made of leaflet midribs

palm after flowering is preferred for this purpose as the tissue is harder at that time. The blow tube is about 150 cm in length and called *sput* (*sumpit* in Malay). Poisonous substances are smeared on the tip for hunting birds and monkeys.

**10.1.2.d *Tagai* (leaflet midrib)**
The *tagai* is a long narrow stick of considerable strength. It is more properly referred to as *tagai da'an balau*. The *tagai* can be used to make a variety of implements.
**Biga:** A fishing implement (Figure 10-4).
**Sapaw:** The sticks are cut to a length of 40 cm or so and bundled together into a broom. Leaflet midribs of the coconut palm (*tagai da'an benyoh*) are also used for this purpose.

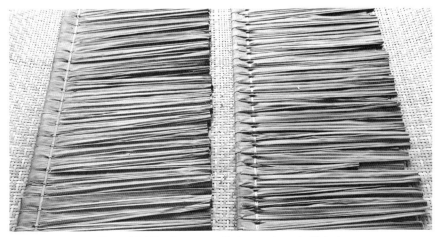

Figure 10-5 Roofing mats called *sapau* (*atap* in Malay)
Two-ply mat (*sapau sikai*) on the left and single-ply mat (*sapau japah japah*) on the right.

### 10.1.2.e *Daun* (leaflet) for roofing

Each leaflet is folded along the midrib, then folded in half, fixed to a shaft called *bakawan* that is laid between the folds, and laced together. This process is repeated to attach leaflets neatly to a 180 cm (about 6 feet) long shaft. It is about 55 cm wide. The resultant roofing mat is called *sapau* in Melanau and *atap* in Malay (Figure 10-5). A roofing mat with single-ply folded leaflets is called *sapau japah japah* or *sapau japah*, and a roofing mat with two-ply folded leaflets is called *sapau sikai*.

The *sapau* are arranged to form a roof. They are traditionally tied to the timber roof frame with string, starting from the bottom end of the roof, but nails are sometimes used these days.

The *sapau* with two-ply leaflets is stronger. The two-ply leaflets are laced together and fixed to the shaft more securely than the single-ply ones. The two leaflets firmly and completely overlap for about the first 15 cm from the shaft. The rest of them only loosely overlap. The top 15 cm are secured firmly to prevent rainwater from seeping through.

The useful life of the *sapau da-un balau* (*atap daun rumbia* in Malay) made of sago palm leaflets is about 7 years for the single-ply type and 10 years for the two-ply type. The selling price is higher for the latter. *Sapau* are also made of nipa palm leaflets (*sapau da-un nyepak,* or *atap daun nyepah* in Malay) but these have a much shorter lifespan of only 3 years or so.

The rachis of a wild palm that resembles the *salak* palm is commonly used as the *sapau* shaft (*bakawan*) although bamboo is used occasionally. This palm is called *lamujan* and bears *salak*-like fruit (*buak*) with mildly acidic taste. The middle segment of the rachis is used as the *bakawan*.

## 10.2 Uses of bark

The bark of the sago palm is used in numerous ways, including as fuel for fire-drying starch in cottage industry-type factories in Malaysia and for wood carvings hung by the front entrance of houses in New Guinea Island. In areas where starch is extracted from the sago palm pith, the remaining bark (Figure 10-6) is used for various purposes.
1. Fuel
2. Building materials: wall, roof, fence and floor (Figure 10-7)

These uses are fundamentally of the bark that is left as waste after starch extraction, not an active exploitation of the characteristics of the sago palm bark. However, it is a preferred flooring material in certain parts of Papua New Guinea for its characteristics such as being softer and more flexible than other available barks. In these areas, the bark of a tree (*Kentiopsis archontophoenix*) called *limbun* in the pidgin language (Tok Pisin) is commonly used for flooring (Mihalic 1971) but the more springy sago palm bark is preferred.

Furthermore, the active utilization of the unused bark is being attempted in

Figure 10-6 Sago palm bark after pith removal

Figure 10-7 Sago palm bark used for flooring (and a sago crushing hammer on it)

areas where large quantities of sago palms are used industrially. Such attempts include processing the sago palm bark into patterned blocks for interior decoration products and applications in wall and fence manufacturing (Aziin and Rahman 2005).

## 10.3 Uses of trunk apex

As mentioned, many parts of the sago palm such as the bark, leaflet, petiole and rachis as well as starch from the trunk are utilized for various purposes. Yet, the trunk apex appears to be of rather limited use. The young, unopened

Figure 10-8 *Songa* shoot of *M. vitiense*
Top left: In Navua on Queen's Road, Viti Levu Island. Top right: At a market in Suva. Center left: Unopened new leaf. Center right: Longitudinal section of an unopened new leaf. Bottom: *Seko* salad in Nadi.

leaf bud wrapped inside the leaf sheath is called cabbage and consumed as a crunchy vegetable similar to bamboo shoot. This edible, so-called 'palm heart' or 'palm cabbage', is slightly sweet and sometimes eaten raw.

The trunk apex of some of the section *Coelococcus* plants found in the South Pacific islands are also used this way. The young unopened leaf bud in the leaf sheath of *M. vitiense* is called *songa* shoot (*songa* means *M. vitiense* in the Fijian language) and in Navua on the island of Viti Levu, the *songa* shoots harvested in the natural stands to the north of Pacific Harbour are sold at places along the arterial Queen's Road (Figure 10-8). They are also available in markets in the capital city of Suva.

*Songa* shoots are used by Indo-Fijians as an ingredient for curries. They are also offered as part of *seko* (young palm shoot) salad at resort hotels in Nadi. In Latin America, the leaf bud of palms such as *Euterpe, Roystonea* and peach palm is called *palmito* (palm shoot in Spanish) and the shoots are widely eaten. Canned products are exported to North America as well. The palmito is generally harvested by felling young trees (aged 2–4 years). In the case of

*M. vitiense*, *songa* shoots appear to be harvested from palms several years after trunk formation judging from the size of the *songa* shoots for sale. The sago palm (*M. sagu*) typically forms a trunk 3–4 years after sucker transplantation, or a bit longer if grown from seed. As *M. vitiense* seems to grow at the same or a slightly slower rate than the sago palm (based on the leaf emergence rate) and propagates only from seed, *songa* shoots are probably harvested from palms that are more than 5–6 years old.

## 10.4 Uses of fruit

The above-ground part of the sago palm (*Metroxylon sagu*) and the section *Coelococcus* plants are highly utilized. Their fruit also have some known uses besides seeding. As discussed in the preceding chapters, the sago palm propagates both by vegetative reproduction through axillary buds at low-level leaf axils, adventitious buds from subterranean stems (suckers) and by seed germination. While it is common to use suckers for new plantings, it is not easy to secure large quantities of suckers suitable for transplantation at large cultivation sites. It is hoped that seed propagation can be improved in order to make up for the shortage of transplantation material. Although the sago palm seed (inside the fruit) has a low germination rate, it may be possible to improve the germination rate by removing seed coat tissues, thereby eliminating germination inhibitors (Ehara et al. 2001). The section *Coelococcus* plants growing in the South Pacific and Micronesia, in contrast, propagate only by seeding (they do not sucker) but they have high germination rates. Some species such as *M. warburgii* found in Vanuatu, Fiji and Samoa even produce viviparous seeds (Ehara et al. 2003d) (Figure 10-9).

Figure 10-9 Germinating fruits of *M. warburgii*
Upolu Island, Samoa.

Among the section *Coelococcus* plants, *M. amicarum* and *M. warburgii* have large seeds (about 9 cm and 6 cm in equatorial diameter respectively) (Figure 10-10), which are called palm ivory. They are used to make

Figure 10-10 *M. amicarum* fruits (left) and longitudinal sections (right)
Chuuk Island, Micronesia.

Figure 10-11 Folk crafts made of the fruit and endosperm of the section *Coelococcus* plants
Endosperm: Left and top center – *M. amicarum* (Pohnpei, Micronesia), bottom center – *M. warburgii* (Port Vila, Vanuatu). Fruit: Right – *M. warburgii* (Port Vila, Vanuatu).

Figure 10-12 Sago palm fruits used in flower arrangement
West Java, Indonesia.

306　　　　　　　　　　　　　　　　　　　　　　　　　　　　　　　　　　　　　　Chapter 10

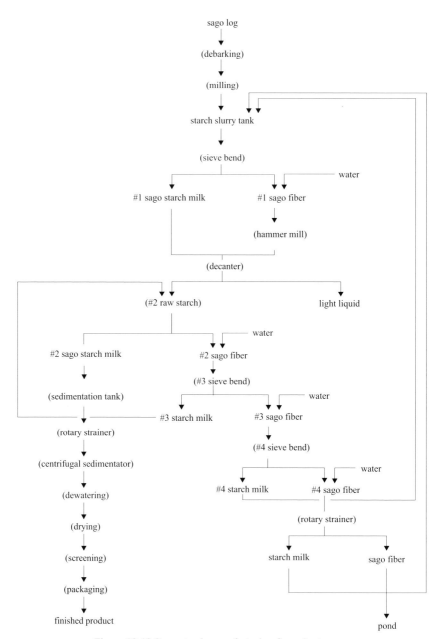

Figure 10-13 Sago starch manufacturing flow chart
Source: Ohno (2004)

buttons and various craft products (Dowe 1989; Ehara et al. 2003d; McClatchey et al. 2006) (Figure 10-11). These large seeds are called palm ivory because the endosperm of the seed of the genus *Metroxylon* plants is composed of reserve nutrients of cellulose and the keratinous endosperm becomes as hard as animal horns or ivory as it matures. The endosperm of *M. amicarum* was once used as a substitute for ivory in making seals and stamps in Micronesia prior to the end of the Second World War. It is now used to make pendant tops and ornaments (Figure 10-11). In Vanuatu, the immature fruits of *M. warburgii* are also used as they are in making seashell necklaces (Figure 10-11).

The fruit of other *Metroxylon* palms are sometimes dried and dyed for use in flower arrangements (Figure 10-12). Sago palm fruit are reportedly used as a herbal medicine for intestinal regulation on Bangka Island in South Sumatra, Indonesia, but this information has not been verified. There is currently very little information about their medicinal effect.

## 10.5 Uses of sago residue after starch extraction

### 10.5.1 Sago palm starch extraction process and residue formation

Collection of starch from the sago palm is carried out by the local people where sago palms grow in the wild (mostly uncultivated) and at starch production factories where sago palms are grown in plantation farms. The basic process is the same in both cases but a large part of the production process in factories is mechanized. Figure 10-13 is a diagram of the starch manufacturing process flow at the factory (Ohno 2004). The sago palm trunk is mostly comprised of the starch accumulating pith and the bark (Figure 10-14). The pith is primarily composed of parenchyma cells in which starch granules are stored. The starch granules grow in diameter as the palm ages (Ogita et al. 1996). A trunk is felled, cross-cut into logs about 1 m long, and debarked. Debarking is mechanized in some factories but the locals often work as subcontractors to perform debarking manually on a 'piece-work' basis for cash income (Figure 10-15). Removing the hard bark with a hand axe is extremely hard work. The soft debarked pith is milled by hammer mills or raspers before being sent into the starch extraction process. The milled pith is soaked in water, filtered and/or centrifuged to separate the starch and residue. After this operation is repeated 3–5 times, sago starch is either machine- or sun-dried. The more times the extraction process is repeated, the higher the starch yield, but the proportion of impurities in the starch (primarily cellulose fibers from the parenchyma) also increases, reducing the starch quality. Since it is impractical to extract all of the starch, a considerable amount of starch remains in the residue (hereinafter called 'sago residue') after starch extraction. Sago residue currently has no

Figure 10-14 Cross section of sago palm pith
The white part is the pith and the outer periphery is the bark.

Figure 10-15 Manual debarking

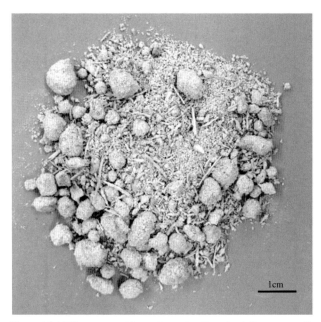

Figure 10-16 Dry sago residues after extracting starch from sago pith

industrial application and mostly goes to waste.

Figure 10-16 shows the sago residue. Dried sago residue forms small cylinders 1–5 mm in length and 0.5–1 mm in width, or spherical aggregates of them. The starch yield is rather low at 30% or less of the pith dry weight. The remaining 70% or so is disposed of as waste. The estimated annual starch production in Sarawak is about 50,000 t (see Table 10-1, which shows the quantity of sago starch exported from the Malaysian state of Sarawak, where sago starch production is carried out on an industrial scale (Department of Agriculture Sarawak 2005)), which means that around 110,000 t of sago

Table 10-1 Sago starch exports from Sarawak State, Malaysia, by destination country

(thousand tons)

| Destination country | Year | | | | |
|---|---|---|---|---|---|
| | 2001 | 2002 | 2003 | 2004 | 2005 |
| China | 0 | 0 | 0.14 | 0.15 | 0 |
| Indonesia | 0.09 | 0.07 | 0.07 | 0.07 | 0 |
| Japan | 8.00 | 8.10 | 7.36 | 8.38 | 12.77 |
| Malaysia (Peninsula) | 15.68 | 18.11 | 18.62 | 24.91 | 23.30 |
| Malaysia (Sabah) | 0.81 | 0.60 | 1.30 | 0.91 | 0.67 |
| Singapore | 4.47 | 5.56 | 4.47 | 5.08 | 5.78 |
| Taiwan | 0 | 0 | 0 | 0 | 0.29 |
| Thailand | 1.45 | 2.00 | 2.88 | 2.81 | 1.62 |
| United States | 0.71 | 0 | 0.04 | 0.13 | 0.29 |
| Vietnam | 0 | 0 | 0.22 | 0.53 | 0.62 |
| Total | 31.21 | 34.44 | 35.10 | 42.90 | 45.33 |

Source: Department of Agriculture, Sarawak (2005)

residue go to waste every year. Whether or not there are sufficient quantities of sago residue to provide an industrial resource, determining its properties as a biomass resource should reveal ways to utilize the sago residue. Although many researchers have been studying sago starch, there have been very few studies of the properties and potential uses of the sago residue, which is produced and disposed of in quantities that are more than double that of sago starch.

## 10.5.2 Chemical and physical properties of sago residue

Table 10-2 shows the chemical composition of the sago residue sampled at a starch manufacturing factory (Nitei Sago Industry Co. Ltd.) in Mukah, Sarawak (Sasaki et al. 2002a). It was found to contain a high level of holocellulose as well as a large amount of unextracted starch. It also had a much lower level of lignin than common arboreous plants. The high starch and cellulose contents suggest a high percentage of glucose as a constituent monosaccharide, which is likely to be the key to utilizing sago residue.

Table 10-2 Chemical composition of sago palm pith and sago residue (%)

|         | Holocellulose | Hemicellulose | Starch | Lignin |
|---------|---------------|---------------|--------|--------|
| Pith    | 14.4          | 12.9          | 54.1   | 1.4    |
| Residue | 20.6          | 19.8          | 30.9   | 5.1    |
| Cedar   | 48.6          | 24.7          | -      | 32.3   |
| Beech   | 53.8          | 30.2          | -      | 23.5   |

Source: Sasaki et al. (2002a)

One possibility for utilizing sago residue with these compositional characteristics is to enhance its functionality through chemical modification. The efficiency of reaction with chemicals is an important factor in chemical modification and is influenced by the degree of powderization. Therefore the grindability of sago residue was tested. The result is shown in Figure 10-17. The sapwood of Japanese cedar, which was used as the control, has a relatively low density among Japanese coniferous trees and is therefore relatively amenable to mechanical processing. The particle size decreased with prolonged grinding in both cedar and sago residue. Cedar particles were not small enough to pass through a screen finer than 100 mesh after 10 minutes or so of grinding whereas sago residue was ground fine enough to pass through a 200-mesh screen after about 2 minutes of grinding. Finer grinding in less time than the control means that less energy is required for grinding. The degree of cellulose crystallization in sago residue was also decreased by fine grinding (Sasaki et al. 2003). Observation of the constituent cells of the sago pith and extraction residue suggested that the extraction residue had high grindability as it was primarily composed of parenchyma cells in comparison with tree species that largely consist of tracheids.

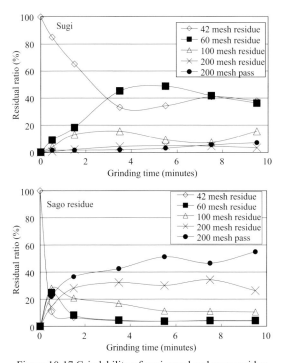

Figure 10-17 Grindability of sugi wood and sago residue

## 10.5.3 Thermo-plasticization by chemical modification

Chemical modification can confer various functions to plant resources depending on the chemicals used. An attempt was made to confer thermal plasticity, i.e., to plasticize the sago residue. The main constituents of plant cell walls include cellulose, hemicellulose and lignin. The most abundant among them is cellulose, which is a linear polymer of linked beta-D-glucose units. In thermoplastic polymeric materials, heat application causes thermal motion of the polymer backbone chain, altering them from a solid to liquid state. However, cellulose is highly stereoregular and its molecular chains (hydroxyl groups in a glucose residue) are held together by hydrogen-bonding. It does not exhibit thermal plasticity even when heated because the thermal motion of the polymer backbone chain is inhibited by the hydrogen bond and pyrolysis occurs at temperatures below the melting point or the glass-transition point (Tg). A number of studies report, however, that it is possible to confer thermal plasticity to plants by substituting nonpolar groups for the hydroxyl groups which are abundant in cellulose. Thermal plasticization by such reactions as acetylation (Shiraishi et al. 1986), lauroylation (Funakoshi et al. 1979) and benzylation (Kiguchi 1990) have been reported.

Thermal plasticization was attempted on sago residue which had been ground to fine particles to reduce cellulose crystallization, i.e., the amount of hydrogen bonding between cellulose chains, by acetylation and lauroylation. While acetylation failed to produce a sufficient level of thermal mobility, lauroylation produced a higher level of reaction than in cedar and the sago residue expressed sufficient thermal softening (Figure 10-18) to form sheets (Watanabe and Ohmi 1997; Ohmi et al. 2003). The lauroylated sago residue sheets prepared by hot pressing were buried in the soil and the temporal change was measured. It was found that their mass had reduced, which pointed to their degradability (Sasaki et al. 2002b).

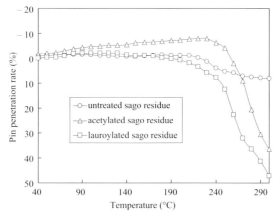

Figure 10-18 Thermal softening curves of untreated, acetylated and lauroylated sago residue
Source: Watanabe and Ohmi (1997) and Sasaki et al. (2003)

## 10.5.4 Preparation of urethane foam from sago residue and its physical properties

Polyurethane resins are produced by a reaction between a polyol and an isocyanate compound. They are classified as either rigid or flexible depending on the combination of the polyol (polymeric alcohol) and the isocyanate compound as well as the type of plasticizing agent used. They are used in a wide range of industries as they can be formed into film, sheet or foam. Polyurethane foam is a low density material containing many voids and thus has many applications such as sponge, cushioning material and shock absorbing material. The main plant constituents of cellulose and hemicellulose might be used as polyols for the abundance of hydroxyl groups in their chemical structure. The use of materials liquefied by the acid catalyzed liquefaction process as polyols has been studied (Shiraishi et al. 1985; Yao et al. 1993). Using renewable resources, including plants, as industrial materials can contribute to a reduction in the consumption of exhaustible resources such as petroleum. Yet, the liquefaction reaction of plants needs to be carried out at relatively high temperatures, which require energy input. A urethane foam production experiment was therefore conducted with a focus on the starch in sago residue because starch can be gelatinized by hot water alone and reacts with isocyanates. The resultant urethane foams were evaluated for their buffering property (Ohmi and Saito 2007). They demonstrated a buffering characteristic (static cushion factor) against a very wide range of compressive stresses according to the sago residue content of each urethane foam (see Figure 10-19). The foams manufactured at an optimum level of sago residue content had low density and exhibited the same level of buffering characteristic as flexible polyurethane foam. The density increased markedly as the sago residue content increased, however, so other applications need to be considered for foams containing high levels of sago residue.

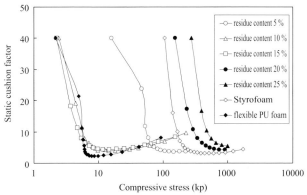

Figure 10-19 Relationship between compressive stress and static cushion factor of foams with various sago residue contents
Source: Ohmi and Saito (2007)

## 10.5.5 Future prospects for sago residue use

The sago palm is a very rare woody plant which accumulates and produces large quantities of starch as a ligneous resource. Nevertheless, the current global area of cultivation of the sago palm as a starch crop is small, limited to Southeast Asia and Melanesia. For the non-starch part of the plant, there are no wide-spread common uses as yet. However, the demand for starches as a biomass resource has been steadily rising for applications in biodegradable plastics and bioethanol. Hence, there is a good chance that the sago palm cultivation area will expand and sago starch production will increase due to the sago palm's high starch productivity and its ability to grow in tropical peat soils in which other crops cannot grow. Hence, the starch production potential of the sago palm is beginning to attract attention from non-food industries. However, there are also growing concerns that the dramatic increase in demand for starches as industrial materials for energy and plastics production is pushing up the prices of starches and other farm products as food resources. As the dietary production and consumption of sago starch increases in the future, it will become increasingly necessary to find beneficial uses for the increasing amounts of waste products such as extraction residue, leaf and bark. Further studies on the biomass resource properties of sago residue and its applications are therefore necessary.

## 10.6 Uses of weevils

### 10.6.1 Sago weevils

Sago weevils refer to coleopteran insects belonging to the genus *Rhynchophorus* which eat into the sago palm pith. Local people living in sago palm growing areas commonly eat the larvae of these insects. They are an important source of protein in some places. Other pests such as *Oryctes rhinoceros* (rhinoceros beetle) and the Passalidae beetles also infest sago palms but they are not called sago beetles. The larvae of sago weevils are called sago grubs or sago worms. As the larvae of the *Rhynchophorus* weevils feed on a majority of palm species and many other plants, they are collectively called the palm weevil. More than one species of *Rhynchophorus* weevils are known to infest the sago palm, including the red palm weevil (*Rhynchophorus ferrugineus* (= *R. signaticollis*)) (Figure 10-20), mostly found in Papua New Guinea, and *R. bilineatus* and *R. vulneratus* (= *R. schach* and *R. pascha*) in tropical and subtropical Asia. Some people consider *R. ferrugineus* on New Guinea Island to be a subspecies called *R. ferrugineus papuanus*.

10 mm

Figure 10-20 *Rhychophorus ferrugineus* (red palm weevil) adult

### 10.6.2 Life cycle

Adult red palm weevils (*R. ferrugineus*) lay eggs in open lesions on the sago trunk. They cannot insert their ovipositors into the pith of sago palm through hard bark unless it is injured. The injured sago palm releases volatile chemicals that are attractive to adult red palm weevils. The release of the chemical reaches a maximum level 4–6 days after injury (Hallett et al. 1993).

The adults are attracted to the lesion by released odors and lay eggs in the pith. The growing point at the crown of the palm tree is often damaged by feeding of rhinoceros beetles and the volatile odors released from the wounds attract adult red palm weevils and induce oviposition. They sometimes lay eggs on the cut end surface of the petiole. The adult weevil uses its mouth to make a hole about 5 mm deep into the palm tissue and then lays a single egg in it. After oviposition, it seals the hole with a pink-colored secretion (Corbett 1932). Adult red palm weevils have strong negative phototaxis, but they are not active at night; they hide themselves in wood chips near cut sago palms. They are active during the day, especially in the morning, when they fly around and lay eggs. Some researchers, however, report that they are active in both the morning and afternoon (Hagley 1965).

The eggs of the red palm weevil are white ovals about 2.5 mm long and 1.1 mm wide and hatch in 3–4 days. The hatched larvae burrow into the sago palm pith to feed. The infested pith turns brown and begins to ferment. The sago palm rarely dies from feeding damage by rhinoceros beetles alone, but when red palm weevils lay eggs in the feeding wounds and the hatched larvae enter the pith and burrow down as they feed, the palm inevitably dies. The larvae take 30–80 days to mature. They molt 6–10 times during this period (Rahalkar et al. 1985). Their body consists of 13 segments which are covered with a very flexible, thick, yellow-white membranous skin except for the brown head and the light brown pronotum. They are legless and move by wriggling. The body is soft, having well-developed fatty tissue. The mature larva reaches about 70 mm long and 20 mm wide at the thickest part of the body (Figure 10-21).

Figure 10-21 Mature larva of *R. ferrugineus*

The mature larva chews and folds the fiber of sago palm to make a cocoon. The cocoon has an oblong shape 50–90 mm long and 20–40 mm wide. The larva does not molt to a pupa immediately after completion of a cocoon. It stays still for 3–4 days as its body shrinks. It spends another 3–7 days in a prepupal state before molting to a pupa. The pupa is about 50 mm long, and is shorter than the larva. The pupal body is light brown overall. The pupal stage lasts 12–33 days and the metamorphosed adult stays in the cocoon for further 4–17 days. When it is sexually mature, it emerges from the cocoon by pushing inside of one end of the cocoon.

Adult weevils are even smaller with a body length of about 40 mm. While those which are found on New Guinea Island tend to have glossy black bodies (Figure 10-20), they are mostly reddish brown elsewhere; hence the name red palm weevil. Adult weevils have well-developed membranous hind wings and are strong flyers. They have a long lifespan, with females living up to 76 days and males 133 days according to the report of Sadakathulla (1991). Adult weevils mate several times. The females begin to oviposit 1–7 days after mating. The average number of eggs laid by a female is said to reach 275.

## *10.6.3 Collecting*

After the pith is removed from the harvested sago palm, some parts of the tree crown and stump are left at the site. Red palm weevils lay eggs on them. Local people come back to the site 2–4 months after logging, and break down the palm remnants with an axe or bush knife to collect larvae. Although the larvae are a byproduct of sago palm harvesting, it is common practice to leave parts of sago palms after logging to attract red palm weevils. In many cases, the ownership of the sago palm remnants is established. This can be considered a primitive form of breeding. Several hundred grams of larvae can be collected from a single tree 2–3 months later. To collect larvae, the bark needs to be removed first, then the pith is hacked away (Figure 10-22) until a browned area appears. Many larvae can be found in and around the discolored area. Although adult weevils are also eaten, it is not easy to catch them in large numbers. They are caught when they fly to palms cut down, or while they are hiding inside starch filtration residues.

Figure 10-22 Hacking a sago trunk away with a bush knife

## 10.6.4 Cooking

People in sago palm growing areas often swallow red palm weevil larvae alive. Children watch adults crushing sago palm trunks left for long time, and scramble to eat the larvae as soon as they are found. They toss them into their mouth and swallow down in a moment (Figure 10-23). They would not be able to taste them but they may like the feeling of the wiggling larva passing down their throat. A majority of adults also swallow the living larva without chewing. The trick of eating a live larva is to crush its head between one's teeth before swallowing it. If one is not quick enough to do so, one may be bitten on the lip or tongue by the larva. The red palm weevil larvae which have been feeding on fresh sago palm pith are clean in the guts and safe to eat raw but those which have been feeding on rotten pith have various microorganisms in the guts and are dangerous to eat uncooked.

Figure 10-23 A boy swallowing a *R. ferrugineus* larva

Larvae are used in dishes such as satay, kebab, braise, stir fry and stew (Figure 10-24).

Figure 10-24 Satay made with *R. ferrugineus* larvae

## 10.6.5 Sago beetle as a product

In areas where red palm weevil larvae are consumed as food, they are considered delicious. This perception creates demand to eat. Local people collect them not only for their own consumption but also for sale at markets and street stalls. They often sell for higher prices than other food products.

## 10.6.6 Nutritional value

The red palm weevil larva contains 73.4% water, 8.2% carbohydrate, 6.7% protein, 11.0% fat and 0.7% minerals. Besides water, the largest component was fat, which was likely to be the source of the larva's faintly sweet taste. The second largest component was carbohydrate, which was likely to have come from the sago palm pith tissue that remained in its guts when the analysis was conducted. The protein content was also high. Among amino acids constituting protein, the percentage of glutamic acid was the highest (14.3 % of all amino acids), followed by aspartic acid (9.1 %) and leucine (9.0 %). Among important essential amino acids, a reasonable amount of threonine (4.9 %) was found but the cystine and tryptophan contents were very low and hence the quality of protein was not particularly good (Mitsuhashi and Sato 1994).

## 10.6.7 Celebration and sago beetle

In the coastal area of Asmat, Papua Province, Indonesia, local people practiced some rituals where large numbers of red palm weevil larvae were consumed (Ponzetta and Paoletti 1997). The exchange of larvae at these rituals symbolized the forming of friendship and played a vital role at feasts held in celebration of reconciliation. As the larvae consecrated at these rituals were considered to be infused with powerful forces, children, the sick, the elderly and pregnant women are not allowed to eat them. Even men were said to refrain from eating them if their wives had given birth recently out of fear that misfortunes might befall their babies if they ate the larvae. Large amounts of weevil larvae were used at three different types of celebratory feasting called *imui* or *imbui*, *an*, and *firauwi*. At *imui*, larvae were exchanged between two men or two women and certain bonds are formed between them. Two such people were considered to be partners linked by a special friendship which should last for the rest of their lives. *An* was held between families or communities to restore a peaceful and amicable relationship that had been lost due to earlier incidents of head hunting. *Firauwi* was also called *basu suangkus*, a ritual to symbolize revenge for men killed in battle.

The Onabasulu people near Mount Bosavi in southwestern Papua New Guinea also gather large numbers of *Rhynchophorus* larvae for celebration. Once every year in June or July, they collect tens of kilograms of larvae, wrap them in banana leaves to make a giant sausage-like food several meters long and steam-roast it in a long house especially constructed for the occasion. The food is distributed to the suppliers of the larvae, their marital relations and families who continue to make merry late into the night (Meyer-Rochow 1973).

Authors:
10.1: Yusuke Goto and Youji Nitta.
10.2: Yukio Toyoda.
10.3 and 10.4: Hiroshi Ehara.
10.5: Masaharu Ohmi
10.6: Jun Mitsuhashi

# 11
# Cultural Anthropological Aspect

## 11.1 Root cropping culture

### 11.1.1 What is root crop farming

The concept of 'root crop farming' has been used to explain farming practices in the areas where sago palms have been utilized by humans. This concept was proposed by Carl Sauer and other scholars and advanced by Sasuke Nakao. Its main feature is the theory that root crop farming originated in Southeast Asia, one of several places of origin of agriculture, a challenge to the then prevailing view that agriculture originated in one area only, Mesopotamia, and spread from there to the rest of the world.

Nakao (1966) proposed four types of agriculture, which have the following characteristics.

1. Southeast Asia – 'Root crop farming culture'
   Domestication of crops such as banana, yam, taro and sugar cane originated in the tropical rainforest region of Southeast Asia.
2. Mesopotamia – 'Mediterranean farming culture'
   The fertile delta was the place of origin of annual winter cropping, mainly of wheat and other grasses.
3. Meso-America – 'New World farming culture'
   This type of farming in cassava and root vegetables originated in Central America and spread to North and South Americas and the surrounds.
4. West Africa – 'Savanna farming culture'
   Cultivation of miscellaneous cereals, including rice, and legumes originated in the sub-Saharan savanna region of Africa known as the Sahel.

Root crop farming culture originated in Southeast Asia and Oceania, relying mainly on vegetative propagation, rather than seeding. The discovery of irrigation channels from circa 9000 BCE in Central Highlands, Papua New Guinea, is considered to be evidence of plant cultivation at that time. It is said to be the oldest type of agriculture among the four types.

Root crop farming contrasts with seed propagation farming in a few aspects. Nakao initially nominated banana, yam and taro as the main root crops and later added sugar cane, sago palm and the breadfruit tree. Starch extraction from sago palms is believed to have played a role in this type of root crop farming.

According to Nakao, root crop farming is a broad concept that entails not only farming agricultural produce but also farming systems, crop processing methods and ways to eat them. The following characteristics of root crop farming were outlined by Nakao.
1. Vegetative propagation: Plants are propagated by means such as root division, tilling and cuttage rather than seeding.
2. Advanced use of polyploids.
3. Absence of legumes and oil crops.
4. Digging stick is the only farming tool.

Based on these characteristics, Nakao hypothesized that root crop farming developed as follows.
1. Gathering wild fruits and roots.
2. Domesticating major crops and forming a complex culture based on root cropping.
3. Establishing swidden agriculture and irrigation facilities for taro cultivation.

Food preparation techniques changed as well. At the early stage, banana and root crops were placed in a hole in the ground and steam-roasted with hot stones. This technique is still used in the Pacific region. Poisonous root crops were probably soaked in water to remove toxins. This technique was applied to starch refining as well.

This technique is important for the study of sago palms. According to Nakao, the technique to extract starch from the sago palm is considered to have developed out of this toxin (or astringency) removal technique.

Nakao believed that root crop farming originated in southern China or some place in the tropical rainforests of Southeast Asia and diffused to the surrounding areas. Three routes were considered:
1. Eastward transmission to Oceania
2. Westward transmission to East Africa and Madagascar Island
3. Northward transmission to East Asia

Areas in which starch from the sago palm is utilized are generally situated within the root crop farming culture zone and share common elements. Some of the common characteristics found in the root crop farming culture zone are discussed below.

## 11.1.2 Characteristics of root crop farming cultures

### 11.1.2.a Multi-variety cultivation

In seed farming, typically, a single variety of a crop is extensively cultivated whereas in root crop farming it is common for many varieties of a crop to be grown simultaneously on the same plot of land. The varieties are typically classified in a very detailed manner. For example, Table 11-1 shows the current status of multi-variety cultivation by the Kwanga people in Sepik, Papua

New Guinea, where root crop farming is still practiced (Toyoda 2003). As depicted in the table, yam is divided into purple yam (*D. alata*) and lesser yam (*D. esculenta*); the purple yam is further divided into 39 varieties, and the lesser yam is divided into a further 38 varieties; there are also 24 varieties of taro, 65 varieties of banana and nine varieties of sago. This schema is of course independent of the taxonomic classification system. It must instead be called a 'folk classification' system, as it works according to the unique criteria of the local people. Such classification is an aspect of local languages and hence varies from one language group to another. This particular classification system is only applicable among about 13,000 Kwanga speakers. They understand these numerous divisions and are more or less able to identify the varieties even by looking at young plants or some part of them. This knowledge is acquired through working in the fields day after day.

Table 11-1 Folk classification of crops in Wanjeaka Village, Papua New Guinea

| Variety | Number of folk varieties | Male | Female |
|---|---|---|---|
| yam (*D. alata*) | 39 | 19 | 20 |
| yam (*D. esculenta*) | 38 | 16 | 22 |
| taro | 24 | 24 | 0 |
| *Metroxylon sagu* Rottb. | 9 | 5 | 4 |
| banana | 65 | 31 | 34 |

Source: Toyoda (2003)

From a scientific perspective, reasons for multi-variety cultivation include continuous food supply achieved by cultivating crops with different growing seasons, and minimizing the risk of total crop loss to pests or natural disasters. However, whether the residents are aware of such reasons is unclear. They often explain that they grow multiple varieties because they want to enjoy different tastes and flavors, or give other reasons.

**11.1.2.b Personification of crops**

In the root crop farming area, crops are often treated as if they were human beings. For example, people are said to address their yams as 'my children' when cultivating them in Papua New Guinea. Some people stroke their yams in the hope that they will grow larger (Kaberry 1941-1942). It is believed that yams grow better when planted together with other crops. For instance, it is recommended that taros be planted with yams because taros are believed to be the children of yams. Yams reportedly 'feel happy' to be with their taro family and therefore grow larger.

This phenomenon is known as crop personification or anthropomorphism (Toyoda 2002). Crops are treated as if they were human. Various examples of personification are observed throughout the root crop farming areas.

The Abelam people in Papua New Guinea, for example, decorate yams as human beings and display them during the yam harvest festival. Assigning gender to each variety may be an aspect of this personification phenomenon. It appears that gender is assigned to relatively important crops and that important crops are personified and assigned a gender accordingly.

There are of course examples of what might be considered to be a personification phenomenon in other types of farming, too, such as belief in the 'spirits' of certain crops in seed farming but the phenomenon seems to be much more prevalent in root crop farming.

## 11.2 Social structure of the 'sago palm culture zone'

Areas in which sago palm starch has traditionally been utilized are found throughout Southeast Asia and part of the South Pacific. There are no clear cultural traits common enough throughout this region to give them the collective name of 'sago culture zone'.

The ways in which sago starch has been used in these traditional sago starch eating areas are varied. It has been eaten as the staple food in some areas while it is only harvested when other crops are in short supply in other areas. Nevertheless, New Guinea Island and the immediately surrounding areas in which sago palm starch has traditionally been a staple food share certain common cultural traits. Some of these common cultural traits are discussed as follows.

In order to use sago starch, particular technologies and techniques to extract starch from the sago palm are needed. These include fine milling techniques

Figure 11-1 Sago extraction work in Papua New Guinea

(hammer-like tools are often used) and starch washing vessels (Figure 11-1), which are unique to the sago starch production process.

Sago starch alone does not provide adequate nutrition. Therefore fish and animals such as wild pigs are used as protein sources. As people who eat sago starch as the staple often live in and around swamps, they tend to catch fish for food. Fishing and hunting tend to supplement the harvesting of sago palm starch. In this sense, their livelihood is closer to hunting and gathering than to farming according to the broad classification of cultural types into hunting and gathering, farming and pastoralism.

Often where sago starch is eaten as the staple, root vegetables such as yam and taro or sugar cane are also cultivated at the same time for food use. This may qualify as farming but it is root crop farming rather than seed farming, which limits the ability to preserve food and thus does not lend itself easily to commodity accumulation and hence wealth accumulation. This is likely to have influenced their social structure.

The traditional social structures found in Melanesia (New Guinea Island and the surrounds, Solomon Islands, Vanuatu, New Caledonia and Fiji) are characterized by an absence of social hierarchy as a general rule. There is no social status difference based on family background in principle. On this point, the region forms a sharp contrast with Polynesia (part of the Pacific area triangulated by New Zealand, Hawaii and Easter Island, including Tonga, Samoa and Tahiti) and Micronesia (north of Melanesia, including Micronesia, Marshall Islands and Kiribati) where communities have chiefdom systems presided over by figures who are called chiefs (headmen). In Melanesia, chiefs do not exist (except Fiji); there are simply some people who have more power than others.

In cultural anthropology and ethnology, the people with more power in Melanesia have been known as 'big men'. Their status is not inherited; they acquire power because their individual qualities are recognized by the others. Means to win power vary from one location to another but the most common is through personal networking and exchanging courtesies. Other means include wealth, combat capability and eloquence, and so on.

The power of the big man does not develop into a large social or political organization because it is built on his personal network. In Polynesia and Micronesia, social stratification has developed and created a distinction between the common people and the chief. When such a system develops further, it assumes the forms of a monarchy as in Tonga or Hawaii before European contact. The influences of the political system extend to a broader area and a larger population. In New Guinea and the surrounding Melanesian region, the basic political unit is a community or a language group with a population on average of a few hundred residents up to just over a thousand people at most (Sahlins 1963).

## 11.3 Social role of sago palm

In areas where sago starch is eaten as the staple food, the sago palm sometimes plays social roles beyond its role as mere food. Many such examples are found in New Guinea Island (Papua New Guinea and the Papua Province of Indonesia). Some of these examples are introduced below to explain the social roles of the sago palm and its relationship with society.

### 11.3.1 Sago as gift

Agricultural crops often play social roles in various societies. For example, certain agricultural crops may be treated as 'goods for gifting' or 'gifts'. Exchanging goods is widely practiced in New Guinea Island and the rest of Melanesia where some crops are treated as gifts. Yam is widely used for this purpose. Sago palm is also used occasionally as goods for gifting.

For example, the groom's family offers some goods (bride price) to the bride's family for matrimony in many cases. While pigs are typically presented as the bride price, bags of sago starch are sometimes gifted at the same time in areas where sago starch is eaten as the staple (Figure 11-2). Sago starch appears to be treated as a symbolic food in this case.

These 'gift' crops often receive special treatment relative to other crops. In Melanesia, yam is the typical example. Yam often plays symbolic roles as a gift

Figure 11-2 Bagged sago palm starch in Papua New Guinea

Figure 11-3 Sago palm starch served at a feast in Papua New Guinea

Figure 11-4 Cooked sago jellies (sago dumplings) in Papua New Guinea

and various ceremonies are performed in relation to yam. In Melanesia, pigs are similarly treated as symbolic gifts. The bride price is typically expressed in the number of pigs. The Arapesh people in Sepik, Papua New Guinea, have a traditional aphorism, 'Your own pigs you may not eat, your own yams you may not eat'. It represents their view that pigs and yams are essentially things they give to others.

In some areas, people assign this characteristic to sago palm and say, 'Your own sago you may not eat'. However, this rule was found only in a few areas on New Guinea Island, and is rarely adhered to even in these areas today.

## 11.3.2 Sago as symbolic food

In areas where sago is the staple food, it is customary to serve sago to invited guests at special occasions such as weddings (Figure 11-3). In Sowom Village on the northern coast of Papua New Guinea, sago jellies (sago dumplings, Figure 11-4) are offered to visitors from other villages at funerals. If the deceased is a young person, his or her close relatives stay at the family home of the deceased for a while (usually three weeks) and a lavish feast is held at the end of the three-week period as an expression of gratitude to the relatives. It is said that one whole sago palm is used to prepare these sago jellies.

In Papua New Guinea, it is customary for close relatives of the deceased to refrain from eating their most favorite food as a mark of mourning. As sago is treated as typical food in areas where sago starch is eaten as the staple, it is often sago that the mourners stop eating. There is no fixed period for stoppage and it is individual choice; it can be a year or sometimes a lifetime.

## 11.3.3 Large-scale trade

The Motu people on the southern coast of Papua New Guinea have traditionally engaged in what is known as Hiri trade with people in the Gulf area to the west to obtain sago. The land inhabited by the Motu had poor soils and little rain and did not produce much agricultural crops. They made unique earthenware pots as their local specialty goods and loaded them on a type of sailboat called *lakatoi* or *lagatoi* to trade for sago. They usually departed for the Gulf about 300 km away in October or November when the southeasterly trade wind blew. They obtained large quantities of sago from their trading partners in exchange for the earthenware and headed home when the northwesterly monsoonal wind began in December or January. As the Motu could not obtain sago in their neighborhood, they resorted to this kind of trade which involved a two to three month-long journey.

Although the traditional form of the Hiri trade is said to have ended in the 1960s, the tradition is celebrated by way of reenactment at the Hiri Moale Festival, which is held annually in the Papua New Guinean capital of Port Moresby (Figure 11-5).

## *11.3.4 Gender of sago*

Sago palms are divided into male or female in some areas of Papua New Guinea. For example, the Kwanga in Papua New Guinea identify nine types of sago palm, treating five of them as male and four as female as in Table 11-2

Figure 11-5 *Lakatoi* boat at Hiri Moale Festival in Papua New Guinea

(Toyoda 2003).

Since this type of categorization is completely independent of biological sexing, it is considered more appropriate to use the term 'gender'. Gender assignment also exists for other crops such as yam, taro and banana. It is easy

Table 11-2 Folk classification of *M. sagu Rottb.* in Wanjeaka Village, Papua New Guinea

| Type | Local name | Gender | Meaning of name |
|---|---|---|---|
| Spiny | 1 naksapmama | f. | very fibrous |
|  | 2 kiermpa | f. | highly water soluble |
|  | 3 minaku | m. | grow tall as a tree |
|  | 4 nakainje | m. | pitpit-like leaves |
|  | 5 nakusia | m. | similar to copra |
|  | 6 nakapsambu | f. | cassowary-like |
|  | 7 nakafija | m. | petioles as white as parrot |
| Spineless | 1 nakrame | m. | prickly |
|  | 2 krumbuwalau | f. | short prickles |

Source: Toyoda (2003)

to assume that in these cases long ones are classified as male and short or round ones are regarded as female in principle. However, this assumption does not fully explain the gender classification in sago palm. Judging from the fact that not only crops but also other plants and inanimate objects such as rocks are gender-classified, it appears that gender is applicable to 'objects' rather than crops according to the basic rule that long objects are male and short or round objects are female. The distinction of male and female seems to exist as one way to perceive objects.

An alternative explanation of the phenomenon of gender classification is the concept of personification or anthropomorphism. Some crops are treated as if they were human beings in some cases. As discussed, people in Papua New Guinea often address yams as 'my children' and stroke them in the hope that they will grow larger. Assigning gender to crops is perhaps best understood in accordance with this concept of 'personification'.

### 11.3.5 Gender-based division of sago starch extraction labor

In many areas where sago starch is the staple food, a man and a woman are paired up to perform sago starch extraction work. The work is often divided between them according to gender. Usually the first half of the sago starch extraction work is mainly performed by the man and the second half by the woman. In many areas on New Guinea Island, the man is in charge from sago palm selection, felling and debarking up to pith milling and the woman handles subsequent stages, including washing the pith in water, precipitating, transporting and storing starch, in general. However, the division is subtly different from one area to another and the woman performs pith crushing in some places.

The sexual division of labor used to be closely adhered to but the lines are blurring in recent years. Although the man often helps the woman with her tasks, he rarely does so when others are looking.

### 11.3.6 Sago cooking as home training

In areas where sago is the staple food, cooking is usually done by women. The most common cooking method of sago is sago jelly making, which is not easy. The ability to make sago jelly tends to be viewed as a prerequisite to entering womanhood.

In Sowom Village in Papua New Guinea, girls of around 10 years old are considered to be of a suitable age to begin to learn how to make sago jelly. Girls spend much time with their mothers and help with dish washing, water fetching, vegetable gathering and cooking. However, making sago jelly is difficult and failure renders the sago inedible and thus wasted. Accordingly, mothers begin to teach their daughters how to cook sago jelly when they turn 10 or so. Most women can make sago jelly correctly by the time they turn 15. The ability to cook sago jelly correctly has a symbolic meaning for the women of Sowom Village, signifying that they are fully grown women.

## 11.4 Mythology surrounding sago palm

Sago starch is the staple food in the low lands on New Guinea Island and its nearby islands today. The term staple here is used to mean that sago starch is objectively the largest source of food for the residents. This does not necessarily mean that the residents themselves regard it as their staple crop. One area which remains heavily reliant upon sago starch as the staple in this sense is the Sepik River Basin in East Sepik Province, Papua New Guinea. This section examines some of the myths and legends involving sago that were collected in this area.

### 11.4.1 Peculiarity of sago palm mythology

Examination of the sago-related myths collected in the Sepik River Basin inevitably reveals a certain peculiarity. Just as in the New Guinean low lands in general, cultivated plants in the Sepik River Basin include root crops such as yam and taro, fruit trees such as coconut palm, sago palm, papaya, pandanus and breadfruit, banana and sugar cane. The type of farming that relies heavily on these cultivated plants is collectively known as root crop farming. The entire South Pacific region, including New Guinea, falls into the root crop farming zone. Of the cultivated plants that are regarded as root crop farming crops, sago palm is the only one to have no traditional origin story (Jensen 1963).[1] Many myths begin with the presupposition that sago palm existed from the very beginning of time, just as heaven and earth and nature itself. This is the case at least in the Sepik River Basin. For example, a story handed down to the Abelam Boiken people says that people did not crush sago in the old days; they just went up to a sago palm, tapped on the trunk and received sago (McElhanon 1962). Exactly how to obtain sago by tapping the trunk is not explained. The Arafundi people who live along the upper reaches of the Karawari River also say, 'Women did not crush sago in the old days; without felling a sago palm, they went to where a sago palm grew, men peeled the bark of the trunk and women scooped up sago milk trickling out of it by hand, collected it in a basket and simply took it home' (Kamimura 1998). It is unknown if starchy sap would really ooze out of the sago trunk. In any case, people evidently perceive as far as these mythological stories imply that sago palms were originally growing spontaneously in the wild. This contrasts sharply with other cultivated plants such as root crops, coconut and banana which have definitive stories of some original or ancient mythological events which gave the crops to mankind. Most sago-related myths tell of either god-like beings who taught people how to obtain or cook sago starch or some mythological events that happened in the process of sago starch preparation. This difference inevitably draws our attention to the peculiar position sago palm occupies in the root crop farming cultures.

The only conceivable reason at this point of time is the unique characteristics of sago palm as a plant. Unlike root crops and banana, there is no need to plant chopped seed tubers or to divide individual plants. Once sago palm shoots are inserted in the swampy land, they do not need to be looked after by man. Unlike other species of the palm family such as coconut palm, sago palm naturally suckers and form clumps. When the author came across a vast wild community of sago palms in the downstream basin of the Sepik River, it was hard to imagine that those dense intergrowths lining the river banks for kilometers would be wild sago palm forests. As far as sago palm is concerned, it is difficult for outsiders to determine whether it is a wild species or a cultivated species. Conversely, it is hard to say if sago palm can be clearly defined as a cultivated plant. The late Sasuke Nakao, Japanese agronomist, coined the term *akebono nōgyō* (prenatal agriculture) for the intermediate stage that could not be called hunting-and-gathering nor crop farming and mentioned sago palm as one of the prenatal agriculture crops (Nakao 1985).[2] While the author is unsure if this concept still has validity, it seems to confirm the difficulty of defining sago palm as a cultivated plant. It is perhaps not surprising that so far no origin stories have been discovered in a clear form for sago palm endowed with these characteristics while the mythological origins of other cultivated plants have been told and passed down.

For this reason or otherwise, it seems that people at least in the Sepik area do not consciously perceive sago palm as a staple crop. The Kwoma people near Ambunti in the upstream Sepik Basin grow yam as their staple crop but sago is their staple food in terms of quantity as they eat far more sago starch than yam.

### *11.4.2 Jensen's two types of cultivation plant origin myths*

There are at this stage no better theories of origin myths of plant cultivation than Jensen's two types: Prometheus and Hainuwele (Jensen 1963). The Prometheus type is named after the ancient Greek myth, which involves some god-like heroes who stole cereal grains from the heavens or another world and gave them to humans. Such crop origin myths are endemic to cereal cultivator cultures according to Jensen. The Hainuwele type is named after a girl born of a banana in the tradition of the Wemale people in western Seram Island and involves some god-like figures (Jensen calls them Dema deities) who were murdered, cut into small pieces, buried in the ground, and turned into plants that were cultivated by humans for the first time. This type of crop origin myth is endemic to root crop farming cultures according to Jensen (1963), who considered these people as archetypal or early cultivators. Jensen believed that the root crop farming culture was an older cultivator culture than the cereal farming culture, and probably the earliest farming culture in human cultural history but the validity of this hypothesis is outside the scope of this discussion. We shall instead simply acknowledge that the crop origin myths endemic to the root crop farming cultures are of the Hainuwele type. The worldview represented by Jensen's Hainuwele type myths is said to be an integrated, coherent worldview comprised not only of the mythological aspect of the creation of cultivated plants out of the pieces of the cut-up body of some murdered demigod, but also meaningfully interconnected elemental characteristics such as killing and procreation, life and death, cultivation plants, and the moon and woman (Obayashi 1977). In the next section we shall see that the sago palm-related myths present such meaningfully interconnected characteristics even though they do not explicitly explain the creation of sago palm out of corpses.

## 11.4.3 Killing and procreation, death and life, or being killed and birth of new life

The frequency with which one encounters mythological stories with complex meanings surrounding death and life, killing and procreation, being killed and new birth in the Sepik region is quite remarkable. Even more remarkable is the matter-of-fact tone in which stories of killing are told, as if it is part of the daily activity of a predatory animal that is not motivated by self-interests or grudges.

The Arafundi people live near the Arafundi River, the uppermost stream of the Karawari River, a major midstream tributary of the Sepik River. The mythology of their culture hero Ape clearly interconnects the beginning of sago cooking and the human reproductive organs, inferring that man was made of the dead and the living. According to the mythology, Ape came to a swamp to see two sisters who were crushing and washing sago there. Ape initially hid behind a sago palm and covered his entire body with sago leaves but he was eventually spotted by the sisters.

> The sisters caught the man Ape and removed the sago leaves covering his body. Underneath was the skin of the sago palm trunk. When they removed it, it separated into the black bark and the white pith. The white pith turned into a white person and the black bark turned into a black person. This is how white people and black people came into the world. (...) The elder sister put sago starch in a mesh bag and carried it on her back. The elder sister walked first, followed by Ape, followed by the younger sister, and the three went home. When they reached their home, they hid Ape under the floor and the sisters silently entered the house. Inside the house was a black dog who was the husband of the sisters. The sisters joined forces to batter the black dog to death and made Ape their husband from then on. They threw the dead dog out of the house. At that point, their parents came home and saw the corpse of the dog. They asked their daughters, 'Why did you kill the dog?' The sisters merely replied, 'Father, Mother, you will see something interesting tonight'. The sisters went outside and brought Ape out from

under the floor. Ape ordered the two, 'In the house, you sit with your right knee up and pass your right arm under the right leg. Lay down sago leaves on the floor, place lumps of sago starch on them, roll starch dumplings with your right hand, wrap them in sago leaves, and bake them by holding them over the fire. When you finish baking, place the sago dumplings on the floor and call out to me. Then I will come into the house'. The sisters did what they were told by Ape and he came into the house, sat down comfortably as the sisters, their parents and brothers watched, and began to eat the sago dumplings. They all felt happy as they watched him. Ape said, 'As you can see, men never cook. Men only eat what women cook for them'. Thus Ape became part of the family. However, his wives did not have a vagina. Their father and brothers did not have a penis. Only Ape had a penis. Ape found an egg of a wild chicken, cracked it in half, attached it to the elder sister's crotch to make a vagina and had intercourse with her. He made vaginas for the younger sister and the brothers' wives and had intercourse also with them. Afterwards, he picked some tubers of wild taro, peeled a piece of skin at the tip of each tuber, made a small hole and attached it to the crotch of the sisters' father and brothers as a penis. Ape told them, 'Have intercourse with your spouses, just the way I did a little while ago'. This is how people learned how to have intercourse. (Recorded by author on 23 August 1990, informant: Yabiki Tapain, Auwim Village)

This story clearly states that the method and manner of cooking sago, table manners for eating sago, and the origin of human genitals and sexual intercourse were taught by a strange-looking sago palm man named Ape. Moreover, the sisters killed the then husband, the black dog, without even hesitating so that they could marry Ape. They did so quite mercilessly and coldheartedly without showing a speck of emotional hesitation. This must be the type of killings Jensen (1966) called the 'primordial violence of nature'. The story also suggests the inevitability of human mortality. This is implied in the part which tells that the bark of the sago palm man Ape became a black person and the pith became a white person. The informant explained that the black person represented the Papuans and the white person represented the Europeans. However, the author thinks that this is a very recent reinterpretation to make the story more interesting. Just as the spirit of the dead named Miminja is painted white in the iconographic representations of the Arafundi people, the white person should be interpreted as the dead and the black person as the living. In other words, the story is very likely suggesting that the existing humans represented by the living and the dead were born out of the sago palm man named Ape. It represents their perception that human-kind consists of living people and dead people. It appears to suggest that human mortality began with this prehistoric event.

Myths with interconnected meanings of life and death, killing and reproduction are also found among the Kwoma people on Washikuku Hill, Upper Sepik River. As mentioned earlier, the Kwoma people rely very heavily on sago starch in their diet even though their staple crop is yam. They naturally have a separate origin myth for yam. A myth surrounding sago is told as follows as part of the mythology passed on in a clan that has the eagle as their totem.

> A giant eagle lived in a huge tree standing beside a spirit house. As the giant eagle had a habit of catching and eating children, people hatched a plan to hide two young brothers inside a *kundu* drum (single-headed hand drum) and send it up to the giant eagle. The younger brother used magic to put the giant eagle to sleep and tried to slash the base of the giant eagle's neck. This is a so-called 'Trojan horse' type narrative. The critically injured giant eagle nonetheless took to the air at dawn.
>
> The giant eagle flew but fell onto a sago palm forest. (...) The two boys, Sambaaku and Kwaiga, went to the place where the giant eagle had fallen. At that place, two girls were about to start washing sago. However, there was no water in the stream to wash sago. The girls went upstream to investigate what was happening. To their amazement, the dead body of a giant eagle was lying across the river and stopping the flow. They moved the body and concealed it under a mass of sago mash (the pith of sago palm trunk crushed with a sago hammer). They plucked some feathers from the giant eagle's head and decorated their own hair. The two brothers, Sambaaku and Kwaiga arrived. They asked the girls, 'Have you seen a giant eagle around here?' The girls replied, 'No, we don't know; we don't know where a giant eagle is'. The brothers asked more sharply, 'You have the giant eagle's feathers on your head; you must have hid its dead body'. The girls gave up and brought the dead giant eagle out of the sago mash. The two brothers told the girls, 'We'll take the two legs of the giant eagle; you can eat the rest of the flesh on its body'. The two girls were called Yuas and Kanuwaya. The name of the stream where the girls were washing sago was Washukapamashuku, which is now Mino Village. Although the two brothers told the girls that they were taking only the two legs back to their village, they secretly returned to where the girls were. The girls had cooked and eaten most of the meat by then. As they couldn't give back what they had eaten, the girls proposed the brothers, 'Let's take the remaining meat and sago starch back to our home together'.
>
> On their way to the girls' home, the brothers asked, 'Do you have areca (betel) nuts?' They replied, 'Yes, we do; there is an areca palm on the side of the road, bearing nuts'. When they reached the areca palm, the

brothers commanded the girls, 'You two climb the tree and pick some nuts for us'. The two girls climbed up the areca palm. Yuas climbed first and Kanuwaya followed. The two brothers stayed on the ground. The brothers inserted a talon of the giant eagle into the areca palm so that it stuck out like a hook. The brothers shouted at the girls, 'Come down quickly, the areca palm is about to fall'. The brothers tricked the girls. The girls were deceived and tried to climb down the tree quickly. Kanuwaya got down first and Yuas followed. They climbed down so fast that their crotches were caught by the talon hook. Kunuwaya's crotch suffered a large laceration as she got down first. It was very painful. Yuas' laceration was small and not so painful as she got down next. As the two girls now had vaginas, the brothers decided to marry them and take them home. The elder brother Sambaaka married Kanuwaya and the younger brother Kwaiga married Yuas. They lived in newly built bush houses. They produced many children and the place became a large village, Nuguruwui Parunjuwui. Thus, our Gushemp clan was formed. (Recorded by the author on 12 August 2003 in Tongshemp Village, informant: Terence Yuwendu Yanbokai)

This myth says nothing about the beginning of sago starch collection or sago cooking, let alone the origin of the sago palm; it merely states that the corpse of the god-like giant eagle was concealed in a pulpy mass of crushed sago by the female sago collectors. However, the episode about two boys who were hidden in a *kundu* drum and sent over to the giant eagle certainly conjures up the image of being 'swallowed up' into a *kundu* in a way. This narrative is frequently used as a mythological explanation of their rite of passage. In fact, the two boys used their wit to kill the giant eagle and then met their future wives, created genitals for them and married them. They were young boys at the beginning but as soon as they killed off the giant eagle they grew up to fully fledged men who were able to turn young girls into marriageable women. In other words, this story about killing the giant eagle was in fact about the rite of passage through which young boys transform into adult men. Mythological features that have interconnected meanings such as killing a divine creature in the form of a giant eagle, the rite of passage, the invention of reproductive organs and marriage are clearly represented in the story. What is the mythological significance of the episode in which the girls were collecting sago when they met the brothers and they hid the dead body of the giant eagle under the heap of mashed sago? Why did the giant eagle have to crash into a sago forest and lie dead to stop the flow of the sago washing water? It appears that the corpse of the giant eagle was transformed into lumps of animal flesh to be eaten by the brothers and the girls as well as a magical tool to create the vagina for the girls by being buried once under a sago mash by the sago collector girls, then dug up and brought out again. Moreover, the story appears to imply that the corpse of the giant eagle also transformed into sago starch by being brought out of the sago mash. The girls were trying to wash sago but they could not extract starch because the stream had been dammed up. Yet, the girls were carrying the flesh of the giant eagle and sago starch when they headed home accompanying the two men. It appears to imply that after being killed and buried inside the sago mash the giant eagle was transformed into edible meat, sago starch and a magic tool. The source of the stream for sago washing was named Washukapamashuku. Since Washukapu is the name of the giant eagle and mashuku means the head, this name means the head of the murdered giant eagle. It appears to suggest that the source of the stream for sago washing is a place named 'giant eagle's head' and that the very source of the river water to extract starch from a sago mash originated in the giant eagle's head.

The fact that the head, right wing, left wing and body of this giant eagle are the main totems of four clans also appears to suggest that the four clans were formed out of the body parts of the giant eagle. Consequently, it seems that the slain and cut up body of the giant eagle produced meat, sago starch, a magical genital-creation tool as well as four clans. The story tells of the rite of passage for boys and marriage which maintain a strong connection to this primordial event. And the snatching and eating of village children by the giant eagle are simply part of the typical behavior of a carnivorous bird of prey. The villagers who were troubled by the giant eagle's behavior and plotted to kill it were also simply neutralizing the threatening predator as part of their hunting behavior; they were not at all motivated by enmity toward it. In short it was a hunt rather than human battle. It certainly appears to represent the worldview that is endemic to an archetypal cultivator culture typified by what Jensen calls the Hainuwele type mythology.

### 11.4.4 The moon, woman and cultivated plants
This section shall look at a sago-related myth about the interconnection between the moon and women. The following myth has been passed down among the aforementioned Kwoma people on Washikuku Hill.

> Long ago, there was no light at night. So, people harvested sago and fished only during the day. One woman went to Lake Membokwao to fish. She caught fish using a net. She saw something shiny in the lake water. She used her net to pull the shining thing, the moon, out of the water. She put the moon in her fish bag and went home. She placed the shining thing in an earthenware pot for storing sago starch (*nogujau*). Her husband usually hunted pigs which came to feed on sago at lightless night but it was difficult. One night, he looked inside a *nogujau* which was full of light. It was the moon. After other men went to sleep, the man took the moon with him, went to a sago palm forest and chased a pig which came to eat sago. He successfully caught

the pig. This man's family got the pig thanks to the moon. Other men's families couldn't do that.

One day, this man and his wife went to crush sago and left their two children at home. The man's cousin came to visit. He asked the two children, 'What are you eating?' The children replied, 'We are eating a pig'. The cousin asked, 'How did you manage to catch the pig? What kind of light does your father carry?' The children replied, 'Our father takes the moon to the sago palm forest'. The cousin asked, 'Where does your father keep the moon?' The children replied, 'The moon is kept in a *nogujau*'. The cousin wanted to have the moon and pressed the children. The children declined to tell him and went up a wooden ladder to hide in the house. The cousin chased them into the house. There were many earthenware pots in the house. The first pot was empty. The second pot was also empty. He opened the last pot and found the moon inside. The cousin looked at the moon and understood how it was possible to find a pig at night. The cousin threw the moon up toward the sky. The moon fell on a tall tree. As the moon got injured, the cousin treated the moon's back with a breadfruit leaf. This is why the moon is still scarred. After treatment, the cousin climbed up on the roof of the house this time and threw the moon up high. The moon rose up on the cloud and did not fall onto the ground this time. Seeing the moon up in the sky, the two children cried out of sadness. They went to look for their parents hunting in the sago forest and cried again when they saw the moon in the sky and in the river water. They held sago dumplings against their bodies and mourned. The children's parents saw that the moon was now in the sky. The moon rose so high that it was out of reach of the man and his wife. This is how the light from the moon became available to everyone at night. (Recorded by author in Tongshemp Village on 2 August 1997; informant: Alan Sasaap Kurawaar)

This myth tells that the moon was collected from the water by a woman. The woman was the owner of the moon in that sense. The water was of the river for sago washing, and the moon was stored in a large earthenware pot for sago starch storage (*nogujau*). It had a close association with sago starch in that sense. Because what is collected from the water of the river for sago washing and stored in the *nogujau* is normally sago starch, it can be said that the moon was assigned an almost equal value to sago starch. The owner of the moon was a woman. This equates the woman with the moon and sago starch. The moon was stolen by a male cousin due to gaffes made by her children and thrown away high into the sky. A very similar myth has been reported from Ali Island on the northern coast of West Sepik Province. According to this story, a married couple living near the Sepik River crushed and washed sago and collected and stored sago starch. When the wife went to their sago starch store, a lump of sago starch had transformed into a large round object shining softly, i.e., the moon (Stokes and Wilson 1987). Although the story's provenance is unreliable, it expressly states that sago starch had transformed into the moon. Moreover, it was the sun which took the moon to the sky and two children who tried to stop it were pasted on the moon (Stokes and Wilson 1987). The snatching of the moon to the sky and the snatching of the children were superimposed. In other words, it was a detachment of the moon and the children from the earth by an act of abduction by the sun. In the Kwoma myth, the moon was thrown into the sky by man. This was also a detachment of the moon from the earth. The fate of the children was not mentioned in that story.

The Kwoma people have another myth about the origin of the moon which has a similar narrative without any direct reference to sago. It says that the children who could not stop the man from throwing the moon into the sky were packed inside a *kundu* drum and thrown in the river. The abandoned children were saved by a river turtle, sheltered by a dog in the forest, and finally adopted by a female bird of paradise. They completed a spirit house and slit drums and lived happily (Recorded on 20 August 2006, informant: Ambrose Maukos). In other words, the abandoned children traveled via the water world and the mountain world to the heavens symbolized by the bird of paradise. The children were detached from their parents and the earth. It can be said that the children were expelled to the sky by man on the earth just as the moon was expelled to the sky by man on the earth. Despite the difference between the Ali Island myth in which the moon and the children were taken to the sky by the sun in the sky and the Kwoma myth in which the moon and the children were expelled to the sky by man on the earth, the common theme tells of the detachment of the moon and the children from the earth forever.

It has become clear that woman, sago starch and the moon have closely interconnected meanings and that the theme of the permanent expulsion of the moon and the children born of the woman to the sky is also closely associated with it. The moon thus ascended to the sky permanently and the existing cosmic order was established. What is the significance of the children who went up to the sky with it? It perhaps means that children are forced to leave their parents, especially their mother, to go on a long journey to adulthood. In that case, it has a connection to an ideological representation underpinning the rite of passage. Or does departure to another world in the sky symbolize dying after all? In that case, the theme must be life and death. The Kwoma myth says that the children were adopted by a female bird of paradise, completed a spirit house and slit drums and lived happily ever after, so the story must mean that the children died as human beings and were reborn as adult men in the world inhabited by the bird of paradise.

The Ali Island myth, in contrast, says that the children were taken to the sky and pasted on the head of the moon. This story alludes to the origin of the dark patches on the moon. It means that the children died, became part of the moon and created the existing cosmic order. It is certainly possible to observe a situation in which a new life or the entire order of human living is created as a result of the death or killing of a being (though not god-like) which characterizes the worldview endemic to what Jensen calls the archetypal cultivator culture (1966, pp. 158).

### 11.4.5 Conclusion

As discussed, no origin myth has been found in the Sepik River region of Papua New Guinea that explains how sago palm came to exist and became available to humans to begin with. Instead, it is suggested that a god-like being (man or animal) was killed in many a folk story, or a normal person was killed in a small number of cases, and this primordial killing incident gave rise to human mortality, sago cookery and eating manners, human reproductive organs or sago starch and edible meat as well as various clans, rites of passage and weddings at some stage in the process of producing sago starch from the sago palm, which already existed as part of nature. A woman obtained the moon while collecting sago starch and hid it in a vessel for storing sago, but an outside party (man or the sun) took it and expelled it to the sky, creating the current state of the moon. During this incident, children die. These characteristics of the sago-related myths found in the Sepik region appear to represent the worldview endemic to the archetypal cultivator cultures which, according to Jensen's theory, created the Hainuwele type myths even though they do not carry Hainuwele type mythological elements in a clear form such as killing a god-like figure, cutting up its body, and the generation of cultivation plants from the dismembered and buried body parts.

**Notes:**
1. The only myth that has been found so far to clearly tell of the origin of sago palm is the Tuwale myth of the Wemale people on Seram Island reported by Jensen. The story says that the two children of Tuwale became sago palms. Conversely, though, this does not clear up doubts about the hypothesis that the Wemale is a typical example of what Jensen calls the archetypal cultivator culture.
2. The late Dr. Sasuke Nakao presented the concept as 'prenatal agriculture' at the Conference on Social and Cultural Change in Papua New Guinea, at the National Museum of Ethnology, Japan, held on 21 December 1985.

Authors:
11.1–3: Yukio Toyoda
11.4: Toru Kamimura

# 12
# Future of Sago Palm in the 21st Century

## 12.1 Sago palm as starch material

The 21st century has been characterized as an age of environmental challenges in which resource depletion and environmental degradation will continue to accelerate if we sit around and do nothing. Background factors behind this include industrial capitalism and population growth. The United Nations (UN) forecasts that the world population will reach 9.1 billion by 2050 (Figure 12-1). Population pressure and inequities in food distribution will lead to food shortages, especially in developing areas. However, the Food and Agriculture Organization (FAO) is mildly optimistic, based on a marked improvement in the global food situation over the last three decades. While the world population increased by more than 70% in that period, the per capita food consumption grew only about 20%. In developing countries, the proportion of the population that is chronically malnourished was almost halved, dropping to as low as 18% in 1995/1997 even though the population almost doubled. However, the number of people on the brink of starvation remains very high: the FAO estimates that 580 million people will suffer from chronic malnutrition in 2015. Yet, the FAO projects that the rate of increase in food production will surpass the rate of population growth. Animal feed will become the most important factor driving global grain production; it will account for 44% of additional demand for grain.

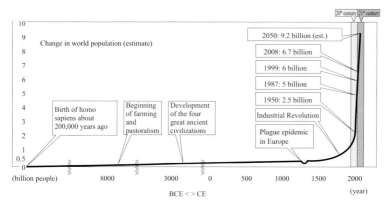

Figure 12-1 Change in world population
Source: www.unfpa.or.jp/p_graph.html

The only way to solve this food shortage is either to increase the yield of grain crops such as wheat, rice and corn, or to increase the acreage of productive agricultural land. As there is virtually no extra arable land left in the world, the starch accumulating sago palm has come to attention as a crop that can be grown on land that is unsuitable for common grain crops. Other starch producing plants include corn, cassava, potato and sweet potato. In 2007, the total global production of corn amounted to 780 million tons (which would produce 550 million tons of starch if all of it were processed into starch), cassava and other root crops totaled 750 million tons (230 million tons of starch), and potato totaled 320 million tons (50 million tons of starch). In comparison, the total production of sago palm yielded less than 1 million tons of starch (Figure 12-2). No definitive statistics are available, partly because the habitat of sago palm is limited to low land swamps in Southeast Asia and the proximity between the sago starch production area and consumption area through small local markets makes verification of production quantities difficult. The other starch accumulating crops such as corn, cassava, potato and sweet potato are cultivated in agricultural fields and therefore in competitive relationships with one another. The optimum cultivation environment for sago palm is wetland. Sago palm is not very selective about soil type and is able to absorb nutrients in the water efficiently. Of course, it grows even better in properly prepared soils complete with beneficial elements. In the low land sago palm habitat of Southeast Asia, there are 23 million hectares of low pH, low bearing-capacity peat soils and 6.7 million hectares of acid sulfate soils which become very acidic when oxidized. It requires a crop that is suited to growing in these types of soil as well as low land swamp conditions. People in

Figure 12-2 Sago starch

Southeast Asia have appreciated the sago palm for a very long time as can be seen in the references to it in *The Travels of Marco Polo*.

Sago palm can be propagated or cultivated by either suckering or seeding. Suckering produces individual plants that express traits identical to those of the mother palm with little variation. Propagation by seed germination, in contrast, can result in phenotypic divergence in the offspring. Suckers are commonly used for cultivation as they grow faster after planting and become harvestable in a shorter time. Once the sago palm forms a clump, the mother palm is harvested and its suckers continue to grow, becoming the mother palm for subsequent generations. Accordingly, there is no need in principle for replanting. This is another advantage of the sago palm over other starch accumulating crops. Recognizing these advantages, a sago palm research organization (The Society of Sago Palm Studies; http://www.bio.mie-u.ac.jp/~ehara/sago/sago-j.html) was formed. Its members, with fields of expertise from basic to applied research, have been endeavoring to publicize the sago palm's potential.

Sago palm is the only acid resistant crop that grows in low land swamps (Yamamoto 1998a). Each sago palm yields 200–300 kg of dry starch. A harvest rate of 100 sago palms per hectare would produce a yield of 20–30 t/ha. Compared with an average yield of 10 t/ha from the other starch accumulating crops, it is obvious that sago palm has a much higher starch accumulating capacity.

Figure 12-3 Sago palm plantation (Mukah, Sarawak, Malaysia)
(Photo: Yusuke Goto, 6 March 1996)

Yet, sago palm lags behind the other starch crops in total production and sales. The reasons for this include its long maturity time (about 10 years), poor access to cultivation areas, lack of an efficient process from palm production to starch extraction, and lack of information (as it has essentially been a local-production-for-local-consumption crop). The sago starch industry cannot operate if the essential raw material for starch production is not delivered to factories on time (Ohno 2003). For the modernization of sago starch production, i.e., sustainable sago palm cultivation and stable starch production, the development of production infrastructure, a reduction in sago log transportation costs and thorough control of sago starch quality (starch content, whiteness etc.) are needed. One sago palm plantation project in the Malaysian state of Sarawak involves the conversion of cleared low land forest sites on well-developed tropical peat (woody peat) soils into sago palm plantation farms (Figure 12-3) (Land Custody and Development Authority 2009). The project was launched in 1987 and scheduled to develop a total of 250,000 ha of land by 2020, including about 14,000 ha of sago palm cultivation (6,300 ha in Mukah, 4,006 ha in Dalat, and 3,640 ha in Sebakong). If this sago palm cultivation area begins to supply sago starch in a stable manner, the level of interest in sago starch is expected to rise (CRAUN Research SDN. BHD 2009).

Sago starch's solubility and heated swelling property are similar to those of sweet potato starch and tapioca (manioca starch). Applications to utilize these characteristics are being considered. Developing new ways to use starch by taking advantage of the characteristics and functionality of sago starch will be essential for expanding sago starch demand. For instance, sago starch is said to be less allergenic (containing fewer allergens) than other starches (Sakai 1999b; Tsuji Anzen Shokuhin 2009). This characteristic is expected to create new demand but more rigorous studies are needed for this purpose (Ueda 1999).

## 12.2 Sago palm as a biofuel

Sago palm has come into the spotlight as a potential source of bio-fuels. The world's fossil energy resources have a limited future, with the confirmed recoverable reserve of crude oil estimated to be between 800 billion and 1,212.9 billion barrels (Takeishi 2004). This gives us a reserve-production ratio of about 45 years. The confirmed recoverable reserve of coal is reported to be 909.1 billion tons with the reserve-production ratio of 155 years. Fossil energy resources are destined to run out eventually. In his State of the Union address in January 2006, President Bush announced a substantial increase in the clean energy research and development budget with a goal of replacing more than 75% of crude oil imports from the Middle East with alternative resources by 2025 which spurred people's interest in biofuels (Figure 12-4).

As Japan imports virtually all fossil energy resources from other countries, what can she do in that regard? In December 2002, the cabinet approved the Biomass Nippon Strategy for the purposes of global warming prevention, the formation of a recycle-based society, strategic industry development and the revitalization of rural communities. The Japanese government adopted biomass utilization as a national policy (MAFF 2009). The comprehensive strategy provided that in order to accelerate the use of biomass-based transportation fuels the government should take initiative in laying out a timetable for their introduction and develop necessary facilities. More specifically, in order to promote the use of locally produced biomass-based transportation fuels:
1. the relevant ministries and agencies cooperate in creating model projects of biomass use;
2. cheaper supply of raw materials and framework for cooperation between concerned parties will be established; and
3. low-cost, high-efficiency production technologies will be developed (ethanol etc. produced from biomass-rich agricultural crops and woody materials).

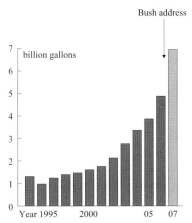

Figure 12-4 Change in ethanol production in the United States
Corn-based ethanol production per year is forecast to double between 2006 and 2016 in the United States.

In March 2006, the government conducted a review of the Biomass Nippon Strategy based on the progress of biomass use and changes in circumstances that took place after the commencement of the strategy such as the Kyoto Protocol, which came into effect in February 2005. It instigated policy measures for the full-scale introduction of locally produced biomass fuels and the acceleration of the development of biomass towns to utilize unused biomass materials such as residual wood materials on the forest floor (Table 12-1).

Table 12-1 Biomass use in Japan

|  | Quantity (1,000 tons) | Utilization ratio (%) | (2010 target %) | Energy potential of unused part (petroleum equivalent) |
|---|---|---|---|---|
| Waste material biomass | 298,000 | 72 | (80) | 530 PJ (4 million kL) |
| Unused biomass | 17,400 | 22 | (25) | |
| Resource crops | None | - | - | 240 PJ (6.2 million kL) |

http://www.maff.go.jp/j/biomass/b_energy/pdf/kakudai01.pdf

Soaring crude oil prices are also stimulating the introduction of transportation biofuel. Let us compare its price competitiveness. In 2009 the ex-refinery price of gasoline was US$ 0.70/l. When gasoline tax is added, the price was US$ 1.28/l. In comparison, the price of Brazilian ethanol (Cost Insurance and Freight (CIF) price where ownership of the goods is transferred from the vendor to the purchaser upon delivery according to the purchaser's instructions) was US$ 1.03/l (including taxes other than gasoline tax). If the cost of raw material is based on feed prices, the price of ethanol from molasses

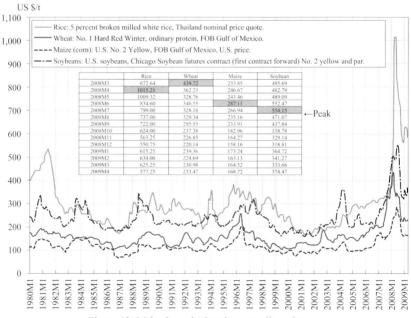

Figure 12-5 Rise in agricultural commodity prices
Note: Monthly data.
Source: IMF Primary Commodity Prices
http://www2.ttcn.ne.jp/honkawa/4710.html

was estimated to be US$ 0.96/l and the price of non-standard wheat ethanol was estimated to be US$ 1.01/l (excluding gasoline tax), both of which are slightly cheaper than gasoline. For bioethanol to compete with gasoline, they need to be exempt from fuel taxes or to be taxed at a reduced rate. Still, these estimates have made plans to manufacture the so-called biofuels from rice, wheat, corn, molasses and beet more realistic. In the United States, corn and many other grains are being converted to ethanol for fuel production (Koizumi 2006), which is driving up the prices of agricultural produce (Figure 12-5) (OECD/FAO 2007). Reduction in surplus agricultural commodities and export subsidies contributes to long-term structural changes in agricultural markets. One of the options is that grains, sugars, oil seeds and vegetable oils are used increasingly for the production of alternative energy, including ethanol and biofuel.

At the Chicago Exchange, the prices of corn and soybean rose to new highs in 10-1/2 and 1-1/2 years respectively after January 2007. The high prices were not caused by shortages due to a crop failure. The US corn harvest in 2006 was the third highest in history, at an estimated 268 million t. Nevertheless, the market rose to the maximum allowable single day gain on 16 January and closed at around US$ 4/bushel (about 25 kg of corn). The amount of corn used for ethanol production was forecast to increase by more than 30% over 2006 to a record 55 million t. This was an increase from only 6% of the total corn production 10 years earlier. The soybean market also traded high, closing at around US$ 7/bushel (about 27 kg of soybean) on 12 January 2007. The prices of all major grains hit record highs influenced by droughts, crop failures, sharply increasing demand for bioethanol in the US and the influx of speculative funds against the backdrop of expanding demand for animal feed in China and other developing countries.

The market peaked in March 2008 for wheat, April for rice, June for corn and July for soybeans. The prices remained high after hitting these peaks but then dropped dramatically during the global financial crisis and economic recession which began in the US, triggered by the collapse of Lehman Brothers in September. Although the prices are still at higher levels than before, people are watching closely at which levels the markets will bottom in the future. The prices began to fall in January 2009 (Google Ads 2009). The rise of the grain prices had an impact on the prices of fat and oil as well as grain-based processed food products such as bread and *natto* (fermented soybean) while the corn price affected the prices of dairy and meat products via the feed price and pushed up the prices of many food products. However, the price of sago starch did not move very much as Japan makes price adjustments (institutional buying and sell-back and producer support) between imported starches and domestic potato starch.

Sago starch is produced from sago palms mostly in Malaysia and Indonesia but accurate production quantities are unavailable as the precise amounts of local consumption are unknown. However, the annual total export of sago starch from Sarawak (Malaysia) has been around 50,000 t since 1992. The price of sago starch has risen eight- or nine-fold over a period of 30 years to 800 ringgit (US$ 239). The annual yield per hectare is 8.6 t for corn, 5 t for rice, 5 t for wheat, 59 t for beet, 62 t for sugar cane and 20 t for sago palm. When ethanol is produced from these crops, the conversion rate per hectare is 3.4 kl for corn, 2.4 kl for rice, 2.0 kl for wheat, 5.9 kl for beet, and 0.65 kl for sugar cane. Only a very small amount of ethanol production from sago palm is carried out at the moment and the conversion rate is unknown but sago palm can potentially produce 8 kl.

## 12.3 Sago palm as a biomass resource

Sago palm has been a very useful biomass resource for a long time. All parts of the sago palm biomass are usable resources. Its leaf has long been used as a highly durable roofing material (Figure 12-6) as it has a flexible structure

Figure 12-6 Biomass use

full of strong fibers and dense mesophyll tissue (Abe 1994). Its bark is sun-dried and used for fencing and matting. It is sometimes charred and used as activated carbon. Sago palm biomass harvested from brackish water areas is often incinerated in the field as it contains a high level of sodium chloride (NaCl), which can be corrosive to some incinerators.

The development of new applications for sago palm biomass may determine the future of sago palm. Attempts are being made to produce biofuel (ethanol) and biodegradable plastics from materials such as sago starch filtration residue, sago palm leaf and bark (Watanabe and Ohmi 1997).

The US President GW Bush ignited the biofuel boom. After he announced, in the 2007 State of the Union address, that 20% of the nation's gasoline consumption would be replaced with renewable energy in the next 10 years, farmers switched from soybean to corn in droves, partly helped by government subsidies. Corn acreage increased, corn for human consumption was sold for fuel, and the prices of both soybean and corn rose sharply. The corn price doubled from US$ 2/bushel in 2006 to around US$ 4 from 2007. The world's bioethanol production is shown in Figure 12-7. There is a need to switch raw materials for biofuel production from food crops such as sugar cane and corn to some currently unused resources. Sago palm provides a large amount of biomass (2 t fresh weight per palm). About 200 kg of starch can be extracted from each sago palm but around 50% of starch still remains in the sago starch filtration residue. It converts into ethanol more readily than other types of biomass. Studies are being carried out on the possibility of producing ethanol from the unused resource of sago residue (Nadhry et al. 2009).

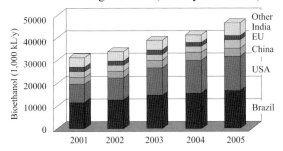

Figure 12-7 Bioethanol production in the world
http://www.maff.go.jp/j/biomass/b_energy/pdf/bea_04.pdf

Japan's plastics production reached 15 million t in 1999 and its domestic consumption exceeded 10 million t. It is estimated that about 10 million tons of plastics are discarded across the country each year. Polyethylene accounts for 29% of the total plastics production, polypropylene 20%, polystyrene 17%, vinyl chloride 13% and others 21% (Figure 12-8). While 21% of discarded plastics are incinerated, 33% end up in landfills and the remaining 46% are reused, only 14% of total plastic waste is reused as raw materials. This situation led to growing calls for plastics which are biodegradable after disposal. Waste generation can be minimized as biodegradable plastics are eventually broken down by water and carbon dioxide when buried in the ground. They are carbon-neutral in that they generate less heat compared with normal plastics when incinerated and that they release previously photosynthesized carbon in the form of carbon dioxide. The use of biomass-based materials can reduce the use of fossil fuel resources. These advantages encouraged the manufacturing of biodegradable plastic. However, market growth is rather slow and erratic due to higher prices than conventional plastics and limited applications (Figure 12-9). Biodegradable plastics are grouped into several types, including biomass-based (plant biomass-based only), microbe-based (polymers produced by

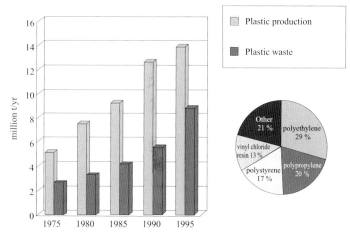

Figure 12-8 Plastics production and waste in Japan

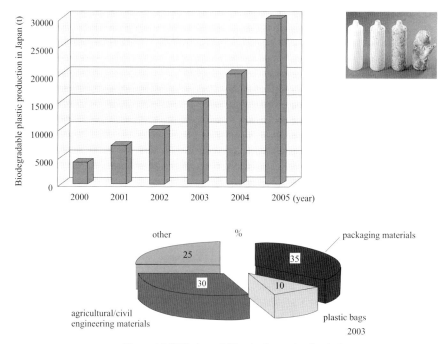

Figure 12-9 Biodegradable plastic production in Japan

microbes), synthetic (petrochemical, e.g. polylactate) and synthetic (plant biomass-based, e.g. starch plus polymers). The synthetic (petrochemical) type such as polybutylene succinate (PBS) has been the mainstream in the past. In recent years, however, manufacturing sago palm biomass-based plastic has been trialed as emphasis has shifted from 'biodegradability in the natural environment' to 'plant biomass-based' (Sasaki et al. 1999; Sasaki et al. 2002a, b). The greatest hurdle to popularization of biodegradable plastics is price. The current biodegradable plastic price of US$ 4.26–6.38/kg is much higher than the price of normal plastics (around US$ 1.60/kg). Lower raw material costs and the introduction of larger manufacturing facilities hold the key to achieve price reduction to the target of US$ 2.13/kg by 2010. Since the price gap is likely to be smaller if the cost of recycling is taken into account, demand is expected to increase further when this comparative assessment approach is adopted more widely (Table 12-2). Demand for biodegradable plastics is expected to come from manufacturing products such as films, sheeting, everyday clothing and miscellaneous goods, containers and foams.

Table 12-2 Fields in which the practical use of biodegradable plastics is expected to expand

| Field | Material | Use |
|---|---|---|
| Used in the environment | Agriculture/forestry/fishery materials. | Mulching film, seedling pots, fishing line, fishing net etc. |
| | Civil engineering/building materials. | Insulation material, non-recoverable formwork for mountain/marine civil engineering work, retaining wall, desert revegetation work, water-holding sheet, sandbag, vegetation reinforcing net/mat etc. |
| | Outdoor leisure goods etc. | Golf tee, fishing (lures etc.). |
| Non-recoverable or non-recyclable after use | Food packaging film, container. | Perishable food tray, instant food container, fast food container, lunch box etc. |
| | Sanitary products. | Disposable diaper, sanitary napkin etc. |
| | Stationery, clothing etc. | Pen case, lead case, razor, tooth brush, glass/cup, garbage bag, drainer, cushion filler, clothing etc. |
| Used for special functionality | Controlled release coating material. | Coating for medicine, agricultural chemicals, fertilizer, seed etc. |
| | Water retentive or absorbing material. | Revegetation material for desert and waste land, compost moistening material. |
| | Biodegradable medical-related products. | Surgical suture thread, fracture fixator, medical film, medical non-woven fabric etc. |
| | Low oxygen permeable or non-absorptive functional packaging material. | Food packaging film, internal coating for drink pack etc. |
| | Low melting point adhesive material. | Adhesives for packaging, book binding and bag making. |

The sago residue after starch extraction still contains a substantial amount of starch, which acts as an effective binder to make highly degradable plastic (Okazaki and Toyota 2003a, b; Okazaki and Toyota 2004). The future is steadily opening up for the utilization of sago palm biomass and future advances in research will accelerate its wider use.

## 12.4 Expectations for sago palm

In order to increase reliability of the process from sago palm production to the manufacturing of sago products, including sago starch, in response to greater expectations for sago palm, it is essential to collect and select sago palm clones to find sago palm types that are optimal for intended uses (Sato 1993). Biotechnology will become one of the driving forces that will change the conventional plant breeding technology dramatically. Once high quality clones are selected, they are expected to perform at a level that defies common sense. Thus, sago palm is considered to have the potential to surpass all other starch crops and change the world's starch supply and demand situation completely.

Author:
Masanori Okazaki

# Appendix: List of the International Sago Symposia and Proceedings

1. Kuching, Malaysia (5–7 July 1976)
   Tan, K. (ed.) (1977), *Sago: '76: Papers of the 1st International Sago Symposium "The Equatorial Swamp as a Natural Resource"*, Kuching, 330 pages.
2. Kuala Lumpur, Malaysia (15–17 November 1979)
   Stanton, W. R. and M. Flach (eds) (1980), *Sago: The Equatorial Swamp as a Natural Resource (Proceedings of the 2nd International Sago Symposium)*, London: Martinus Nijhoff, 244 pages.
3. Tokyo, Japan (20–23 May 1985)
   Yamada, N. and K. Kainuma (eds) (1986), *Sago: '85: Proceedings of the 3rd International Sago Symposium*, Tokyo: The Sago Palm Research Fund, 233 pages.
4. Kuching, Malaysia (6–9 August 1990)
   Ng, T. T., Y. L. Tie and H. S. Kueh (eds) (1991), *Towards Greater Advancement of the Sago Industry in the '90s: Proceedings of the 4th International Sago Symposium*, Kuching: Lee Ming Press, 225 pages.
5. Songkhla, Hat Yai, Thailand (27–29 January 1994)
   Subhadrabandhu, S. and S. Sdoodee (eds) (1995), *Fifth International Sago Symposium*: Acta Horticulturae, No. 389, 278 pages.
6. Pekanbaru, Indonesia (9–12 December 1996)
   Jose, C. and A. Rasyad (eds) (1998), *The Future Source of Food and Feed: Proceedings of the 6th International Sago Symposium*, Pekanbaru: Riau University Training Center, 270 pages.
7. Port Moresby, Papua New Guinea (27–29 June 2001)
   Sago as Food and Renewable Resource for the New Millennium (The 7th International Sago Symposium). (No published proceedings)
8. Tsukuba, Japan (15–17 October 2001)
   Kainuma, K., M. Okazaki, Y. Toyoda and J. E. Cecil (eds) (2002), *New Frontiers of Sago Palm Studies: Proceedings of the International Symposium on SAGO (SAGO 2001): A New Bridge Linking South and North*, Tokyo: Universal Academy Press, 388 pages.
9. Jayapura, Indonesia (4–6 August 2005)
   Karafir, Y. P., F. S. Jong and V. E. Fere (eds) (2005), *Sago Palm Development and Utilization: Proceedings of the 8th International Sago Symposium*, Manokwari: Universitas Negeri Papua Press, 266 pages.
10. Ormoc, The Philippines (19–21 July 2007)
    Toyoda, Y., M. Okazaki, M. Quevedo and J. Bacusmo (eds) (2008), Sago: *Its Potential in Food and Industry: Proceedings of the 9th International Sago Symposium*, Tokyo: TUAT Press, 238 pages.

11. Bogor, Indonesia (29-30 October 2011)
12. Manokwari, West Papua Province, Indonesia (6-8 November 2013)

# Bibliography

Abbas, B., M. H. Bintoro, H. Sudarsono, M. Surahman and H. Ehara (2006) Haplotype diversity of sago palm in Papua based on chloroplast DNA. *In*: Sago Palm Development and Utilization: The 8th International Sago Symposium. (Karafir, Y. P., F. S. Jong and V. E. Fere eds.) Universitas Negeri Papua Press (Manokwari) 135–148.

Abbas, B., H. Ehara, M. H. Bintoro, H. Sudarsono and M. Surahman (2008) Genetic diversity of sago palm (*Metroxylon sagu*) in Indonesia, based on genes encoding the biosynthesis of waxy starch. *In*: Sago: Its Potential in Food and Industry: Proceedings of the 9th International Sago Symposium. (Toyoda, Y., M. Okazaki, M. Quevedo and J. Bacusmo eds.) TUAT Press (Tokyo) 35–44.

Abe, N. (1994) Durability of leaves of sago palm, nipa palm and coconut palm, and their lamina anatomy. Sago Palm 2: 7–12. (in Japanese)

Akiyama, T. (1966) New Encyclopedia of French Cooking. Yuki Shobo (Tokyo). (in Japanese)

Alang, Z. C. and B. Kirshnapillay (1986) Studies on the growth and development of embryos of the sago palms (*Metroxylon* spp.) in vivo and in vitro. *In*: Sago '85: Proceedings of the 3rd International Sago Symposium. (Yamada, Y. and K. Kainuma eds.) The Sago Palm Research Fund (Tokyo) 121–129.

Anderson, J. A. R. (1961) The Ecology and Forest Types of the Peat Swamp Forests of Sarawak and Brunei in Relation to Their Silviculture. Ph. D. thesis, University of Edinburgh.

Ando, H., D. Hirabayashi, K. Kakuda, A. Watanabe, F. S. Jong and B. H. Puruwant (2007) Effect of chemical fertilizer application on the growth and nutrient contents in leaflet of sago palm at the rosette stage. Japanese Journal of Tropical Agriculture 51: 102–108. (in Japanese)

Andriesse, J. P. (1972) The Soils of West-Sarawak (East Malaysia), with Soil Map Memoir 1 and 2: Soil Survey Division Research Branch/ Department of Agriculture, Sarawak, East Malaysia.

Asakura, I. (1984) Paper-making industry and starches. *In*: Starch Science Handbook. (Nakamura, M. and S. Suzuki eds.) Asakura Shoten (Tokyo) 579–582. (in Japanese)

Aziin, K. A., and A. Rahman (2005) Utilizing sago (*Metroxylon* spp) bark waste for value added products. *In*: Eco Design, 4th International Symposium on Environmentally Conscious Design and Inverse Manufacturing, 102–106.

Baker, J. B., T. A. Henderson and J. Dransfield (2000) Molecular phylogenetics of Calamus (Palmae) and related rattan genera based on 5S nrDNA spacer sequence data. Molecular Phylogenetics and Evolution 14: 218–231.

Barie, B. (2001) Improvement of nutritive quality of crops by-products using bioprocess technique and their uses for animals. E-seminar by International Organisation for Biotechnology and Bioengineering.

Barrau, J. (1959) The sago palm and other food plants of marsh dwellers in the South Pacific Islands. Economic Botany 13: 151–159.

Beccari, O. (1918) Asiatic palms – Lepidocaryeae. Annals of the Royal Botanical Garden, Calcutta 12: 156–195.

Bellwood, P. (1985) Prehistory of the Indo-Malaysian Archipelago. Academic Press (London).

Bintoro, H. M. H. (1999) Pemberdagaan tan arno sagu sebagai pengan alternatit dan bahan baku agroindustri yang potensial dalam rangk Ketahanan pangan Nasior Orasi Ilmiah Guru Besa Tetap Ilmu Tanaman Perkebunan Fkultas Pertarian, Institut Pertanian Bogor. Bogor, 11 Sept. 1999. (in Indonesian)

Bintoro, M. H. (2008) Bercocok tanam sagu. IPB Press (Bogor). (in Indonesian)

Bleeker, P. (1983) Soils of Papua New Guinea. Australian National University Press (Canberra).

Bourke, R. M. and V. Vlassak (2004) Estimates of Food Crop Production in Papua New Guinea. The Australian National University (Canberra). (http://rspas.anu.edu.au/lmg/pubs/estimates_food_crop.pdf)

Bowen, H. J. M. (1979) Environmental Chemistry of the Elements. Academic Press (London).

BPPD TKII INHL-UNRI-FAPETA (1996) Studi inventarisasi dan identifikasi potensi areal pengembangan sagu di kabupaten Indragiri Hilir. Badan Perencanaan Pembangunan Daerah Tingkat II Indragiri Hilir dan Facultas Pertanian Universitas Riau, Pekanbaru. Laporan Awal.

BPPT (1982) Hasil surveipotensi sagu di Kepulanuan Maluku (Bagian I). Kerjasama BPP Teknologi dengan Unversitas Pattimura.

Bujang, K. B. and F. B. Ahmad (2000a) Production and utilisation of sago starch in Malaysia. *In*: Sago 2000: Proceeding of the International Sago Seminar. (Bintro, H. M. H., Suwardi, Sulistiono, M. Kamal, K. Setiawan, and S. Hadi eds.) UPT Pelatihan Bahasa–IPB (Bogor) 1–8.

Bujang, K. B. and F. B. Ahmad (2000b) Production and utilization of sago starch in Malaysia. Sago Communication 11: 1–6.

Burnet, M., P. J. Lafontaine and A. D. Hanson (1995) Assay, purification, and partial characterization of choline monooxygenase from spinach. Plant Physiology 108: 581–588.

Cabalion, P. (1989) Metroxylon, Vanuatu palm. *In*: Palms of the South-West Pacific/ (Dowe, J. L. ed.) Palms and Cycad Societies of Australia (Milton) 178–180.

Chinen, Y., A. Miyazaki, S. Hamada, T. Yoshida, Y. B. Pasolon, Y. Yamamoto and F. S. Jong (2003). Changes of root amount by age in sago palm. Japanese Journal of Tropical Agriculture 47 (Supplement 1): 21–22. (in Japanese)

Corbett, G. H. (1932) Insects of Coconuts in Malaya. Dept. Agr. General, Ser. No. 10, Caxton Press (Kuala Lumpur).

CRAUN Research SDN. BHD. (2009) http://www.craunresearch.com.my/HTML/Info%20On%20Sago/Info %20On%20Sago_%20Introduction.htm.

Daita, T. (1961) Water-soluble polymers in the textile industry. *In*: Water-Soluble Polymers, enlarged edition (Nakamura, M. ed.) Kagaku Kogyo Sha (Tokyo) 167–169. (In Japanese)

Darmoyuwono, K. (1984) Application of remote sensing inventory and mapping of sago palm distribution. The expert consultation on the development of sago palm and product (Jakarta), 24 January.

De la Fuente, J. M., V. Ramirez-Rodriguez, J. L. Cabrera-Ponce and L. Herrera-Estrella (1997) Aluminum tolerance in transgenic plants by alteration of citrate synthesis. Science 276: 1566–1568.

Dengler, N. G., R. E. Dengler and D. R. Kaplan (1982) The mechanism of plication inception in palm leaves: histogenetic observations on the pinnate leaf of Chrysalidocarpus lutescens. Canadian Journal of Botany 60: 2976–2998.

Department of Agriculture Sarawak (2005) Agricultural Statistics of Sarawak, 26–29.

Dowe, J. L. (1989) Palms of the South-West Pacific: their origin, distribution and description. *In*: Palms of the South-West Pacific. (Dowe, J. L. ed.) Palms and Cycad Societies of Australia (Milton) 1–154.

Driesen, P. M. (1980) Peat soils. *In*: Problem Soils: Their Reclamation and Management (Technical paper 12). International Soil Reference and Information Centre (Wageningen) 49–53.

Ehara, H., C. Mizota, S. Susanto, S. Hirose and T. Matsuno (1995a) Sago palm production in Eastern Indonesia: Variation in morphological characteristics and growing environment. Japanese Journal of Tropical Agriculture 39 (Ext 1): 11–12. (in Japanese)

Ehara, H., C. Mizota, S. Susanto, S. Hirose and T. Matsuno (1995b) Sago palm production in Eastern Indonesia: Variation in starch yield and soil environment. Japanese Journal of Tropical Agriculture 39 (Ext 2): 45–46. (in Japanese)

Ehara, H. (1997) Sago production in Riau province, Indonesia: Morphological characteristics of sago palm and the current state and challenges of the sago starch industry in Bengkalis regency. Sago Palm 5: 24–27.

Ehara, H., S. Kosaka, T. Hattori and O. Morita (1997) Screening of primers for RAPD analysis of spiny and spineless sago palm in Indonesia. Sago Palm 5: 17–20.

Ehara, H. (1998) Differentiation of sago palm species and ecospieces (forma), 1. Variation in external morphology, 2. Sago in North Sulawesi, Indonesia). *In*: A study of the identification of species and the relationship between starch productivity and growth environments in sago palm growing areas. (Yamamoto, Y. ed.) Toyota Foundation research output report, 17–19. (in Japanese)

Ehara, H., C. Komada and O. Morita (1998) Germination characteristics of sago palm seeds and spine emergence in seedlings produced from spineless palm seeds. Principes 42: 212–217.

Ehara, H. and C. Mizota (1999) Relationship between variation in sago palm starch productivity, genetic background and growing environment in western Indonesia. 1998 JSPS tropical bio-resources research fund research outputs report.

Ehara, H., S. Susanto, C. Mizota, S. Hirose and T. Matsuno (2000) Sago palm (*Metroxylon sagu*, Arecaceae) production in the Eastern archipelago of Indonesia: Variation in morphological characteristics and pith-dry matter yield. Economic Botany 54: 197–206.

Ehara, H., O. Morita, C. Komada and M. Goto (2001) Effect of physical treatment and presence of the pericarp and sarcotesta on seed germination in sago palm (*Metroxylon sagu* Rottb.). Seed Science and Technology 29: 83–90.

Ehara, H., S. Kosaka, N. Shimura, D. Matoyama, O. Morita, C. Mizota, H. Naito, S. Susanto, M. H. Bintoro and Y. Yamamoto (2002) Genetic variation of sago palm (*Metroxylon* sagu Rottb.) in the Malay Archipelago. *In*: New Frontiers of Sago Palm Studies: Proceedings of the International Symposium on Sago. (Kainuma, K., M. Okazaki, Y. Toyoda and J. E. Cecil eds.) Universal Academy Press (Tokyo) 93–100.

Ehara, H., S. Kosaka, N. Shimura, D. Matoyama, O. Morita, H. Naito, C. Mizota, S. Susanto, M. H. Bintoro and Y. Yamamoto (2003a) Relationship between geographical distribution and genetic distance of Sago Palms in the Malay Archipelago. Sago Palm 11: 8–13.

Ehara, H., M. Matsui and H. Naito (2003b) Absorption and translocation of Na+ in sago palm under NaCl treatments. Sago Palm 11: 35–36. (in Japanese)

Ehara, H., H. Naito, C. Mizota and P. Ala (2003c) Agronomic features of *Metroxylon* palms growing on Gaua in the Banks Islands, Vanuatu. Sago Palm 11: 14–17.

Ehara, H., H. Naito, C. Mizota and P. Ala (2003d) Distribution, growth environment and utilization of *Metroxylon* palms in Vanuatu. Sago Palm 10: 64–72.

Ehara, H. (2005) Report of the 8th International Sago Symposium (EISS2005). Japanese Journal of Tropical Agriculture 49: 386–387. (in Japanese)

Ehara, H., H. Naito and C. Mizota (2005) Environmental factors limiting sago production and genetic variation in *Metroxylon sagu* Rottb. *In*: Sago Palm Development and Utilization: Proceeding of the 8th International Sago Symposium. (Karafir, Y. P., F. S. Jong and V. E. Fere eds.) Universitas Negeri Papua Press (Manokwari) 93–103.

Ehara, H. (2006a) Geographical distribution and specification of *Metroxylon* palms. Japanese Journal of Tropical Agriculture 50: 229–233. (in Japanese)

Ehara, H. (2006b) Diversity of economic plants. *In*: Sustainable Crop Production. (Morita, S., H. Daimon and J. Abe eds.) Asakura Shoten (Tokyo) 25–28. (in Japanese)

Ehara, H. (2006c) Sago palm seed anatomy and germination process. Sago Palm 14: 38–41. (in Japanese)

Ehara H., M. M. Harley, W. J. Baker, J. Dransfield, H. Naito and C. Mizota (2006a) Flower and pollen morphology of spiny and spineless sago palm in Indonesia. Japanese Journal of Tropical Agriculture 50: 121–126.

Ehara, H., M. Matsui and H. Naito (2006b) Avoidance mechanism of salt stress in sago palm (*Metroxylon sagu* Rottb.). Japanese Journal of Tropical Agriculture 50: 36–41.

Ehara, H., H. Naito, A. J. P. Tarimo, M. H. Bintsro and T. Y. Takamara (2006c) Introduction of sago palm seeds and seedlings into Tanzania. Sago Palm 14: 65–71.

Ehara, H., H. Shibata, H. Naito,T. Mishima and P. Ala (2007) Na+ and K+ concentrations in different plant parts and physiological features of *Metroxylon warburgii* Becc. under salt stress. Japanese Journal of Tropical Agriculture 51: 160–168.

Ehara, H., H. Shibata, W. Prathumyot, H. Naito and H. Miyake (2008a) Absorption and distribution of Na+, Cl- and some other ions and physiological characteristics of sago palm under salt stress. Tropical Agriculture and Development 52: 7–16.

Ehara, H., H. Shibata, W. Prathumyot, H. Naito, T. Mishima, M. Tuiwawa, A. Naikatini and I. Rounds (2008b) Absorption and distribution of Na+ and some ions in seedlings of *Metroxylon vitiense* H. Wendl. ex Benth. and Hook. F. under salt stress. Tropical Agriculture and Development 52: 17–26.

Elbeltagy, A., K. Nishioka, T. Sato, H. Suzuki, B. Ye, T. Hamada, T. Isawa, H. Mitsui and K. Minamisawa (2001) Endophytic colonization and in planta nitrogen fixation by a *Herbaspirillum* sp. isolated from wild rice species. Applied and Environmental Microbiology 67: 5285–5293.

Ellen, R. F. (1977) The Place of Sago in the Subsistence Economics of Seram. *In*: Sago '76: Papers of the 1st International Sago Symposium "The Equatorial Swamp as a Natural Resource". (Tan, K. ed.) University of Malaya (Kuala Lumpur) 105–111.

Ellen, R. F. (1979) Sago subsistence and the trade in spices: A provisional model of ecological succession and imbalance in Molluccan history. In: Social and Ecological Systems. (Burnham, P. C. and R. F. Ellen eds.) Academic Press (London) 43–74.

Ellen, R. F. (2004) Processing *Metroxylon sagu* Rottboell (Arecaceae) as a technological complex: A case study from south central Seram, Indonesia. Economic Botany 58: 50–74.

Ellen, R. F. (2006) Local knowledge and management of sago palm (*Metroxylon sagu* Rottboell) diversity in South Central Seram, Maluku, Eastern Indonesia. Journal of Ethnobiology 26: 258–298.

Epstein, J. A. and M. Lewin (1962) Kinetics of the oxidation of cotton with hypochlorite in the pH range 5–10. Journal of Polymer Science 58: 991–1008.

Evans, L. T. (1996) Crop Evolution, Adaptation and Yield. Cambridge University Press (New York) 288–289.

Ezaki, B., R. C. Gardner, Y. Ezaki and H. Matsumoto (2000) Expression of aluminum-induced genes in transgenic arabidopsis plants can ameliorate aluminum stress and/or oxidative stress. Plant Physiology 122: 657–666.

FAO (2002) FAOSTAT: http://faostat.org/site/567/default.aspx#ancor.

FAO (2006) World Reference Base for Soil Resources 2006: A Framework for International Classification, Correlation and Communication. FAO (Rome).

Felenstein, J. (2001) PHYILIP, ver. 3.6. University of Washington (Seattle).

Flach, M. (1977) Yield potential of the sago palm and its realization. *In*: Sago '76: Papers of the 1st International Sago Symposium "The Equatorial Swamp as a Natural Resource". (Tan, K. ed.) University of Malaya (Kuala Lumpur) 157–177.

Flach, M. (1980) Comparative ecology of the main moisture-rich starchy staples. *In*: Sago: The Equatorial Swamp as a Natural Resource (Proceedings of the 2nd International Sago Symposium). (Stanton, W. R. and M. Flach eds.) Martinus Nijhoff (London) 110–127.

Flach, M. (1981) Sago palm resources in the Northeastern part of the Sepik River Basin. Report of a survey. Energy Planning Unit, Dep. of Minerals & Energy (Konedobu) 5–11.

Flach, M. (1983) The Sago Palm. FAO Plant Production and Protection Paper 47, FAO (Rome).

Flach, M. (1984) The Sago Palm Domestication Exploitation and Products: FAO Sponsored Expert Consultation on the Sago Palm and Its Products. FAO (Rome).

Flach, M. (1997) Sago Palm, *Metroxylon sagu* Rottb. *In*: Promoting the Conservation and Use of Underutilized and Neglected Crops, 13. International Plant Genetic Resources Institute (Rome).

Flach, M., F. J. G. Cnoops and G. C. van Roekel-Jansen (1977) Tolerance to salinity and flooding of young sago palm seedlings. *In*: Sago '76: Papers of the 1st International Sago Symposium "The Equatorial Swamp as a Natural Resource". (Tan, K. ed.) University of Malaya (Kuala Lumpur) 190–195.

Flach, M., K. den Braber, M. J. J. Fredrix, E. M. Monster and G. A. M. van Hasselt (1986a) Temperature and relative humidity requirements of young sago palm seedlings. *In*: Sago '85: Proceedings of the 3rd International Sago Symposium. (Yamada, N. and K. Kainuma eds.) The Sago Palm Research Fund (Tokyo) 139–143.

Flach, M., D. W. G. van Kraalingen, and G. Simbardjo (1986b) Evaluation of present and potential production of natural sago palm stands. *In*: Sago '85: Proceedings of the 3rd International Sago Symposium. (Yamada, N. and K. Kainuma eds.) The Sago Palm Research Fund (Tokyo) 86–93.

Flach, M. and D. L. Schuiling (1989) Revival of an ancient starch crop: A review of the agronomy of the sago palm. Agroforestry Systems 7: 259–281.

Flach, M. and D. L. Schuiling (1991) Growth and yield of sago palms in relation to their nutritional needs. *In*: Towards Greater Advancement of the Sago Industry in the '90s: Proceedings of the 4th International Sago Symposium. (Ng, T. T., Y. L. Tie and H. S. Kueh eds.) Lee Ming Press (Kuching) 103–110.

Fong, S. S., A. J. Khan, M. Mohamed and A. M. Dos Mohamed (2005) The relationship between peat soil characteristics and the growth of sago palm (Metroxylon sagu). Sago Palm 13: 9–16.

Fujii, S., S. Kishihara and M. Komoto (1986a) Studies on improvement of sago starch quality. *In*: Sago '85: Proceedings of the 3rd International Sago Symposium. (Yamada, N. and K. Kainuma eds.) The Sago Palm Research Fund (Tokyo) 186–192.

Fujii, S., S. Kishihara, H. Tamaki and M. Komoto (1986b) Studies on improvement of quality of sago starch, part II: Effect of the manufacturing condition on the quality of sago starch. The Science Reports of Faculty of Agriculture, Kobe University 17: 97–106.

Fukui, H. (1984) Utilization of Southeast Asian low land swamps. Japanese Journal of Southeast Asian Studies 21: 409–436.

Funakawa, S., K. Yonebayashi, F. S. Jong and E. C. Oi-Khun (1996) Nutritional environment of tropical peat soils in Sarawak, Malaysia based on soil solution composition. Soil Science and Plant Nutrition 42: 833–843.

Funakoshi, H., N. Shiraishi, M. Norimoto, T. Aoki, H. Hayashi and T. Yokota (1979) Study on the thermoplasticization of wood. Holzforshung 33: 157–166.

Furukawa, H. (1986) Agricultural landscape of low land swamps on the Batang Hari River: Pt 2, development of agricultural landscape. Japanese Journal of Southeast Asian Studies 24: 65–105.

Furukawa, N., K. Inubushi, M. Ali, A. M. Itang and H. Tsuruta (2005) Effect of changing groundwater levels caused by land-use changes on greenhouse gas fluxes from tropical peat lands. Nutrient Cycling in Agroecosystems 71: 81–91.

Furukawa, J. and J. F. Ma (2006) Application of proteomics to the study of plant mineral stress 3. Impact of genomics on the study of plant nutrition. Japanese Journal of Soil Science and Plant Nutrition 77: 109–114. (in Japanese)

Girija, C., B. N. Smith and P. M. Swamy (2002) Interactive effects of sodium chloride and calcium chloride on the accumulations of praline and glycinebetaine in peanut (*Arachis hypogaea* L). Environmental and Experimental Botany 47: 1–10.

Google Ads (2009) Records of social data. http://www2.ttcn.ne.jp/honkawa/index.html (in Japanese)

Goto, Y., Y. Yamamoto, T. Yoshida, L. Hilary and F. S. Jong (1994) A cultivation physiological study of sago palm in Sarawak state, Report No. 4 vascular bundle distribution in pith cross section. Tropical Agriculture 38 (Suppl. 1): 37–38.

Goto, Y. (1996) Strike direction of vascular bundle in sago palm trunk). *In*: JSPS tropical bio-resources research fund project, Kochi University and Tohoku University field study report (Yamamoto, Y. ed.) 37–46.

Goto, Y., Y. Nitta, K. Kakuda, T. Yoshida and Y. Yamamoto (1998) Differentiation and growth of suckers in sago palms (*Metroxylon sagus* Rottb.) Japanese Journal of Crop Science 67 (Suppl. 1): 212–213.

Goto, Y. and S. Nakamura (2004) Forms of sago palm leaf. Sago Palm 12: 24–27.

Groves, M. (1972) Hiri. *In*: Encyclopaedia of Papua and New Guinea, Vol. 1. (Ryan, P. ed.) Melbourne University Press (Carlton) 523–527.

Hagley, E. A. C. (1965) On the life history and habits of the palm weevil *Rhynchophorus palmarum*. Annals of the Entomological Society of America 58: 22–28.

Haji, A., K. Inubushi, Y. Furukawa, E. Purnomo, M. Rasmadi and H. Tsuruta (2005) Greenhouse gas emissions from tropical peatlands of Kalimantan, Indonesia. Nutrient Cycling in Agroecosystems 71: 73–80.

Hallett, R. H., G. Gries, R. Gries, J. H. Borden, E. Czyzewska, A.C.Oehlschlager, H.D. Pierce, N. P. D. Angerilli and A. Rauf (1993) Aggregation pheromones of two Asian palm weevils, *Rhynchophorus ferrugineus* and *R. vulneratus*. Naturwissenschaften 80: 328–331.

Hamanishi, T., T. Hatta, F. S. Jong, S. Takahashi and K. Kainuma (1999) Physicochemical properties of starches obtained from various parts of sago palm trunks at different growth starges. Journal of Applied Glycoscience 46: 39–48. (in Japanese)

Hamanishi, T., T. Hatta, F. S. Jong, K. Kainuma and S. Takahashi (2000) The relative crystallinity, structure and gelatinization properties of sago starches at different growth stages. Journal of Applied Glycoscience 47: 335–341.

Hamanishi, T. (2002) Studies on physicochemical properties of sago starch obtained at different growth stages of the trunk and cooking characteristics of sago starch. Kyoritsu Women's University Ph. D. thesis, 73, 95. (in Japanese)

Hamanishi, T., K. Hirao, Y. Nishizawa, H. Sorimachi, K. Kainuma and S. Takahashi (2002a) Physicochemical properties of sago starch compared with various commercial starches. In: New Frontiers of Sago Palm Studies: Proceedings of the International Symposium on Sago (Kainuma, K., M. Okazaki, Y. Toyoda and J. E. Cecil eds.) Universal Academy Press (Tokyo) 289–292.

Hamanishi, T., N. Matsunaga, K. Hirao, K. Kainuma and S. Takahashi (2002b) Cooking and processing properties of the traditional Japanese confection, kudzumushbiyokan, made from sago starch. Journal of Cookery Science of Japan 35: 287–296. (in Japanese)

Hamanishi, T., N. Matsunaga, K. Hirao, K. Kainuma and S. Takahashi (2002c) The cooking and processing properties of Japanese traditional confectionery made of sago starch: Effect of addition of trehalose and silk fibroin. In: New Frontiers of Sago Palm Studies: Proceedings of the International Symposium on Sago (Kainuma, K., M. Okazaki, Y. Toyoda and J. E. Cecil eds.) Universal Academy Press (Tokyo) 261–264.

Hamanishi, T., K. Hirao, A. Miyazaki, Petrus, F. S. Jong, Y. Yamamoto, T. Yoshida and S. Takahashi (2006) Physicochemical properties of the starches extracted from the sago palm varieties. Proceedings of the 15th Conference of the Society of Sago Palm Studies, 13–16. (in Japanese)

Hamanishi, T., K. Hirao, A. Miyazaki, Y. Yamamoto, T. Yoshida and S. Takahashi (2007) Properties and classification of the starches extracted from the sago palm varieties. Proceedings of the 16th Conference of the Society of Sago Palm Studies, 29–32. (in Japanese)

Haryanto, B. and P. Budhi (1987) Budidaya dan Pengorahan. Penerbt Kanisius (Jogjakarta). (in Indonesian)

Haryanto, B. and P. Pangloli (1994) Potensi dan pemanfaatan sagu. Penerbt Kanisius (Jogjakarta). (in Indonesia)

Haryanto, B. and Suharjito (1996) Model perkebuan inti Rkyat (PIR) sebagai salah satu altanatif Pengembangan sagu. Prosiding Simposium National Sagu III, Dekan Baru. (in Indonesian)

Hashimoto, K., Y. Sasaki, K. Kakuda, A. Watanabe, F. S. Jong and H. Ando (2006) Relationship of sago palm and groundwater in tropical peat soil. Proceeding of the 15th Conference of the Society of Sago Palm Studies, 25–26. (in Japanese)

Haska, N. (2001) Comparison of productivity and properties of the starches from several tropical palms. *In*: Diversity and Optimum Utilization of Biological Resources in the Torrid and Subtropical Zones: Proceedings of the International Sago Symposium (in honor of Prof. Ayaaki Ishizaki's retirement). (Ogata, S., K. Furukawa, K. Sonomoto and G. Kobayashi eds.) Kyushu University Press (Fukuoka) 13–16.

Hassan, A. H. (2002) Agronomic practice in cultivating the sago palm, *Metroxylon sagu* Rottb.: the Sarawak experience. I. *In*: New Frontiers of Sago Palm Studies: Proceedings of the International Symposium on Sago (Kainuma, K., M. Okazaki, Y. Toyoda and J. E. Cecil eds.) Universal Academy Press (Tokyo) 3–7.

Hatta, T., S. Nemoto, T. Hamanishi, K. Yamamoto, S. Takahashi and K. Kainuma (2002) Uppermost surface structure of sago starch granules. *In*: New Frontiers of Sago Palm Studies: Proceedings of the International Symposium on Sago (Kainuma, K., M. Okazaki, Y. Toyoda and J. E. Cecil eds.) Universal Academy Press (Tokyo) 349–354.

Henanto, H. (1992) Sago palm distribution in Irian Jaya Province. Symposium Sagu Nasional (Ambon), 12–13 October.

Hill, R. D. (1977) Rice in Malaya: A Study in Historical Geography. Oxford University Press (Oxford).

Hirao, K., I. Nishioka and S. Takahashi (1989) Studies on the cooking of pearl starch. Part 1. Cooking conditions of tapioca pearls. Journal of Home Economics of Japan 40: 363–371. (in Japanese)

Hirao, K. and S. Takahashi (1996) Studies on cooking of pearl-type starch. Part 3. Cooking method of sago pearls. Sago Palm 4: 14–20. (in Japanese)

Hirao, K., Y. Igarashi and S. Takahashi (1998) Cooking and processing quality of sago starch gel. Part 3. Effects of ingredients ratio of soybean protein isolate and soybean oil on the rheological properties of starch gel. Sago Palm 6: 1–9. (in Japanese)

Hirao, K. (2001) Studies on improvement of serum and liver lipids by sago starch and its physicochemical properties. Iwate University Graduate School Ph. D. thesis, 3–5. (in Japanese)

Hirao, K., T. Hamanishi, Y. Igarashi and S. Takahashi (2002) Effect of the ingredient ratio of sago starch, soybean protein isolate and soybean oil on the physical properties and sensory attributes of blancmange. Journal of Home Economics of Japan 53: 659–669. (in Japanese)

Hirao, K., T. Watanabe and S. Takahashi (2003) Effects of added soybean protein isolate and soymilk powder on the physical properties and sensory evaluation of a blancmange type of starch gel (Part 2) Effects of adding cocoa and powdered green tea. Journal of Home Economics of Japan 54: 469–476 (in Japanese)

Hirao, K., K. Kainuma and S. Takahashi (2004a) Effects of addition of silkfibroin gel on gelatinized properties of starch. Proceedings of the 2004 Conference of the Japanese Society of Applied Glycoscience, 21. (in Japanese)

Hirao, K., K. Kanamori, Y. Yoneyama and S. Takahashi (2004b) Physical and sensory properties of biscuits containing partially replaced sago starch. Journal of Home Economics of Japan 55: 715–723. (in Japanese)

Hirao, K., F. Takei, Y. Yoneyama and S. Takahashi (2005) Effect of adding egg yolk powder on the physiochemical properties of sago starch. Journal of Home Economics of Japan 56: 49–54. (in Japanese)

Hirao, K., T. Hamanishi, H. Sorimachi, Y. Yamamoto, A. Miyazaki, F. S. Jong, T. Yoshida and S. Takahashi (2006) Physical properties and utilization of the sago starches extracted from 8 varieties at harvesting stage. Journal of Applied Glycoscience 53: 49.

Hirao, K., H. Tanaka, S. Konoo, T. Hamanishi and S. Takahashi (2008) The present condition of sago starch using method and processed foods in Riau, Indonesia. Proceedings of the 17th Conference of the Society of Sago Palm Studies, 75–78. (in Japanese)

Horiuchi, M. (1998) Papua New Guinea team dispatch plan. Highland aquaculture development project JICA summary report 1–7.

Ichige, H. and M. Ishikawa (1984) The investigation of food words that old encyclopedias contain. Part 2. The Words of various potatoes, starch, seeds, beans, vegetables, fruits, mushrooms and seaweeds. Journal of Home Economics of Japan 35: 736–746. (in Japanese)

Igura, M., M. Okazaki, S. D. Kimura, K. Toyota, M. Ohmi, T. Kuwabara, and M. Syuno (2007) Tensile strength characteristics of biodegradable plastics made from sago starch extraction residue. Sago Palm 15: 1–8.

IPCC (Intergovernmental Panel on Climate Change) (2007) Climate Change 2007: The Physical Scientific Basis. (Solomon, S., D. Qin, M. Manning, M. Marquis, K. Averyt, M. M. B. Tignor, H. Leroy Miller, Jr. and Z. Chen eds.) Cambridge University Press.

Irawan, A. F., H. M. Bintro, Y. Yamamoto, K. Saitoh and F. S. Jong (2005) Effects of sucker weight on the vegetative growth of sago palm (*Metroxylon sagu* Rottb.) during the nursery period. Shikoku Journal of Crop Science 42: 44–45.

Irawan, A. F., Y. Yamamoto, A. Miyazaki and F. S. Jong (2009a) Characteristics of suckers from sago palm (*Metroxylon sagu* Rottb.) grown in different soil types in Tebing Tinggi Island, Riau, Indonesia. Tropical Agriculture and Development 53: 103–111.

Irawan, A. F., Y. Yamamoto, A. Miyazaki, T. Yoshida and F. S. Jong (2009b) Characteristics of sago palm (*Metroxylon sagu* Rottb.) suckers from various mother palms at different growth stages in Tebing Tinggi Island, Riau, Indonesia. Tropical Agriculture and Development 53: 1–6.

Ishizaki, F., K. Sonomoto, G. Kobayashi, S. Sirisansaneeyakul, S. Karnchatawee, S. Radtong, C. Siripatana, D. Uttapap, S. Tripetchkul, P. Mekvichitsaeng and K. Bujang (2002) Microbial conversion of tropical plant biomass into chemical industrial materials and new biofuel production: International joint research grant project (NEDO grant) output report (Ref. 99GP1), New Energy and Industrial Technology Development Organization. (in Japanese)

Istalaksana, P., Y. Gandhi, P. Hadi, A. Rochani, K. Mbaubedari and S. Bachri (2006) Conversion of natural sago forest into sustainable sago plantation at Masirei district, Waropen, Papua, Indonesia: Feasibility study. *In*: Sago Palm Development and Utilization: Proceedings of the 8th International Sago Symposium. (Karafir, Y. P., F. S. Jong and V. E. Fere eds.) Universitas Negeri Papua Press (Manokwari) 65–77.

Iuchi, S., H. Koyama, A. Iuchi, Y. Kobayashi, S. Kitabayashi, Y. Kobayashi, T. Ikka, T. Hirayama, K. Shinozaki and M. Kobayashi (2007) Zinc finger protein STOP1 is critical for proton tolerance in Arabidopsis and coregulates a key gene in aluminum tolerance. Proceedings of the National Academy of Science of the United States of America 104: 9900–9905.

Jabatan Pertanian Negri Johor (1994) Statistik Pertanian Negeri Johor *1994*. Batu Pahat: Jabatan Pertanian Negri Johor. (in Indonesian)

Jalil, M., N. Hj and J. Bahari (1991) The performance of sago palms on river alluvial clay soils of Peninsular Malaysia. *In*: Towards Greater Advancement of the Sago Industry in the '90s: Proceedings of the 4th International Sago Symposium. (Ng, T. T., Y. L. Tie and H. S. Kueh eds.) Lee Ming Press (Kuching) 114–121.

Jaman, O. H. (1985) The study of sago seed germination. Proceedings of the 22nd Research Officers' Conference, Department of Agriculture, Sarawak, Malaysia (Kuching) 69–78.

Jane, J., T. Kasemsuwan, S. Leas, H. Zobel and F. Robyt (1994) Anthology of starch granule morphology by scanning electron microscopy. Starch 46: 121–129.

Japan Bioplastic Association (2006a) Types and characteristics. *In*: Introduction to biodegradable plastic technology. Omu Sha (Tokyo) 5–18. (In Japanese)

Japan Bioplastic Association (2006b) Manufacturing and synthesizing methods. *In*: Introduction to biodegradable plastic technology. Omu Sha (Tokyo) 45–47. (In Japanese)

Japan Food Industry Association (1991) Research project report on demand for food processing development (sago starch industry) in Indonesia. Shokuhin Sangyo Senta (Tokyo).

Japan-Papua New Guinea Goodwill Society (1984) Report of research on utilization of tropical plant resources in Papua New Guinea. (in Japanese)

Japan-Papua New Guinea Goodwill Society (1985) Report of research on utilization of tropical plant resources in Papua New Guinea. (in Japanese)

Jensen, Ad. E. (1963) Prometheus- und Hainuvere-Mythologem. Anthropos 58: 145–186.

Jensen, A. ([1966] 1974) Slaughtered goddess. (Anthropology seminar 2), translated by T. Obayashi et al., Kobundo (Tokyo). (in Japanese)

JICA (1981a) Report on primary basic research for palms development cooperation in the Malay Peninsula, Malaysia. JICA (Tokyo).

JICA (1981b) Report on primary basic research for sago palms development cooperation in Malaysia (Sabah state) and Indonesia (South Kalimantan province). JICA (Tokyo).

JICA (1981c) Report on primary basic research for sago palms development cooperation in Papua New Guinea. JICA (Tokyo).

Johnson, R. M. and W. D. Raymond (1956) Sources of starch in colonial territories I: The sago palm. Colonial Plant and Animal Products 6: 20–32.

Jones, D. L. (1995) Palms throughout the World. Smithsonian Institution Press (Washington D. C.).

Jong, F. S. (1991) A preliminary study on the phyllotaxy of sago palms in Sarawak. *In*: Towards Greater Advancement of the Sago Industry in the '90s: Proceedings of the 4th International Sago Symposium. (Ng, T. T., Y. L. Tie and H. S. Kueh eds.) Lee Ming Press (Kuching) 69–73.

Jong, F. S. (1995a) Research for the development of sago palm (*Metroxylon sagu* Rottb.) cultivation in Sarawak, Malaysia. Ph. D. thesis of Agricultural University, Wageningen, The Netherlands.

Jong, F. S. (1995b) Distribution and variation in the starch content of sago palm (*Metroxylon sagu* Rottb.) at different growth stages. Sago Palm 3: 45–54.

Jong, F. S. and M. Flach (1995) The sustainability of sago palm (*Meteroxylon sagu*) cultivation on deep peat in Sarawak. Sago Palm 3: 13–20.

Jong F. S. (2001) Sago production in Tebinggi Sub-district, Riau, Indonesia. Sago Palm 9: 9–15.

Jong, F. S. (2002a) The rehabilitation of natural sago forest as sustainable sago plantations: A shortcut to sago plantations. *In*: New Frontiers of Sago Palm Studies: Proceedings of the International Symposium on Sago. (Kainuma, K., M. Okazaki, Y. Toyoda and J. E. Cecil eds.) Universal Academy Press (Tokyo) 61–67.

Jong, F. S. (2002b) Commercial sago palm cultivation on deep peat in Riau, Indonesia. *In*: New Frontiers of Sago Palm Studies: Proceedings of the International Symposium on Sago. (Kainuma, K., M. Okazaki, Y. Toyoda and J. E. Cecil eds.) Universal Academy Press (Tokyo) 251–254.

Jong, F. S. (2006) Technical recommendations for the establishment of a commercial sago palm (*Meteroxylon sagu* Rottb.) plantation. Japanese Journal of Tropical Agriculture 50: 224–228.

Jong, F. S., A. Watanabe, D. Hirabayashi, S. Matsuda, B. H. Puruwanto, K. Kakuda and H. Ando (2006) Growth performance of sago palms (*Metroxylon sagu* Rottb.) in peat of different depth and soil water table. Sago Palm 14: 59–64.

Jong, F. S., A. Watanabe, Y. Sasaki, K. Kakuda and H. Ando (2007) A study on the growth response of young sago palms to the omission of N, P, and K in culture solution. *In*: Sago: Its Potential in Food and Industry: Proceedings of the 9th International Sago Symposium. (Toyoda, Y., M. Okazaki, M. Quevedo and J. Bacusmo eds.) TUAT Press (Tokyo) 103–112.

Josue, A. R. and Okazaki, M. (1988) Stands of sago palms in northern Mindanao, Philippines. Sago Palm 6: 24–27.

Jourdan, C. and H. Rey (1997) Architecture and development of the oil-palm (*Elaeis guineensis* Jacq.) root system. Plant and Soil 189: 33–48.

Kaberry, P. M. (1941–2) The Abelam tribe, Sepik district, New Guinea: A preliminary report. Oceania 11: 233–258, 345–367.

Kagawa, Y. (2001) Standard Tables of Food Composition in Japan (5th edition). Jyosieiyouidaigaku Syuppan (Tokyo). (in Japanese)

Kainuma, K., T. Oda, H. Fukino, K. Tanida and S. Suzuki (1968) Tracing the gelatinization phenomena of starch granule by photopastegraphy: Part. 2, Determination of gelatinization temperature of starch granules by photopastegraphy. Journal of the Japanese Society of Starch Science 16: 54–60. (in Japanese)

Kainuma, K. (1977a) Present status of starch utilization in Japan. *In*: Sago '76: Papers of the 1st International Sago Symposium "The Equatorial Swamp as a Natural Resource". (Tan, K. ed.) University of Malaya (Kuala Lumpur) 224–239.

Kainuma, K. (1977b) The outline and future outlook of the 1st International Sago Symposium. The Food Industry July: 37–42. (in Japanese)

Kainuma, K. (1981) Sago palm in Borneo Island and Papua New Genea: Part 1. Basic studies in Borneo Island. Monthly report of the National Food Research Institute 135: 5–12. (in Japanese).

Kainuma, K., A. Matsunaga, M. Itakawa and S. Kobayashi (1981) New Enzyme System -β-amylase and-pullulnase –to determine the degree of gelatinization and retrogradation of starches or starch foods. Journal of the Japanese Society of Starch Science 28: 235–240. (in Japanese)

Kainuma, K. (1982) Utilization of sago palms in Sarawak, South Kalimantan and Papua New Guinea. Japanese Journal of Tropical Agriculture 26: 177–186.

Kainuma, K., H. Ishigami and S. Kobayashi (1985) Isolation of a novel raw starch digesting amylase from a strain of black mold – *Chalara paradoxa*. Journal of the Japanese Society of Starch Science 32: 136–141.

Kainuma, K. (1986a) *Chalara paradoxa* raw starch digesting amylase obtained from sago palm. *In*: Sago '85: Proceedings of the 3rd International Sago Symposium. (Yamada, N. and K. Kainuma eds.) The Sago Palm Research Fund (Tokyo) 217– 222.

Kainuma, K. (1986b) Cation-exchanged starch. In Methods in starch and related carbohydrates. (Nakamura, M. and K. Kainuma eds.) Gakkai Shuppan Center (Tokyo) 284–287. (in Japanese)

Kakuda, K., H. Ando, T. Yoshida, Y. Yamamoto, Y. Nitta, H. Ehara, Y. Goto and B. H. Puruwanto (2000) Soil characteristics in sago plam grown area; Factors associated with fate of inorganic nitrogen in soil. Sago Palm 8: 9-16. (in Japanese)

Kakuda, K., A. Watanabe, H. Ando and F. S. Jong (2005) Effects of fertilizer application on the root and aboveground biomass of sago palm (*Metroxylon sagu* Rottb.) cultivated in peat soil. Japanese Journal of Tropical Agriculture 49: 264–269.

Kamimura, T. (1998) Folklore on sago utilization: 'Sago adaptation' in the upper Karawari East Sepik province, Papua New Guinea. Sago Palm 6: 10–23. (in Japanese)

Kaneko, T., M. Okazaki, N. Kasai, C. Yamaguchi and A. H. Hassan (1996) Growth and biomass of sago palm (*Metroxylon sagu*) on shallow peat soils of Dalat District, Sarawak. Sago Palm 4: 21–24.

Kaplan, D. R., N. G. Dengler and R. E. Dengler (1982) The mechanism of plication inception in palm leaves: Histogenetic observations on the pinnate leaf of *Chrysalidocarpus lutescens*. Canadian Journal of Botany 60: 2976–2998.

Kasuya, N. (1996) Sago root studies in peat soil of Sarawak. Sago Palm 4: 6–13.

Kato, K. (2002) Chemical stress. *In*: Encyclopedia of Plant Nutrition and fertilizer. Asakura Shoten (Tokyo) 290–294. (in Japanese)

Kawahigashi, M., H. Sumida, K. Yamamoto, H. Tanaka and C. Kumada (2003) Chemical properties of tropical peat soils and peat soil solutions in sago palm plantation. Sago Palm 10: 55–63.

Kawakami, I. (1975) Optical and electron microscopic images: Starch morphology. Ishiyaku Shuppan (Tokyo). (in Japanese)

Kawasaki, M. (1999) A histological and cytological study of reserve synthesis and accumulation in vegetative storage organs in root and tuber crops, Ph. D. thesis of United Graduate School of Agricultural Science, Tokyo University of Agriculture and Technology. (in Japanese)

Kawasaki, M., T. Matsuda and Y. Nitta (1999) Electron microscopy of plastid-amyloplast system involved in starch synthesis and accumulation in potato tuber (*Solanum tuberosum* L.). Japanese Journal of Crop Science 68: 266–274. (in Japanese)

Kelvim, L. E. T., Y. L. Tie and S. C. Y. Patricia (1991) Starch yield determination of sago palm: A comparative study. *In*: Towards Greater Advancement of the Sago Industry in the '90s: Proceedings of the 4th International Sago Symposium. (Ng, T. T., Y. L. Tie and H. S. Kueh eds.) Lee Ming Press (Kuching) 137–141.

Kertopermono, A. P. (1996) Inventory and evaluation of sago palm (*Metroxylon* sp.). *In*: Sago: The Future Source of Food and Feed: Proceedings of the 6th International Sago Symposium. (Jose Chistine, C. and A. Rasyad eds.) Riau University Training Center (Pekanbaru) 52–62.

Kiew, R. (1977) Taxonomy, ecology and biology of sago palms in Malaysia and Sarawak. *In*: Sago '76: Papers of the 1st International Sago Symposium "The Equatorial Swamp as a Natural Resource". (Tan, K. ed.) University of Malaya (Kuala Lumpur) 147–154.

Kiguchi, M., (1990) Chemical modification of wood surfaces by etherification I. Manufacture of surface hot-melted wood by etherification. Mokuzai gakkaishi 36: 651–658. (in Japanese)

Kimura, N. (1979) Pests of sago palm and their control. Japanese Journal of Tropical Agriculture 23: 142–148. (in Japanese)

Kimura, S. D. and M. Okazaki (2006a) Sago and Taro Growth and Production in the Sago/Taro Intercropping Systems of Leyte with Special Reference to Nitrogen. Tokyo University of Agriculture and Technology.

Kimura, S. D. and M. Okazaki (2006b) Sago Growth in Sago-Taro Intercropping System and Sago Starch Extraction of Two Sago Varieties. Tokyo University of Agriculture and Technology.

Konoo, S., T. Yamamoto, Y. Kurashige and H. Utushibara (1997) Elasticity of fish cake added with sago starch to Alaska pollock fish paste. Journal of Seafood Paste Product Technology Research 22: 203–212. (in Japanese)

Konoo, S. (1998) Studies on the application of chemically modified starches to seafood paste products. National Fisheries University Graduate School Ph. D. thesis, 43–56.

Kjar, A. S., A. S. Barfod, C. B. Asumussen and O. Seberg (2004) Investigation of genetic and morphological variation in the sago palm (*Metroxylon sagu*; Arecaceae) in Papua New Guinea. Annals of Botany 94: 109–117.

Koizumi, T. (2006) U.S. ethanol policy: Impacts on corn market. Journal of Agricultural Policy Research 11: 53–72. (in Japanese)

Komada, C., H. Ehara, O. Morita and M. Goto (1998) External and internal factors affecting seed germination properties of sago palm. Newsletter of the Tokai branch of the Crop Science Society of Japan125: 13–14. (in Japanese)

Koyama, H. (2002) Tolerance mechanisms. In Encyclopedia of Plant Nutrition and fertilizer. Asakura Shoten (Tokyo) 337–341. (in Japanese)

Kueh, H. S., Y. L. Tie, E. Robert, C. M. Ung and Hj. Osman (1991) The feasibility of plantation production of sago (*Metroxylon sagu*) on an organic soil in Sarawak. *In*: Towards Greater Advancement of the Sago Industry in the '90s: Proceedings of the 4th International Sago Symposium. (Ng, T. T., Y. L. Tie and H. S. Kueh eds.) Lee Ming Press (Kuching) 127–136.

Kueh, H. S. (1995) The effects of soil applied NPK fertilizers on the growth of the sago palm (*Metroxylon sagu*, Rottb.) on undrained deep peat. Acta Horticulturae 389: 67–76.

Kyuma, K. (1986a) Soils developed from sediments in mangroves: Acid sulfate soil. In Lowland swamps in Southeast Asia. (Tropical Agricultural Research Center ed.) Norin Tokei Kyokai (Tokyo) 56–79. (in Japanese)

Kyuma, K. (1986b) Organic soils in wetland forests: Tropical peat soils. In Lowland swamps in Southeast Asia. (Tropical Agricultural Research Center ed.) Norin Tokei Kyokai (Tokyo) 79–103. (in Japanese)

Laiho, R., T. Sallantaus and J. Laine (1999) The effect of forestry drainage on vertical distributions of major plant nutrients in peat soils. Plant and Soil 207: 169–181.

Land Custody and Development Authority (2009) Sago development. http://www.pelita.gov.my/ sago_development.html.

Luhulima, F., S. A. Karyono, Y. Abdullah and D. Dampa (2006) Feasibility study of natural sago forest for the establishment of commercial sago plantations in South Sorong, Irian Jaya Barat, Indonesia. *In*: Sago Palm Development and Utilization: Proceeding of the 8th International Sago Symposium. (Karafir, Y. P., F. S. Jong and V. E. Fere eds.) Universitas Negeri Papua Press (Manokwari) 7–64.

Ma, J. F., P. R. Ryan and E. Delhaize (2001) Aluminum tolerance in plants and the complexing role of organic acids. Trends in Plant Science 6: 273–278.

Maamun, Y. and I. G. P. Sarasutha (1987) Prospects for sago palm in Indonesia: South Sulawesi case study. Indonesion Agricultural Research and Development Journal 9: 52–56.

Maeda, K., Y. Yamamoto and N. Uchida (1992) Kochi University and Kobe University sago palm field study report, 1991 JSPS tropical bio-resources research fund research output report. (in Japanese)

Maeda, K. (1998) Tropical starch resource crop: cassava and sago palm, Japanese Journal of Tropical Agriculture 42 (Ext 2): 75–80. (in Japanese)

MAFF (1991) Technologies for efficient large scale production of useful microbes and enzymes. *In*: Biomass conversion project. Korin (Tokyo) 493–517. (in Japanese)

MAFF (2002) Comprehensive list of starch demand and supply. Ministry of Agriculture, Forestry and Fishery. (in Japanese)

MAFF (2009), 'Baiomasu Nippon sōgō senryaku (National biomass strategy)',http://www.maff.go.jp/j/biomass/index.html, http://www.maff.go.jp/j/biomass/b_energy/pdf/kakudai01.pdf.

Magat, S. S. and R. Z. Margate (1988) The Nutritional Deficiencies and Fertilization of Coconut in the Philippines. Philippine Coconut Authority, R & D Techn. (Report No.2).

Manan, S and S. Supangkat (1984) Management of sago forest in Indonesia. *In*: The Development of the Sago Palm and its Products. Report of the FAO/BPPT Teknologi Consultation (Jakarta).

Martikainen, P. J., H. Nykanen, J. Alm and J. Silvola (1995) Changes in fluxes of carbon dioxide, methane and nitrous oxide due to forest drainage of mire sites of different trophy. Plant and Soil 168/169: 571–577.

Maruyama, T., N. Suzuki, Y. Kawachi, S. Fujimaki, T. Tanno, E. Miwa and K. Higuchi (2008) Kinetics analysis of suppression mechanism in the translocation of Na+ to aboveground at the shoot base of reed grass. Abstract of the 54th annual meeting of Japanese Society of Soil Science and Plant Nutrition, 82.

Masuda, M. (1991) Past, present and future of sago as foods. Sago Study 2: 6–7. (in Japanese)

Matanubun, H. (2004) Diversity of sago palm based on taxonomy in Central Sentani District, Jayapura Regency, Papua Province, Indonesia. Abstract of the Sixth New Guinea Biology Conference, The State University of Papua, Manokwari, Indonesia.

Matanubun, H. and L. Maturbongs (2006) Sago palm potential, biodiversity socio-cultural considerations for industrial sago development in Papua, Indonesia. *In*: Sago Palm Development and Utilization: Proceedings of the 8th International Sago Symposium. (Karafir, Y. P., F. S. Jong and V. E. Fere eds.) Universitas Negeri Papua Press (Manokwari) 41–54.

Matoh, T. (1991) Salt-tolerance mechanisms of higher plants. Plant Cell Technology 3: 268–272. (in Japanese)

Matoh, T. (1999) What are halophytes. Iden 53: 54–57. (in Japanese)

Matoh, T. (2000) 1. Response of crops to salt stress – its development mechanism, tolerance and acclimatization: Salt tolerance mechanism of hylophytes. Agriculture and Horticulture 75: 783–786. (in Japanese)

Matoh, T. (2002) Salt stress. *In*: Encyclopedia of Plant Nutrition and Fertilizer. Asakura Shoten (Tokyo) 319–321. (in Japanese)

Matoh, T., J. Watanabe and E. Takahashi (1987) Sodium, potassium, chloride, and betaine concentrations in isolated vacuoles from salt-grown Atriplex gmelini leaves. Plant Physiology 84: 173–177.

Matsumoto, M., M. Osaki, T. Nuyim, A. Jongskul, P. Eam-On, Y. Kitaya, M. Urayama, T. Watanabe, T. Kawamura, T. Nakamura, C. Nilnond, T. Shinano and T. Tadano (1998) Nutritional characteristics of sago palm and oil palm in tropical peat soil. Journal of Plant Nutrition 21: 1819–1841.

McClatchey, W. C. (1998) A new species of *Metroxylon* (Arecaceae) from Western Samoa. Novon 8: 252–258.

McClatchey, W. C. (1999) Phylogenetic analysis of morphological characteristics of *Metroxylon* section *Coelococcus* (Palmae) and resulting implications for studies of other Calamoideae Genera. Memoirs of the New York Botanical Garden 83: 285–306.

McClatchey, W. C., H. I. Manner and C. R. Elevitch (2006) *Metroxylon amicarum, M. paulcoxii, M. sagu, M. salomonense, M. vitiense*, and, *M. warburgii* (sago palm), Arecaceae (palm family). (http://www.agroforestry. net/tti/Metroxylon-sagopalm.pdf).

McElhanon, K. A. (ed.) (1974) Legends from Papua New Guinea. Ukarumpa, Summer Institute of Linguistics (Papua New Guinea).

Melling, L., R. Hatano and K. J. Goh (2005a) Soil CO2 flux from three ecosystems in tropical peatland of Sarawak, Malaysia. Tellus 57B: 1–11.

Melling, L., R. Hatano and K. J. Goh (2005b) Methane fluxes from three ecosystems in tropical peatland of Sarawak, Malaysia. Soil Biology and Biochemistry 37: 1445–1453.

Meyer-Rochow, V. B. (1973) Edible insects in three different ethnic groups of Papua and New Guinea. The American Journal of Clinical Nutrition 26: 673–677.

Mihalic, F. (1971) The Jacaranda Dictionary and Grammar of Melanesian Pidgin. The Jacaranda Press (Port Moresby).

Mikuni, K., M. Monma and K. Kainuma (1987) Alcohol fermentation of corn starch digested by Chalara *paradoxa* amylase without cooking. Biotechnology and Bioengineering 29: 729–732.

Ministry of Economy, Trade and Industry (2005 )Yearbook of paper, printing, plastics products and rubber products statistics. Research and Statistics Department, Economic and Industrial Policy Bureau, Ministry of Economy, Trade and Industry 112–113.

Mitsuhashi, J. and H. Sato (1994) Investigation on the edible sago weevils in Papua New Guinea. Sago Palm 2: 13–20.

Mitsuhashi, J. and S. Kawai (1999) A study of pests in sago palm plantations (A preliminary). Proceeding of the 8th conference of the Society of Sago Palm Studies, 1–4.

Miyamoto, E., S. Matusda, H. Ando, K. Kakuda, F. S. Jong and A. Watanabe (2009) Effect of sago palm (*Metroxylon sagu* Rottb.) cultivation on the chemical properties of soil and water in tropical peat soil ecosystem. Nutrient Cycling in Agroecosystems 85: 157–167.

Miyazaki, A., T. Yoshida, Y. Chinen, S. Hamada, Y. Yamamoto, Y. B. Pasolon and F. S. Jong (2003) Changes of root amount by age in sago palm (*Metroxylon sagu* Rottb.). Proceedings of the 12th Conference of the Society for the Studies of Sago Palm Studies, 5–10. (in Japanese)

Miyazaki, A., F. S. Jong, Petrus, Y. Yamamoto, T. Yoshida, Y. B. Pasolon, H. Matanubun, F. S. Rembon and J. Limbongan (2006) Starch extraction in several sago palm varieties grown near Jayapura, Papua State, Indonesia. Proceedings of the 15th Conference of the Society of Sago Palm Studies, 9–12. (in Japanese)

Miyazaki, A., Y. Yamamoto, K. Omori, H. Pranamuda, R. S. Gusti, Y. B. Pasolon and J. Limbongan (2007) Leaf photosynthetic rate in sago palms (*Metroxylon sagu* Rottb.) grown under field conditions in Indonesia. Japanese Journal of Tropical Agriculture 51: 54–58.

Miyazaki, A., T. Yoshida, I. Yanagidate, Y. Chinen, S. Hamada, Y. Yamamoto, Y. B. Pasolon, F. S. Rembon and F. S. Jong (2008) Root development with age in sago palm (*Metroxylon sagu* Rottb.) by trench method. Abstract of the 103rd Academic Meeting of Japanese Society for Tropical Agriculture 1 (Ext 1): 23–24 (in Japanese)

Miyazaki, T. (1999) Cooking characteristics of sago starch: Its physical properties by sago starch types and the breaking properties of sago noodles. Kyoritsu Women's University Faculty of Home Economics 1999 graduation thesis.

Mizuma, S., Y. Nitta, T. Matsuda, Y. Yamamoto, T. Yoshida and A. Miyazaki (2007) Starch accumulation of sago palm grown around lake Sentani, near Jayapura of the Papua province, Indonesia. Japanese Journal of Crop Science 76 (Ext 1): 360–361.

Monma, M., Y. Yamamoto, N. Kagei and K. Kainuma (1989) Raw starch digestion by alfa-amylase and glucoamylase from Chalara paradoxa. Staerke/ Starch 41: 382–385.

Morimoto, H. (1979) Grains. *In*: Animal Feed Science (Morimoto, H. ed.) Yohkendo (Tokyo) 81–105. (in Japanese).

Munns, R. (2001) Avenues for increasing salt tolerance of crops. *In*: Plant Nutrition: Food Security and Sustainability of Agro-Ecosystems. (Horst, W. J., M. K. Schenk, A. Burkert, N. Claassen, H. Flessa, W. B. Frommer, H. E. Goldbach, H. -W. Olfs, V. Romheld, B. Sattelmacher, U. Schmidhalter, S. Schubert, N. von Wiren, L. Wittenmayer eds.) Kluwer Academic Publishers (Dordrecht) 370–371.

Murayama, S. (1995) Degradation of tropical peat in Malaysia. Expert bulletin for International Cooperation of Agriculture and Forestry 15: 13–33. (in Japanese)

Nadhry, N., M. Igura and M. Okazaki (2009) Conversion of glucans in sago starch extraction residue to ethanol: Saccharification by acids and enzymes. Proceedings of the 18th Conference of the Society of Sago Palm Studies, 6–12.

Nagato, I. and H. Shimoda (1979) The present state of sago production and its future. Japanese Journal of Tropical Agriculture 23: 160–168. (in Japanese)

Naito, H., H. Ehara and T. Mizota (2000) Production ecology and genetic background of sago palm in Indonesia. 1999 JSPS tropical bio-resources research fund research output report. (in Japanese)

Nakamura, S., Y. Goto and Y. Nitta (2000) Morphology of leaf and leaf area in sago palm (*Metroxylon sagu* Rottb.). Sago Palm 8: 21–23. (in Japanese)

Nakamura, S., Y. Nitta and Y. Goto (2004a) Leaf characteristics and shape of sago palm (*Metroxylon sagu* Rottb.) for developing a method of estimating leaf area. Plant Production Science 7: 198–203.

Nakamura, S., M. Watanabe, Juliarni, Y. Nitta and Y. Goto (2004b) Elongation and thickening of sago palm trunks. Proceedings of the 13th Conference of the Society of Sago Palm Studies, 9–12. (in Japanese)

Nakamura, S., Y. Nitta, M. Watanabe and Y. Goto (2005) Analysis of leaflet shape and area for improvement of leaf area estimation method for sago palm (*Metroxylon sagu* Rottb.). Plant Production Science 8: 27–31.

Nakamura, S., Y. Nitta, M. Watanabe and Y. Goto (2008) Stem formation of suckers in sago palm (*Metroxylon sagu* Rottb.). Proceedings of the 17th Conference of the Society of Sago Palm Studies, 9–12. (in Japanese)

Nakamura, S., Y. Nitta, M. Watanabe and Y. Goto (2009a) A method for estimating sago palm (*Metroxylon sagu* Rottb.) leaf area after trunk formation. Plant Production Science 12: 58–62.

Nakamura, S., Y. Nitta, M. Watanabe, T. Nakamura and Y. Goto (2009b) Effect of sucker control on the leaf emergence of sago palm (*Metroxylon sagu* Rottb.) before trunk formation. Japanese Journal of Crop Science 78 (Suppl. 1): 46–47. (in Japanese)

Nakanishi, H. (2005) Mangrove forests. *In*: Illustrated Encyclopedia of Japanese Vegetation. (Fukushima, T. and T. Iwase eds.) Asakura Shoten (Tokyo) 22–23. (in Japanese)

Nakao, S. (1966) The origin of domesticated plants and agriculture. Iwanami Shoten (Tokyo). (in Japanese).

Nakao, S. (1983) Southeast Asian agriculture and wheat. *In*: Search for the origin of agriculture-based culture in Japan. (Sasaki T. ed.) Nihon Hoso Kyokai (Tokyo) 149–151. (in Japanese).

Nakao, S. (1985) Prenatal agriculture. Presentation at the National Museum of Ethnology "Social and cultural changes in Papua New Guinea" joint research conference (Osaka). (in Japanese).

Nei, M. and W. H. Li (1979) Mathematical model for studying genetic variation in terms of restriction endonucleases. Proceedings of the National Academy of Science of the United States of America 76: 5269–5273.

Nei, M., J. C. Stephens and N. Saitou (1985) Methods for computing the standard errors of branching points in an evolutionary tree and their application to molecular date from humans and apes. Molecular Biology and Evolution 2: 66–85.

Nishimura, Y. (1995) Agriculture in the villages of Southeast Sulawesi, Indonesia – Part 2: The role of sago exploitation in household and village exonomy. Sago Palm 3: 62–71. (in Japanese)

Nishimura, Y. and T. M. Laufa (2002) A study of traditional methods used for sago starch extraction in Asia and the Pacific. *In*: New Frontiers of Sago Palm Studies: Proceedings of the International Symposium on Sago. (Kainuma, K., M. Okazaki, Y. Toyoda and J. E. Cecil eds.) Universal Academy Press (Tokyo) 211–218.

Nishimura, Y. (2008) Present sago palm situation in Mindanao, Philippines. Proceeding of the 17th Conference of the Society of Sago Palm Studies, 31–36. (in Japanese)

Nitta, Y. (1998) Formation of root primordia in monocotyledon. *In*: Encyclopedia of Plant Root (Editorial Committee of Encyclopedia of Plant Root ed.) Asakura Shoten (Tokyo) 26–28. (in Japanese)

Nitta, Y. and Y. Goto (1998) Differentiation and growth of sucker, Internal and outside structures. *In*: A study of the environments identification of species and the relationship between starch productivity and growth environments in sago palm growing areas, Toyota Foundation research output report (Yamamoto, Y. ed.) 63–91. (in Japanese)

Nitta, Y., T. Matsuda, M. Endo, Y. Goto, S. Nakamura, T. Yoshida and Y. Yamamoto (2000a) Scanning electron microscopy of starch accumulation in ground parenchyma of sago palm. Proceedings of the 9th Conference of the Society of Sago Palm Studies, 58–63. (in Japanese)

Nitta, Y., T. Yoshida, Y. Yamamoto and F. S. Jong (2000b) Effects of micronutrients on the growth and yield of sago palm grown under deep peat soil in the tropics. Proceedings of the 9th Conference of the Society of Sago Palm Studies, 54–57. (in Japanese)

Nitta, Y., Y. Goto, K. Kakuda, H. Ehara, H. Ando, T. Yoshida, Y. Yamamoto, T. Matsuda, F. S. Jong and A. H. Hassan (2002a) Morphological and anatomical observations of adventitious and lateral roots of sago palm. Plant Production Science 5: 139–145.

Nitta, Y., M. Honda, S. Nakamura, Y. Goto and T. Matsuda (2002b) Scanning electron microscopic observation on starch accumulation in the ground parenchyma tissues of sago stem – Feature of plastid-amyloplast system in the apex and its basal portions. Proceedings of the 11th Conference of the Society of Sago Palm Studies, 51–54. (in Japanese)

Nitta, Y., T. Yoshida, Y. Yamamoto, F. S. Jong, S. Nakamura, Y. Goto and T. Matsuda (2003) Partial-dead leaf appearance and their recovery by micronutrients application of sago palm before trunk formation stage. Abstract of the 12th conference of Japanese Society of Sago Palm Studies, 16–17. (in Japanese)

Nitta, Y., R. Miura, T. Matsuda, S. Nakamura, Y. Goto and M. Watanabe (2004) Internal structure of sago palm leaf. Abstract of the 13th Conference of Japanese Society of Sago Palm Studies, 5–8. (in Japanese)

Nitta, Y. and T. Matsuda (2005) Structure and morphology of sago palm root. Sago Palm 13: 16–19. (in Japanese)

Nitta, Y., T. Matsuda, R. Miura, S. Nakamura, Y. Goto and M. Watanabe (2005a) Anatomical leaf structure related to photosynthetic and conductive activities of sago palm. *In*: Sago Palm Development and Utilization: Proceedings of the 8th International Sago Symposium. (Karafir, Y. P., F. S. Jong and V. E. Fere eds.) Universitas Negeri Papua Press (Manokwari) 105–112.

Nitta, Y., Warashina, S., Matsuda, T., Yamamoto, Y., Yoshida, T. and Miyazaki, A. (2005b) Varietal differences in amyloplast accumulation of sago palms grown around Lake Sentani near Jayapura, Indonesia – Electron microscopic study. Abstract of the 14th Conference of Japanese Society of Sago Palm Studies, 16–18. (in Japanese)

Nitta, Y., T. Nakayama and T. Matsuda (2006) Structure and function of intercellular spaces in the stem of sago palm – Electron microscopic study. Proceedings of the 15th Conference of the Society of Sago Palm Studies, 21–24. (in Japanese)

Obayashi, T. (1977a) Foreword by the translator. *In*: Slaughtered goddess. (Jensen, Ad. E. [1966]) Kobundo (Tokyo). (in Japanese)

Obayashi, T. (1977b) The origin of mortuary practices. Kadokawa Shoten (Tokyo). (in Japanese).

OECD/FAO (2007) OECD-FAO Agricultural Outlook 2007–2016. p.88, http://www.oecd.org/ dataoecd/6/10/38893266.pdf.

Ogita, S., T. Kubo, M. Takeuchi, C. Yamaguchi and M. Okazaki (1996) Accumulations and distribution of starch in sago palm (*Metroxylon sagu*) stems. Sago Palm 4: 1–5. (in Japanese)

Ogura, T. (1984a) Other major applications: Corrugated cardboard. *In*: Starch science handbook. (Nakamura, M. and Suzuki S. eds.) Asakura Shoten (Tokyo) 587–589. (In Japanese).

Ogura, T. (1984b) An overview of chemically modified starches. *In*: Starch science handbook. (Nakamura, M. and Suzuki S. eds.) Asakura Shoten (Tokyo) 596–598. (In Japanese).

Ogura, T. (1984c) Dextrin. *In*: Starch science handbook. (Nakamura, M. and Suzuki S. eds) Asakura Shoten (Tokyo) 498–500. (In Japanese).

Ohmi, M., H. Inomata, S. Sasaki, H. Tominaga and K. Fukuda (2003) Lauroylation of sago residue at normal temperature and characteristics of plastic sheets prepared from lauroylated sago residue. Sago Palm 11: 1 7.

Ohmi, M. and A. Saito (2007) Properties of polyurethane foam prepared from sago residue 2. Effect of sago residue contents on mechanical properties of foam. Proceedings of the 16th Conference of the Society of Sago Palm Studies, 53–56. (in Japanese)

Ohno, A. (2003) Sago starch. *In*: Encyclopedia of Starch. (Fuwa, E., T. Komaki, S. Hizukuri and K. Kainuma eds.) Asakura Shoten (Tokyo) 379–387.

Ohno, A. (2004) Sago starch processing. Sago Palm 12: 28–31. (in Japanese)

Ohtsuka, R. (1977) The sago eater's adaptation in the Oriomo Plateau, Papua New Guinea. *In*: Sago '76: Papers of the 1st International Sago Symposium "The Equatorial Swamp as a Natural Resource". (Tan, K. ed.) University of Malaya (Kuala Lumpur) 96–104.

Ohtsuka, R. (1983) Oriomo Papuans: Ecology of Sago Eaters in Lowland, Papua.University of Tokyo Press.

Ohya, C. and S. Takahashi (1987) Application of sago starch to baked foods. Journal of Cookery Science of Japan 20: 362–370. (in Japanese)

Ohya, C., S. Takahashi and T. Watanabe (1990) Texture evaluation of fenpi made of sago starch by physical and sensory method. Journal of Cookery Science of Japan 23: 293–301. (in Japanese)

Okazaki, M. (1998) Sago Study. Tokyo University of Agriculture and Technology.

Okazaki, M. (2000) Sago Study in Mindanao. Tokyo University of Agriculture and Technology.

Okazaki, M. (2004) Sago Study in Panay and Leyte. Tokyo University of Agriculture and Technology.

Okazaki, M. and C. Yamaguchi (2002) The non-molecular nitrogen balance in an experimental sago garden at Dalat, Sarawak. *In*: New Frontiers of Sago Palm Studies: Proceedings of the International Symposium on Sago (Kainuma, K., M. Okazaki, Y. Toyoda and J. E. Cecil eds.) Universal Academy Press (Tokyo) 297–302.

Okazaki, M. and K. Toyota (2003a) Sago Study in Leyte. Tokyo University of Agriculture and Technology.

Okazaki, M. and K. Toyota (2003b) Sago Study in Cebu and Leyte. Tokyo University of Agriculture and Technology.

Okazaki, M., K. Toyota and S. D. Kimura (2005) Sago Project in Leyte. Tokyo University of Agriculture and Technology.

Okazaki, M. (2006) Sago Palm protects global warming. *In*: Creating a new system informed by living organisms. (Editorial Committee for Creating a New System Informed by Living Organisms) Hakuyu Sha (Tokyo) 49–53.

Okazaki, M., K. Toyoda, S. D. Kimura, S. Matsumura, M. Yoshikawa, T. Hamanishi and A. M. Mariscal (2007) Ecological distribution and characteristics of sago palm in the Philippines. Proceedings of the 16th Conference of the Society of Sago Palm Studies, 1–3.

Omori, K., Y. Yamamoto, F. S. Jong and T. Wenston (2000a) Relationship between leaf weight, length or width and leaf area in sago palm (*Metroxylon sagu* Rottb.). Japanese Journal of Tropical Agriculture 44 (Ext 1): 15–16. (in Japanese)

Omori, K., Y. Yamamoto, T. Yoshida, A. Miyazaki and Y. B. Pasolon (2000b) Differences of maximum leaflet characters of sago palm (*Metroxylon sagu* Rottb.) in varieties, palm ages and leaf positions. Proceedings of the 9th Conference of the Society of Sago Palm Studies, 31–38. (in Japanese)

Omori, K., Y. Yamamoto, Y. Nitta, T. Yoshida, K. Kakuda and F. S. Jong (2000c) Stomatal density of sago palm (*Metroxylon sagu* Rottb.) with special reference to positional differences in leaflets and leaves, and change by palm age. Sago Palm 8: 2–8.

Omori, K. (2001) Agro-physiological studies on sago palm (*Metroxylon sagu* Rottb.). Master's thesis for the Subtropical Agriculture Course 2000, Graduate School in Agricultural Science, Kochi University. (in Japanese)

Omori, K., Y. Yamamoto, F. S. Jong, T. Wenston, A. Miyazaki and T. Yoshida (2002) Changes in some characteristics of sago palm sucker growth in water and after transplanting. *In*: New Frontiers of Sago Palm Studies: Proceedings of the International Symposium on Sago. (Kainuma, K., M. Okazaki, Y. Toyoda and J. E. Cecil eds.) Universal Academy Press (Tokyo) 265–269.

Osozawa, K. (1982) On completion of a sago palm study in Mukah, Sarawak state. Noko no Gijutsu 5: 73–84. (in Japanese)

Osozawa, K. (1988) The extensive nature of sago palm forest management: A case study of sago palm producing communities in South Sulawesi Province, Indonesia. Noko no Gijutsu 11: 101–117. (in Japanese)

Osozawa, K. (1990) Sago palm and sago production in South Sulawesi: Essay on tropical low land development. (in Japanese)

Othman, A. R. (1991) Sago: A minor crop in peninsular Malaysia. *In*: Towards Greater Advancement of the Sago Industry in the '90s: Proceedings of the 4th International Sago Symposium. (Ng, T. T., Y. L. Tie and H. S. Kueh eds.) Lee Ming Press (Kuching) 17–21.

Parthasarathy, M. V. (1980) Mature phloem of perennial monocotyledons. Berichte Deutsche Botanische Gesellschaft 93: 57–70.

Ponzetta, M. T. and M. G. Paoletti (1997) Insects as food of the Irian Jaya populations. Ecology of Food and Nutrition 36: 321–346.

Power, A. (2002) Commercialization of sago in Papua New Guinea: PNG–World leader in sago in the 21st century. *In*: New Frontiers of Sago Palm Studies: Proceedings of the International Symposium on Sago. (Kainuma, K., M. Okazaki, Y. Toyoda and J. E. Cecil eds.) Universal Academy Press (Tokyo) 159–165.

Puchongkavarin, H., S. Shonbsngob, T. Nuyim, P. Luangpituksa and S. Varavinit (2000) Production of alkaline noodles produced from the partial substitution of wheat flour with sago starch. Sago Communication 11: 7–14.

Puruwanto, B. H., K. Kakuda, H. Ando, F. S. Jong, Y. Yamamoto, A. Watanabe and T. Yoshida (2002) Nutrient availability and response of sago palm (*Metroxylon sagu* Rottb.) to controlled release N fertilizer on coastal lowland peat in the tropics. Soil Science and Plant Nutrition 48: 529–537.

Rahalkar, G. W., M. R. Harwalkar, H. D. Dananavare, A. J. Tamhankar and K. Shantkram (1985) Rhynchophorus ferrugineus. *In*: Handbook of Insect Rearing Vol. 1. (Singh, P. and R. F. Moore eds.) Elsevier Science (Amsterdam) 279–286.

Rasyad, S. and K. Wasito (1986) The potential of sago palm in Maluku (Indonesia). *In*: Sago '85: Proceedings of the 3rd International Sago Symposium. (Yamada, N. and K. Kainuma eds.) The Sago Palm Research Fund (Tokyo) 1–6.

Rauh, W. (1999) Encyclopedia of Plant Morphology. Translated by Nakamura, S. and H. Tobe, Asakura Shoten Tokyo).

Rauwerdink, J. B. (1986) An essay on *Metroxylon*, the sago palm. Principes 30: 165–180.

Renwari, H. T., H. Matanubun and A. Barahima (1998) Identification, collection, and evaluation of sago palm cultivars in Irian Jaya for supporting commercial and plantation sago palm in Indonesia. Competitive Research Grant Report. (in Indonesian).

Richards, P. W. (1978) Tropical rainforest: An ecological study. Translated by Uematsu, S. and T. Kira, Kyoritsu Shuppan (Tokyo).

Sadakathulla, S. (1991) Management of red palm weevil, *Rhynchophorus ferrugineus* F. in coconut plantations. Planter 67: 415–419.

Sahamat, A. C. (2007) Commercial potential of sago in Malaysia. Abstract of the 9th International Sago Symposium, Leyte, Philippines.

Sahlins, M. (1963) Poor man, rich man, big-man, chief: Political types in Melanesia and Polynesia. Comparative Studies in Society and History 5: 285–300.

Saitoh, K., M. H. Bintro, F. S. Jong, H. Hazairin, J. Louw and N. Sugiyama (2004) Studies on the starch productivity of sago palm in Riau, West Kalimantan and Irian Jaya, Indonesia. Japanese Journal of Tropical Agriculture 48 (Ext 2): 1–2.

Sakai, K. (1999a) Sweet Potato. Hosei University Press (Tokyo).

Sakai, K. (1999b) The true state of food allergy and diet. *In*: Food allergy guidebook. (Ueda, N. ed.) Nihon Hyōron Sha (Tokyo) 3–12. (in Japanese)

Sasaki, S., C. Yamaguchi, H. Tanaka, M. Ohmi and, H. Tominaga (1999) Thermoplasticization of sago residue by estirification with plant oil. Sago Palm 7: 1–7.

Sasaki, S., M. Ohmi, H. Tominaga and K. Fukuda (2002a) Component analysis of sago waste after starch extraction. *In*: New Frontier in Sago Palm Studies: Proceedings of the International Symposium on Sago. (Kainuma, K., M. Okazaki, Y. Toyoda and J. E. Cecil eds) Universal Academy Press (Tokyo) 331–335.

Sasaki, S., M. Ohmi, H. Tominaga and K. Fukuda (2002b) Degradability of the plastic sheet prepared from esterificated sago residue. Sago Palm 10: 1–6.

Sasaki, S., M. Ohmi, H. Tominaga and K. Fukuda (2003) Characteristics of sago residue as a lignocellulosic resource. I. anatomical and physicochemical properties. Sago Palm 10: 73–78.

Sasaki, Y., H. Ando, A. Watanabe, K. Kakuda, F. S. Jong and L. Jamallam (2007) Effects of groundwater level and fertilizer application in sustainable sago palm cultivation in tropical peat. JSPS report. (in Japanese)

Sasaoka, M. (2006) The meaning of sago palm ownership: A monograph on sago eaters in the highlands of Seram Island. Japanese Journal of Southeast Asian Studies 44: 105–144. (in Japanese)

Sasaoka, M. (2007) Effect of "sago-based vegeculture" on the forest ladscape: A case study of Munusela village in central Seram, Eastern Indonesia. Sago Palm 15: 16–28. (in Japanese)

Sastrapradja, S. (1986) Seedling variation in *Metroxylon sagu* Rottb. *In*: Sago '85: Proceedings of the 3rd International Sago Symposium. (Yamada, N. and K. Kainuma eds.) The Sago Palm Research Fund (Tokyo) 117–120.

Sato, T. (1967) Palms in Southeast Asia. Japanese Journal of Southeast Asian Studies 5: 229–275. (in Japanese)

Sato, T., T. Yamaguchi and T. Takamura (1979) Cultivation, harvesting and processing of sago palm. Japanese Journal of Tropical Agriculture 23: 130–136. (in Japanese)

Sato, T. (1986) Starch resource plant of the equatorial high rainfall zone – sago palm. Annual Report of Advances in Agronomy 33: 1–5.(in Japanese)

Sato, T. (1993) Expectation for sago palm as a crop in the 21st century, especially from the view of agronomy. Sago Palm 1: 8–19. (in Japanese)

Schoch, T. J. (1967) Dextrin. In: Starch: Chemistry and Technology, Vol. II. (Whistler, R. L. and E. F. Pashall eds.) Academic Press (New York) 404–408.

Schuiling, D. L. and M. Flach (1985) Guidelines for the Cultivation of Sago Palm. Agricultural University (Wageningen).

Schuiling, D. L. (2006) Traditional starch extraction from the trunk of sago palm (*Metroxylon sagu* Rottb.) in West Seram (Maluku, Indonesia). In: Sago Palm Development and Utilization: Proceedings of the 8th International Sago Symposium. (Karafir, Y. P., F. S. Jong and V. E. Fere eds.) Universitas Negeri Papua Press (Manokwari) 189–200.

Scott, I. M. (1985) The Soil of Central Sarawak Lowlands, East Malaysia. Soil Survey Division Research Branch, Department of Agriculture, Sarawak, East Malaysia.

Secretariat of Directorate General of Estates (2006) Tree Crop Estate Statistics of Indonesia 2004–2006 Sago. (in Indonesian)

Shimizu, T. (2001) Root-related terminology. In: Illustrated dictionary of botanical terms. Yasaka Shobo (Tokyo) 233–249. (in Japanese)

Shimoda, M. (2005) Aquatic plants. In: Illustrated Encyclopedia of Japanese Vegetation. (Fukushima, T. and T. Iwase eds.) Asakura Shoten (Tokyo) 52–53. (in Japanese)

Shimoda, H. and A. P. Power (1986) Investigation into development and utilization of sago palm forest in the East Sepik region, Papua New Guinea. In: Sago '85: Proceedings of the 3rd International Sago Symposium. (Yamada, N. and K. Kainuma eds.) The Sago Palm Research Fund (Tokyo) 94–104.

Shimoda, H. and A. P. Power (1990) Present condition of sago palm forest and its starch productivity in East Sepik province, Papua New Guinea 1. Outline of survey area and natural environmental conditions of sago palm forest. Japanese Journal of Tropical Agriculture 34: 292–301. (in Japanese)

Shimoda, H. and A. P. Power (1992a) Present condition of sago palm forest and its starch productivity in East Sepik province, Papua New Guinea 2. Varieties of sago palm and their distribution. Japanese Journal of Tropical Agriculture 36: 227–233. (in Japanese)

Shimoda, H. and A. P. Power (1992b) Present condition of sago palm forest and its starch productivity in East Sepik province, Papua New Guinea 3. Growth habit of sago palm (1). Japanese Journal of Tropical Agriculture 36: 242–250. (in Japanese)

Shimoda, H., K. Saito and A. P. Power (1994) Investigation studies on the starch productivity of sago palm: A case study in Sepik basin, Papua New Guinea. Sago Palm 2: 1–6. (in Japanese)

Shinozaki, K. (1995) Molecular mechanism of signal transduction. *In*: Genetic expression in plants. (Nagata, T. and H. Uchimiya eds.) Kodan Sha (Tokyo) 124–136. (in Japanese)

Shiraishi, N., S. Onodera, M. Ohtani and T. Masumoto (1985) Dissolution of etherified or esterified wood into polyhydric alcohols or bisphenol A and their application in preparing wooden polymeric materials. Mokuzai Gakkaishi 31: 418–420. (In Japanese)

Shiraishi, N., and M. Yoshioka (1986) Plasticization of wood by acetylation with trifluoroacetic acid pretreatment. Sen'i Gakkaishi 42: T346–T355. (in Japanese)

Shoji, S. (1976) Peat soil. Urban Kubota 13: 14–15. (in Japanese)

Shrestha, A., K. Toyota, M. Okazaki, Y. Suga, M. A. Quevedo, A. B. Loreto and A. M. Mariscal (2007) Enhancement of nitrogen-fixing activity of Enterobacteriaceae strains isolated from sago palm (*Metroxylon sagu*) by microbial interaction with non-nitrogen fixers. Microbes and Environments 22: 59–70.

Shrestha, A., K. Toyota, Y. Nakano, M. Okazaki, M. Quevedo and E. I. Abayon (2006) Nitrogen fixing activity in different parts of sago palm (*Metroxylon sagu*) and characterization of aerobic nitrogen fixing bacteria colonizing sago palm. Sago Palm 14: 20–32.

Sim, E. S. and M. I. Ahmed (1978) Variation of flour yield in the sago palm. Malaysian Agriculture Journal 51: 351–358.

Sim, E. S. and M. I. Ahmed (1990) Leaf nutrient variation in sago palms. *In*: Towards Greater Advancement of the Sago Industry in the '90s: Proceedings of the 4th International Sago Symposium. (Ng, T. T., Y. L. Tie and H. S. Kueh eds.) Lee Ming Press (Kuching) 94–102.

Sim, S. F., A. J. Khan, M. Mohamed and A. M. D. Mohamed (2005) The relationship between peat soil characteristics and the growth of sago palm (*Metroxylon sagu*). Sago Palm 13: 9–16.

Society of Sago Palm Studies (2009) http://www.bio.mie-u.ac.jp/~ehara/sago/sago-j.html.

Soedewo, D. and B. Haryanto (1983) Prospek pengembangan daya guna sagu sebagai bahan industri. Seri Monitorin Strategis perkembangan IPTEK No. Monstra/6/1983, Biro Koorinasi dan Kebijaksanaan Ilmiah-LIPI. (in Indonesian)

Soekarto, S. T. and S. Wiyandi (1983) Prospek pengembangan sagu sebagai bahan pangan di Indoensia. Seri Monitoring Strategis Perkembangan IPTEK No. Monstra/4/1983, Biro Koordinasi dan kebijaksanaan Ilmiah-LIPI. (in Indonesian)

Soerjono, R. (1980) Potency of sago as a food-energy source in Indonesia. *In*: Sago: The Equatorial Swamp as a Natural Resource: Proceedings of the 2nd International Sago Symposium. (Stanton, W. R. and M. Flach eds.) Martinus Nijhoff (London) 35–42.

Stokes, D. S. and B. K. Wilson (eds.) (1987) Folk stories from Papua New Guinea. Miraisha (Tokyo). (in Japanese).

Takahashi, E. (1991) Mechanism of salt injury and salt tolerance in plants. *In*: Soil salinity and agriculture. Hakuyu Sha (Tokyo) 123–154. (in Japanese)

Takahashi, R. (1984a) Textile industry and starch. *In*: Starch science handbook. (Nakamura, M. and Suzuki S. eds) Asakura Shoten (Tokyo) 575–578. (In Japanese)

Takahashi, R. (1984b) Oxydized starch. *In*: Starch science handbook. (Nakamura, M. and Suzuki S. eds.) Asakura Shoten (Tokyo) 501–503. (In Japanese)

Takahashi, S., H. Kitahara and K. Kainuma (1981) Chemical and physical properties of starches from mung bean and sago. Journal of the Japanese Society of Starch Science 28: 151–159. (in Japanese)

Takahashi, S. and T. Watanabe (1983) Effect of soybean protein on gelatinization of starch. Bulletin of the Faculty of Home Economics, Kyoritsu Women's University 29: 127–140. (in Japanese)

Takahashi, S., R. Kobayashi, T. Watanabe and K. Kainuma (1983) Effects of addition of soybean protein on gelatinization and retrogradation of starch. Journal of food science and technology 30: 276–282. (in Japanese)

Takahashi, S., K. Hirao and T. Watanabe (1985a) Effect of soybean protein on gelatinization of starch (Pt. 2). Bulletin of the Faculty of Home Economics, Kyoritsu Women's University 31: 32–42. (in Japanese)

Takahashi, S., K. Hirao, A. Kawabata and M. Nakamura (1985b) Effect of preparation methods of starches from mung beans and broad beans and preparation method of noodles on the physico-chemical properties of harusame noodles. Journal of the Japanese Society of Starch Science 30: 257–266. (in Japanese)

Takahashi, S. (1986) Some useful properties of sago starch in cookery science. *In*: Sago '85: Proceedings of the 3rd International Sago Symposium. (Yamada, N. and K. Kainuma eds.) The Sago Palm Research Fund (Tokyo) 208–216.

Takahashi, S., K. Hirao and T. Watanabe (1986) Effect of added soybean protein on physico-chemical properties of starch noodles (Harusame). Japanese Society of Starch Science 33: 15–24 (in Japanese).

Takahashi, S., K. Hirao, R. Kobayashi, A. Kawabata and M. Nakamura (1987) The degree of gelatinization and texture during the preparation of harusame noodles. Journal of the Japanese Society of Starch Science 34: 21–30. (in Japanese)

Takahashi, S. and K. Kainuma (1989) Palm trees accumulating starch in their trunks: Properties of sago starch. Dietary Scientific Research 10: 13–21. (in Japanese)

Takahashi, S. and K. Hirao (1992) Food-cultural studies on cooking and processing properties of sago starch. Bulletin of the Faculty of Home Economics, Kyoritsu Women's University 38: 17–23. (in Japanese)

Takahashi, S. and K. Hirao (1993) Physico-chemical properties of starch noodles (Harusame) made from sago and warm water treated potato starch. Bulletin of the Faculty of Home Economics Kyoritsu Women's University 39: 103–108. (in Japanese)

Takahashi, S. and K. Hirao (1994) Studies on the physical and the chemical properties of sago starch, and its performance when used in the preparation of Japanese sweets. Bulletin of the Faculty of Home Economics, Kyoritsu Women's University 40: 59–64. (in Japanese)

Takahashi, S., K. Hirao and K. Kainuma (1995) Physicochemical properties and cooking quality of sago starch. Sago Palm 3: 72–82. (in Japanese)

Takahashi, S. and K. Kainuma (2006) Palm trees accumulating starch in their trunks: Physicochemical, cooking and processing properties of sago starch. Japanese Journal of Tropical Agriculture 50: 238–243. (in Japanese)

Takamura, T. and E. Yuda (1985) Sago palm (*Metroxylon* spp.) distribution and utilization 1: Ambon Island and northern Sulawesi Island, Indonesia. Proceedings of the 57th Annual Meeting of Japanese Society for Tropical Agriculture, 66–67.

Takamura, T. (1990) Recent research activities and the problems on sago palm. Japanese Journal of Tropical Agriculture 34: 51–58. (in Japanese)

Takamura, T. (1995) Agronomic problems of sago palms with special reference to the possibilities of introduction to new areas. Sago Palm 3: 26–32. (in Japanese)

Takaya, Y. (1983) Sago production in south Sulawesi. Japanese Journal of Southeast Asian Studies 21: 235–260. (in Japanese)

Takaya, Y. and A. Poniman (1986) Traditional life and its transformation among the Melayu people on the east coast of Sumatra. Japanese Journal of Southeast Asian studies 24: 263–288. (in Japanese)

Takeishi, R. (2004) Oil reserve estimation and hurdles for production expansion in the Middle Eastern oil producing countries. Gendai no chuto 36: 2–35. (in Japanese)

Tan, K. (1986) Plantation sago in the Batu Pahat floodplain. *In*: Sago '85: Proceedings of the 3rd International Sago Symposium. (Yamada, N. and K. Kainuma eds.) The Sago Palm Research Fund (Tokyo) 65–70.

Tarimo, A. J. P., H. Ehara, H. Naito, M. H. Bintoro and T. Y. Takamura (2006) Sago palm (Metroxylon sagu Rottb.) cultivation trial in Tanzania, Africa. *In*: Sago Palm Development and Utilization: Proceedings of the 8th International Sago Symposium. (Karafir, Y. P., F. S. Jong and V. E. Fere eds.) Universitas Negeri Papua Press (Manokwari) 123–134.

Teramoto, Y., and F. Matsumoto (1966) Studies on the cooking quality of various starch (Pt.1) Kudzu dumpling. Journal of Home Economics of Japan 17: 384–388. (in Japanese)

Tie, Y. L., Hj. O. Jaman and H. S. Kueh (1987) Performance of sago (*Metroxylon sagu*) on deep peat. Proceedings of the 24th Research Officer's Conference, Department of Agriculture, Sarawak, 105–118.

Tie, Y. L. and E. T. Kelvin Lim (1991) The current status and future prospects of harvestable sago palms in Sarawak. *In*: Towards Greater Advancement of the Sago Industry in the '90s: Proceedings of the 4th International Sago Symposium. (Ng, T. T., Y. L. Tie and H. S. Kueh eds.) Lee Ming Press (Kuching) 11–16.

Tie, Y. L., K. S. Loi and E. T. Kelvin Lim, (1991) The geographical distribution of sago (*Metroxylon* spp.) and the dominant sago-growing soils in Sarawak. In: Towards Greater Advancement of the Sago Industry in the '90s: Proceedings of the 4th International Sago Symposium. (Ng, T. T., Y. L. Tie and H. S. Kueh eds.) Lee Ming Press (Kuching) 36–45.

Tomlinson, P. B. (1970) The flowering in *Metroxylon* (the sago palm). Principes 15: 49–62.

Tomlinson, P. B. (1990) The Structural Biology of Palms. Clarendon Press (New York).

Toyoda, Y. (1997) Some cultural and social problems in the implementation of agricultural development projects in the Sepik area, Papua New Guinea. Japanese Journal of Tropical Agriculture 41: 27–36. (in Japanese)

Toyoda, Y. (2002) Socio-economic and anthropological studies regarding sago palm growing areas. *In*: New Frontiers of Sago Palm Studies: Proceedings of the International Symposium on Sago. (Kainuma, K., M. Okazaki, Y. Toyoda and J. E. Cecil eds.) Universal Academy Press (Tokyo) 15–23.

Toyoda, Y. (2003) Rationales for multivariety cultivation in the Sepik region, Papua New Guinea. *In*: Root vegetables and man: Root crop farming that supported human survival. (Yoshida, S., M. Hotta and M. Into eds.) Heibonsha (Tokyo) 95–111. (in Japanese)

Toyoda, Y., R. Todo and H. Toyohara (2005) Sago as food in the Sepik area, Papua New Guinea. Sago Palm 12: 1–11.

Toyoda, Y. (2006) Multicropping in Sago (*Metroxylon sagu* Rottb.) growing areas of Papua New Guinea. *In*: Sago Palm Development and Utilization: Proceeding of the 8th International Sago Symposium. (Karafir, Y. P., F. S. Jong and V. E. Fere eds.) Universitas Negeri Papua Press (Manokwari) 209–216.

Toyohara, H., M. Amano and T. Konishi (1994) Extracting, cooking and eating starch from sago palm in Papua New Guinea. *In*: Report on Overseas Scientific Studies Commemorating a Centenary of Tokyo University of Agriculture, 136–142.

Tsuji Anzen Shokuhin (2009) Saku saku ko (Sago flour). http://www.a-soken.com/item/NO–212314.html.

Uchida, N., S. Kobayashi, T. Yasuda and T. Yamaguchi (1990) Photosynthetic characteristics of sago palm, *Metroxylon rumphii* Martius. Japanese Journal of Tropical Agriculture 34: 176–180. (in Japanese)

Ueda, N. (1999) Factors for the onset of food allergies. *In*: Food Allergy Guidebook. (Ueda, N. ed.) Nihon Hyoron Sha (Tokyo) 35–44. (in Japanese)

Ulijaszek, S. J. (1991) Traditional methods of sago palm management in the Purai Delta of Papua New Guinea. *In*: Towards Greater Advancement of the Sago Industry in the '90s: Proceedings of the 4th International Sago Symposium. (Ng, T. T., Y. L. Tie and H. S. Kueh eds.) Lee Ming Press (Kuching) 122–126.

Umemura, Y. (1984) Potato, its relation with human beings. Kokinsyoin (Tokyo). (in Japanese)

United States of America Department of Agriculture (1975) Soil Taxonomy. Washington.

Utami, N. (1986) Penyerbukan pada sagu (*M. sagu*). Berita Biologi 3: 229–231. (in Indonesian)

Van Breemen, N. (1980) Acid sulphate soils. *In*: Problem Soils: Their Reclamation and Management, Technical Paper 12. International Soil Reference and Information Centre (Wageningen) 53–57.

Van Kraalingen, D. W. G. (1983) Investigation on Sago Palm in East Sepik District, PNG, Konedobu. Report of Department of Minerals and Energy.

Van Kraalingen, D. W. G. (1984) Some Observation on Sago Palm Growth in Sepik River Basin, Papua New Guinea. Konedobu, Report of Department of Minerals and Energy.

Van Kraalingen, D. W. G. (1986) Starch content of sago palm trunks in relation to morphological characters and ecological conditions. *In*: The Development of the Sago Palm and its Products. Report of the FAO/BPPT Consultation (Jakarta) 105–111.

Wada, K. (1999) Mechanisms of salt tolerance. Iden 53: 58–62. (in Japanese)

Wagatsuma, K. (1994) The present method of manufacturing sago starch and its industrial use. Bioscience, Biotechnology, and Biochemistry 68: 844–848. (in Japanese)

Wagatsuma, T. (2002) Aluminum stress. *In*: Encyclopedia of Plant Nutrition and Fertilizer. Asakura Shoten (Tokyo) 332–337.

Waisel, Y. (1972) Biology of Halophytes. New York: Academic Press.

Warashina, S., Y. Nitta, T. Matsuda, T. Nakayama and Y. Sasaki (2007) Formation portion of intercellular spaces and feature of starch accumulation in sago palm (*Metroxylon sagu* Rottb). Japanese Journal of Crop Science 76 (Ext 1): 356–357.

Watanabe, A., K. Kakuda, B. H. Purwanto, F. S. Jong and H. Ando (2008) Effect of sago palm (*Metroxylon sagu* Rottb.) plantation on $CH_4$ and $CO_2$ fluxes from a tropical peat soil. Sago Palm 16: 10–15.

Watanabe, A., B. H. Purwanto, H. Ando, K. Kakuda and F. S. Jong (2009) Methane and $CO_2$ fluxes from an Indonesian peatland used for sago palm (*Metroxylon sagu* Rottb.) cultivation: Effects of fertilizer and groundwater level management. Agriculture, Ecosystems and Environment 134: 14–18.

Watanabe, H. (1984) Exploitation of swamp forest and cultivation of sago palm. Japanese Journal of Tropical Agriculture 28: 134–140. (in Japanese)

Watanabe, M., S. Nakamura, Juliarni, Y. Nitta and Y. Goto (2004) Characteristics of leaves in sago palm before trunk formation. Proceedings of the 13th Conference of the Society of Sago Palm Studies, 1–4. (in Japanese)

Watanabe, M., S. Nakamura, Y. Nitta, Y. Yamamoto and Y. Goto (2008) The course of vascular bundles in the stem of sago palm after trunk formation. Proceedings of the 17th Conference of the Society of Sago Palm Studies, 79–82. (in Japanese)

Watanabe, T. and M. Ohmi (1997) Thermoplasiticization of sago palm by acetylation. Sago Palm 5: 10–16.

Westphal, E. and P. C. M. Jensen (1989) Plant Resources of South-East Asia: A Selection. Pudoc (Wageningen).

Whitemore, T. C. (1973) On the Solomon's sago palm. Principes 17: 46–48.

Widjono, A., Y. Mokay, Aminsnaipa, H. Lakuy, A. Rouw, A. Resubun, dan P. Wihyawari (2000) Jenis-jenis Sagu Beberapa Daerah Papua. Badan Penelitian dan Pengembangan Pertanian. Pusat Penilitian Sosial Ekonomi, Bogor. Proyek Penelitian Sistem Usaha Tani Irian Jaya/Sustainable Agriculture Development Project) P2SUT/SADP. (in Indonesian)

Wina, E., A. J. Evans and J. B. Lowry (1986) The composition of pith from the sago palms *Metroxylon sagu* and *Arenga pinnata*. Journal of the Science of Food and Agriculture 37: 352–358.

Wright, L., B. Boundy, B. Perlack, S. Davis, B. Saulsbury (2006) Ethanol overview. *In*: Biomass Energy Data Book, Edition 1. U. S. Department of Energy (Washington D. C.) 20–44.

Wttewaal, B. W. G. (1954) Report on the possibilities for a mechanized sago operation at Tarof'. [cited from Flach (1980)]

Yamada, I. and J. Akamine (2002) Sustainable utilization of the upper mountain sago species (*Eugeisona utilis* and *Arenga nudulatifolia*) in the central mountain range of East Kalimantan, Indonesia. *In*: New Frontiers of Sago Palm Studies: Proceedings of the International Symposium on Sago (Kainuma, K., M. Okazaki, Y. Toyoda and J. E. Cecil eds.) Universal Academy Press (Tokyo) 237–244.

Yamaguchi, C., M. Okazaki and T. Kaneko (1994) Sago palm growing on tropical peat soil in Sarawak, with special reference to copper and zinc. Sago Palm 2: 21–30.

Yamaguchi, C., M. Okazaki, T. Kaneko, K. Yonebayashi and A. H. Hassan (1997) Comparative studies on sago palm growth in deep and shallow peat soils in Sarawak. Sago Palm 5: 1–9.

Yamaguchi, C., M. Okazaki and A. H. Hassan (1998) The behavior of various elements in tropical swamp forest and sago plantation. Japanese Journal of Forest Environment 40: 33–42

Yamamoto, Y. (ed.) (1997) Kochi University and Tohoku University Field Study Report (1996 JSPS Tropical Bio-Resources Research Fund Project). (in Japanese)

Yamamoto, Y. (1998a) Sago palm. Tropical Agriculture Series, Tropical Crop Manual No. 25, Tokyo: Association for International Cooperation of Agriculture and Forestry. (in Japanese)

Yamamoto, Y. (1998b) VII. Sago palm growth and starch yield. *In*: A study of the identification of species and the relationship between starch productivity and growth environments in sago palm growing areas. (Yamamoto, Y. ed.) Toyota Foundation Research Output Report, 135–142. (in Japanese)

Yamamoto, Y. (1999) The present condition of sago palm cultivation and utilization in Ambon, Seram and Southeast Area in Sulawesi Island in Indonesia. Japanese Journal of Tropical Agriculture 43: 206–212. (in Japanese)

Yamamoto, Y. (2005) On attending the 8th International Sago Symposium. International Cooperation in Agriculture and Forestry 28: 21–26. (in Japanese)

Yamamoto, Y. (2006a) Starch productivity of sago palm (*Metroxylon sagu* Rottb.) in Indonesia and Malaysia. Japanese Journal of Tropical Agriculture 50: 234–237. (in Japanese)

Yamamoto, Y. (2006b) Visiting to see sago palm, a tropical starch crop. Kurashi to Nogyo 20: 5–7. (in Japanese)

Yamamoto, Y. (2006c) Biodiversity and starch productivity of several sago palm varieties in Indonesia. International workshop on Domestication, super-domestication and gigantism: Human manipulation of plant genomes for increasing crop yield (OECD and NIAS Sponsored workshop). (Tsukuba), Program and Abstract.

Yamamoto, Y. (2009) Strategy for bio-fuel, community to world. Research for Tropical Agriculture 2: 104–109. (in Japanese)

Yamamoto, Y., K. Omori, T. Yoshida, Y. Nitta, Y. B. Pasolon and A. Miyazaki (2000) Growth characteristics and starch productivity of three varieties of sago palms in Southeast Sulawesi, Indonesia. Proceedings of the 9th Conference of the Society of Sago Palm Studies, 15–22. (in Japanese)

Yamamoto, Y., K. Omori, Y. Nitta, K. Kakuda, Y. B. Pasolon, R. S. Gusti, A. Miyazaki and T. Yoshida (2002a) Changes of leaf characters in sago palms after trunk formation. Proceedings of the 11th Conference of the Society of Sago Palm Studies, 43–46. (in Japanese)

Yamamoto, Y., K. Omori, Y. Nitta, K. Kakuda, Y. B. Pasolon, R. S. Gusti, A. Miyazaki and T. Yoshida (2002b) Changes of dry weight and dry matter percentage in each organ or part in sago palms after trunk formation. Proceedings of the 11th Conference of the Society of Sago Palm Studies, 47–50. (in Japanese)

Yamamoto, Y., Y. Chinen, T. Yoshida, A. Miyazaki, S. Hamada, T. Wenstone and F. S. Jong (2003a) Sago palms (*Metroxylon sagu* Rottb.) growing on the seashore in Tebing Tinggi Is., Indonesia. Proceedings of the 12th Conference of the Society of Sago Palm Studies, 62–66. (in Japanese)

Yamamoto, Y., T. Yoshida, Y. Goto, F. S. Jong and L. B. Hilary (2003b) Starch accumulation process in the pith of sago palm (*Metroxylon sagu* Rottnb.). Japanese Journal of Tropical Agriculture 47: 124–131. (in Japanese)

Yamamoto, Y., T. Yoshida, Y. Goto, Y. Nitta, K. Kakuda, F. S. Jong, L. B. Hilary and A. H. Hassan (2003c) Differences in growth and starch yield of sago palms (*Metroxylon sagu* Rottb.) among soil types in Sarawak, Malaysia. Japanese Journal of Tropical Agriculture 47: 250–259. (in Japanese)

Yamamoto, Y., T. Yoshida, K. Yamashita, A. Miyazaki, T. Wenstone and F. S. Jong (2004a), Differences in growth and mineral element content in each part of sago palm (*Metroxylon sagu* Rottb.) growing at different locations from the seacoast. Proceedings of the 13th Conference of the Society of Sago Palm Studies, 48–52. (in Japanese)

Yamamoto, Y., T. Yoshida, F. S. Jong, Y. B. Pasolon, H. Matanubun and A. Miyazaki (2004b) Studies on the sago palm (*Metronxylon sagu* Rottb.) varieties in Irian Jaya, Indonesia. Proceedings of the 13th Conference of the Society of Sago Palm Studies, 43–47. (in Japanese)

Yamamoto, Y., A. Miyazaki, T. Yoshida, K. Ōmori, Y. Chinen, T. Wenston, Gunawan and F. S. Jong (2005a) Growth performance of sago palm suckers after transplanting in the plantation at Tebing Tinggi Island, Riau Province, Indonesia. Shikoku Journal of Crop Science 42: 46–47. (in Japanese)

Yamamoto, Y., T. Yoshida, A. Miyazaki, F. S. Jong, Y. B. Pasolon and H. Matanubun (2005b) Studies on starch productivity and the related characters of the sago palm (*Metroxylon sagu* Rottb.) varieties in Irian Jaya, Indonesia. Proceedings of the 14th Conference of the Society of Sago Palm Studies, 8–13. (in Japanese)

Yamamoto, Y., T. Yoshida, A. Miyazaki, F. S. Jong, Y. B. Pasolon and H. Matanubun (2005c) Biodiversity and productivity of several sago palm varieties in Indonesia. *In*: Sago Palm Development and Utilization: Proceedings of the 8th International Sago Symposium. (Karafir, Y. P., F. S. Jong and V. E. Fere eds.) Universitas Negeri Papua Press (Manokwari) 35–40.

Yamamoto, Y., T. Yoshida, A. Miyazaki, F. S. Jong, Y. B. Pasolon and H. Matanubun (2005d) Studies on growth characteristics and starch productivity of sago palm (*Metoxylon sagu* Rottb.) varieties in Irian Jaya, Indonesia. Japanese Journal of Tropical Agriculture 49 (Ext 2), 1–2. (in Japanese)

Yamamoto, Y., K. Katayama, T. Yoshida, A. Miyzaki, F. S. Jong, Y. B. Pasolon, H. Matsunubun, F. S. Rembon, Nicholus and J. Limbongan (2006a) Changes of leaf characters with ages in two sago palm varieties grown near Jayapura, Papua state, Indonesia. Proceedings of the 15th Conference of the Society of Sago Palm Studies, 1–4. (in Japanese)

Yamamoto, Y., K. Katayama, T. Yoshida, A. Miyzaki, F. S. Jong, Y. B. Pasolon, H. Matsunubun, F. S. Rembon, Nicholus and J. Limbongan (2006b) Starch accumulation process in two sago palm varieties grown near Jayapura, Papua state, Indonesia. Proceedings of the 15th Conference of the Society of Sago Palm Studies, 5–8. (in Japanese)

Yamamoto, Y., K. Omori, Y. Nitta, A. Miyazaki, F. S. Jong and T. Wenston (2007a) Efficiency of starch extraction from the pith of sago palm: A case study of the traditional method in Tebing Tinggi Island, Riau, Indonesia. Sago Palm 15: 9–15.

Yamamoto, Y., I. Yanagidate, T. Yoshida, A. Miyazaki, Y. B. Pasolon, S. Darmawanto, J. Limbongan, F. S. Jong, A. F. Irawan and F. S. Rembon (2007b) Leaf characteristics of sago palm varieties grown near Jayapura, Papua Province, Indonesia. Proceedings of the 16th Conference of the Society of Sago Palm Studies, 12–16. (in Japanese)

Yamamoto, Y., K. Omori, T. Yoshida, A. Miyazaki and F. S. Jong (2008a) The annual production of sago (*Metroxylon sagu* Rottb.) starch per hectare. *In*: Sago: Its Potential in Food and Industry: Proceedings of the 9th International Sago Symposium. (Toyoda, Y., M. Okazaki, M. Quevedo and J. Bacusmo eds.) TUAT Press (Tokyo) 95–101.

Yamamoto, Y., T. Yoshida, I. Yanagidate, F. S. Jong, Y. B. Pasolon, A. Miyazaki, T. Hamanishi and K. Hirao (2008b) Studies on growth characteristics and starch productivity of sago palm (*Metroxylon sagu* Rottb.) varieties in Seram Island, Maluku Province, Indonesia. Research for Tropical Agriculture 1 (Ext 1): 19–20. (in Japanese)

Yamamoto, Y., T. Yoshida, I. Yanagidate, F. S. Jong, Y. B. Pasolon, A. Miyazaki, T. Hamanishi and K. Hirao (2008c) Sago palm and its utilization in Ambon and Seram Island, Maluku Province, Indonesia. Proceedings of the 17th Conference of the Society of Sago Palm Studies, 17–22. (in Japanese)

Yamamoto, Y., K. Omori, A. Miyazaki and T. Yoshida (2010) Changes in the composition and content of sugars in the pith during the growth of sago palm. Sago Palm 18: 41–43.

Yanagidate, I., Y. Yamamoto, T. Yoshida, A. Miyazaki, Y. B. Pasolon, S. Darmawanto, J. Limbongan, F. S. Jong, A. F. Irawan and F. S. Rembon (2007) Characteristics of growth and starch productivity of so-called wild type sago palm "Manno" grown near Jayapura, Papua Province, Indonesia. Proceedings of the 16th Conference of the Society of Sago Palm Studies, 8–11. (in Japanese)

Yanagidate, I., Y. Yamamoto, T. Yoshida, H. Pranamuda, S. A. Yusuf, U. E. Suryadi, I. Yuliati, A. Miyazaki and N. Haska (2008) Growth characteristics and starch productivity of sago palm (*Metroxylon sagu* Rottb.) and fishtail palm (*Caryota mitis* Lour.) grown at Pontianak and Singkawang, west Kalimantan province, Indonesia. Proceedings of the 17th Conference of the Society of Sago Palm Studies, 13–16. (in Japanese)

Yanagidate, I., F. S. Rembon, T. Yoshida, Y. Yamamoto, Y. B. Pasolon, F. S. Jong, A. F. Irawan and A. Miyazaki (2009) Studies on trunk density and prediction of starch productivity of sago palm (*Metroxylon sagu* Rottb.): A case study of a cultivated sago palm garden near Kendari, Southeast Sulawesi Province, Indonesia. Sago Palm 17: 1–8.

Yanbuaban, M., M. Osaki, T. Nuyim, J. Onthong and T. Watanabe (2007) Sago (*Meteroxylon sagu* Rottb.) growth is affected by weeds in a tropical peat swamp in Thailand. Soil Science and Plant Nutrition 53: 267–277.

Yao, Y., M. Yoshioka and N. Shiraishi (1993) Combined liquefaction of wood and starch in a polyethylene glycol/glycerin blended solvent. Mokuzai Gakkaishi 39: 930–938. (in Japanese)

Yatsugi, T. (1987a) Tapioka starch. *In*: Starch Science Handbook. (Nakamura, M. and S. Suzuki eds.) Asakura Shoten (Tokyo) 396–403. (in Japanese)

Yatsugi, T. (1987b) Sago starch. *In*: Starch Science Handbook. (Nakamura, M. and S. Suzuki eds.) Asakura Shoten (Tokyo) 404–410. (in Japanese)

Yomiuri Shimbun (2007) Expanding demand for biofuel raw materials, 22 January 2007.

Yoneta, R., M. Okazaki and Y. Yano (2006) Response of sago palm (*Metroxylon sagu* Rottb.) to NaCl stress. Sago Palm 14: 10–19.

Yoneta, R., M. Okazaki, K. Toyota, Y. Yano and A. P. Power (2003) Glycine betaine concentrations in sago palm (*Metroxylon sagu*) under different salt stress. Proceedings of the 12th Conference of the Society of Sago Palm Studies, 18–22. (in Japanese)

Yoneta, R., M. Okazaki, Y. Yano and A. P. Power (2004) Possibility of osmotic pressure regulation using K+ in sago palm (*Metroxylon sagu*). Proceedings of the 13th Conference of the Society of Sago Palm Studies, 17–22. (in Japanese)

Yoshida, S. (1980) Folk classification of the sago palm (*Metroxylon* spp.) among the Galeda. Senri Ethnological Studies 7: 109–117.

Yoshida, S. (1994) Low temperature-induced cytoplasmic acidosis in cultured mung bean (*Vigna radiata* [L.] Wilczek) cells. Plant Physiology 104: 1131–1138.

Yoshida, S. (1977) Folk classification of sago palm. Plant and culture 20: 50–57. (in Japanese)

Yoshida, S. (1985) Preliminary report of folk classification on Iwam agriculture, East Sepik province, Papua New Guinea. Bulletin of National Museum of Ethnology 10: 615–680. (in Japanese)

Yoshida, S. (2002) Physical stress. *In*: Encyclopedia of Plant Nutrition and Fertilizer. Asakura Shoten (Tokyo) 278–281. (in Japanese)

Yuda, E., H. Yamashita, K. Watanabe, H. Ehara and T. Takamura (1985) Biology of sago palm and its cultivation and use in South Sulawesi. Proceedings of the 57th Annual Meeting of Japanese Society for Tropical Agriculture, 4–5. (in Japanese)

Zenkoku Komugi Ko Bunri Kako Kyokai (2009) http://www15.ocn.ne.jp/~zenbun/panfu.pdf#search= 'コムギ澱粉'.

Zimmermann, M. H. and P. B. Tomlinson (1972) The vascular system of monocotyledonous stems. Botanical Gazette 133: 141–155.

Zwollo, M. (1950) Report on Investigations into sago production at Inanwatan. [cited from Flach (1980)]

# Index

16S rDNA gene sequence 152

acetylated starch 279
acetylation 311, 312
acetylene reducing activity 155
acid stress 54
acid sulfate soil 50
adhesiveness 264
adventitious root primordia, 63
adventitious root 61, 114
adzuki bean 270
albuminous seed 90
α-amylase 293
altitudes 221
aluminum ion tolerance 54
aluminum stress 143, 144
aluminum tolerance 144
amylase 293
amylopectin 253
amyloplast 88, 210
amylose 253
animal feed 284
annual starch production 227
anther 89
anthropomorphism 323, 330
*Apis dorsata* 123
apoplastic ion transport 140
*Arecaceae* 1
Asmat 318
*Azospirillum* 151

*Bacillus* 152
bark 79, 301
bearing capacity, 49
benzylation 311
β-1,3-glucan 144
β-amylase-pullulanase (BAP) 253, 263

β-amylase-pullulanase method 256
big men 325
biodegradable plastic 281, 348, 349
bioethanol 282, 284, 347
biofuel 344
biomass 134, 346
Biomass Nippon Strategy 345
blancmange 253, 270
bracken 253
bracken starch 268
brackish water 59
bracteal leaf 118
bracteole 119
branch 12
bread and muffin 270
bride price 328
bud differentiation 97
*Burkholderia* 152

$C_3$ plant 129
callose 144
caloric production 231
caloric productivity 230
carbon assimilation 130
carbon dioxide 282
carbon neutral 283
carboxymethyl starch 280
carrier 143
cassava 342
cat clay 53
cell membrane 145
cell wall 145
cellulose 279, 295, 311, 312
*Chalara paradoxa* 292
*Chalara paradoxa* PNG-80 293
channel 143
chiefdom 325
Chinese vermicelli 271

chipping axe  235
chloroplasts  45
chronic malnutrition  341
citric acid  144
classification  3
clump  36
$CO_2$ compensation point  45
cold stress  44
coleoptile  102
compatible solutes  143
composition of sugar  207
conductivity  59
cooking  260
cooking properties  259, 261
cord-like tissue  67
Coriolis force  46
corn  253, 342
cortex  64
creeping sucker  78
cultivated stand  224
cultivated variety  216
cultivation management  225
cuticular transpiration  128

debarking  243
deep peat soil  173, 204
degradable plastic  352
degree of gelatinization  256
dextrin  261, 275
diagram  291
diameters of adventitious root
    primordia  63
diffusion  240
dissolved organic carbon (DOC)  196
dissolved organic matte  196
distribution  241
DOM  196
drying stress  44
dry matter percentage  203
dry matter production  133
dry matter weight  134
dynamic viscoelasticity  257, 258

early maturing variety  217
embryo  101
emergence of new leaves  175
emergence rate  106
emergence rate of new leaves  175
endodermis  140
endosperm  12, 92, 125
ensiform leaf  66, 76, 96
*Enterobacter*  151, 152
entisols  41
epiblast  102
epidermis  64, 140
establishment rate  168
exocarp  90
exodermis  64
extraction method  200
extruded pellet  284, 287

factory  245
family  1
feeding attractant  289
*fen pi*  269
ferrihydrite  55
ferrous / ferric iron  55
fertilizer losses  182
fibrous root system  114
fibrous tissue  84
Fiji / Fijian  303, 304, 325
first-order floral axis  118
flooring  301
flour  236
flower bud  88
flower bud development  118
flower bud initiation  118
flowering  121
flowering period  99
flowering stage  88
flowering-fruiting stage  98
fluvisols  41
foams  351
folk variety  7, 199, 216, 261
Food and Agriculture Organization
    341

food resources  264
fossil fuel  349
*Frankia*  150
free-living nitrogen fixing bacteria  151
freezing stress, 44
frequency distributions  227
fructose  207
fruit  3, 12, 88, 125, 304
fruit ripening stage  99
fruiting  124

gasoline tax  346
gel  253
gender  329
genetic distance  18
genetic variation  18
genus *Metroxylon*  2
geographical distribution  18
germination  12, 99
germination rate  166
gley horizon  55
gleysols  41
glucoamylase  293
glucose  207, 273
glycine betaine  143
goethite  53, 55
*gomadofu*  270
greenhouse gas  182, 192, 194, 282
ground parenchyma  87
groundwater  57, 163
groundwater level  177, 182–184, 191
groundwater table  176
growth stage  96, 259
grub  246

Hainuwele  332
Hainuwele type  332, 333, 340
halophytes  141, 147
hardness  92, 257
harvestable sago palm  227
helophytes  138
hematite  56
hemicellulose  154, 295
hermaphrodite flowers  122
high-yielding ability  232
Hiri  328
histosols  41
horny endosperm  92
host plant  152
hydrogen ion tolerance  54
hydrophytes  138
hydroxyethyl starch  280
hyperosmotic stress  44

inceptisols  41
indigenous bacteria  154
inflorescence  iii, 3, 99, 118
INHUTANI  250
intensive cultivation system  160
intercellular space  64, 87
international market  294
internode  76, 110, 111
internode diameter  110, 111, 113
internode elongation  111, 113
internode length  111
inverted V shape  66
ion pump  143
ionic stress  140

jarosite  53
Jensen  332
JICA  248
J shape  172

*keropo*  264
*kerupuk sagu*  265
*Klebsiella*  152
kudzu  253
kudzu starch  270
*Kue(h) bangkit*  265
*Kue(h) pisang*  265
*kurupun*  264
*kuzu mushiyokan*, 270
*kuzukiri*  253, 269

*kuzuzakura* 253
Kwoma 335, 337
Kyoto Protocol to the United Nations Framework Convention on Climate Change (Kyoto Protocol) 282, 283, 345

*lagatoi* 328
landfill 349
large vascular bundles 73
lateral roots 63
lauroylation 311, 312
LCDA 247
leaching 192
leaching losses 163
leaf 66
leaf area 68, 105, 133
leaf area estimation 69
leaf blade 105, 106, 109
leaf development 104
leaf emergence rate 104, 167
leaf formation 107
leaf formation process 108
leaf length 109
leaf longevity 129
leaf primodia 109
leaf scars 118
leaf sheath 3, 66, 298, 299
leaf sheath length 66
leaflet 7, 66, 105, 133
leaflet area 68
leguminous tree 150
*lempeng* 264
life cycle 93
light saturation point 45
log 186, 187, 235
log raft 187
low environmental impact 191
L shape 172
lysigenous aerenchyma 64, 139, 140

maghemite 55
major starch crop 230

Malay Archipelago 17
male (staminate) flower 120
male flower 89
malic acid 144
manioca 344
Marco Polo 259, 343
maximum photosynthesis rate 45
Mediterranean farming culture 321
medulla 79
Melanau 297
mesocarp 90
mesophyll cell 141
mesophyll tissue 73, 348
mesophytes 147
methane 192, 193
methane flux 195
methane transfer 195
methanogens 192
*Metroxylon* 1
Micronesia 304, 307
mid-level axil iii
*mie sagu* 265
milling 243
mineral soil 173, 179, 204, 219
mineral soils 41
modified sago starch 261
Modified starch 276
molasses 346
monocarpic (hapaxanthic) 93
monoecious plant, 120
multistage hydrocyclone 244
mung bean 253
mythology 331

Nakao, Sasuke 321, 332
native starch 275
natural stands 222
needle leaf, 66
New Caledonia 325
New Guinea 316
New World farming culture 321
nipa 147
nitrogen 50

nitrogen fixing bacteria  150
node  76, 110
noodle  253
nursery  165

operculum  90, 101
optimum groundwater level  184
optimum photosynthetic temperature  45
organic acid  144
osmolytes  148
osmosis  138
osmotic stress  140
oxalic acid  144
oxidized starch  275
oxygen partial pressure  153

palisade parenchyma  73
palm height  96
palm plantation  160
*Pantoea*  152
*papeda*  264
paper  275
Papua New Guinea  318
parenchyma, 210, 295
peat soil  46, 163, 177, 179-184, 187, 192, 219
pellet  287
perfect (hermaphrodite) flower, 89, 120
pericarp  90
pericycle  150
personification  323, 330
petiole  3, 7, 66, 106, 109
petiole length  66
phloem fibers  84
photosynthesis  129
photosynthetic rate  132
photosynthetic system I  44
phyllotaxis  69
physical properties  257
physiochemical characteristics  253, 254

pie filling  253, 270
pinnate compound leaves  66, 107
pistillode  89
pith  79, 235, 254
plantation  160, 162, 163, 187, 194
planting density  222
planting pattern  157
plastics  281, 349
pollens  89
pollination, 122, 124
pollinators  122
polylactic acid (PLA)  281, 282
polyphenol  246
potassium  148, 286
potato  253, 342
pre-emergent leaves  108
proliferation of amyloplast  213
proline  143
Prometheus  332
Prometheus type  332
proton ATPase  143
proton pump  143
pyrite  50

rachillae  12, 118
rachis  7, 66
rachises  vii
raft  157, 168, 172, 173, 175, 177, 187
*randang*  264
Rapid Visco Analyzer (RVA)  253
rasper  236
reed grass  142
refining  244
relative humidity  46
reproductive growth  117
rhizome  164
rhizosphere  152
*Rhynchophorus*  314
roofing material  348
root crop farming 321, 331
root cropping culture  321
root development  114

root system 114
rosette stage 95, 104, 224
rotary screen 244
rupture force 263

sago biscuit 271
sago pearl 260
sago residue 307
sago starch 254
sago starch factories 294
*sagomochi* 268
salt stress 44, 128, 138
Samoa 304
Sarawak 254, 309, 310
savanna climate 43
savanna farming culture 321
scale 90
scanning electron microscope 73
scar 57
schizogenous aerenchyma 64
sclerenchyma 64
screw press 244
seawater invasion 221
second-order floral axis 118
section *Coelococcus* 2
section *Metroxylon* 2
seed 12, 90, 100
seed coat 90
semi-cultivated stands 224
sensory evaluation 263, 269
separation 244
Sepik 322, 331, 333, 340
sexual division of labor 330
shade 184
shading 174
shading in the clump 220
shallow peat soil 204
shape retention 263
*sinoli* 264
sizing agents 273
small vascular bundles 73
sodium concentration 149
sodium exclusion 149

*Sohun* 265
soil taxonomy 41
soil temperature 194
solar radiation 45
Solomon Islands 325
solubility 253, 255
South Pacific 303, 304
soybean 347
soy protein 263
soy protein isolate 264
spear leaf 66
species 1
specific gravity 49
spikes 118
spine 3, 7
spineless sago 256
spiny sago 256
spongy parenchyma 73
springiness 264
stablishment rate 176
stamens 89
staminate flower 99, 121
staple food 264
starch 2, 84
starch accumulation process 200
starch content 201
starch density 201
starch derivative 275
starch extraction 235
starch granules 88, 210, 254
starch production 290
starch productivity 199, 290
starch resource crop 230
starch sol 253
starch yield 137, 160, 259
starch yield per area 222
starch yield per palm 204
starch yield per year 204
Stein Hall process 276
stele 64
stem 75
sterile 89
stomata 71

stomatal density  71, 130
stomatal transpiration  128
subfamily  1
sucker  33, 76, 157–159, 163–165
sucker control  95
sucker management  178, 179
sucrose  207
sugar cane  273
sugar content  12
sugar palm  256
Sumatra  264, 307
super decanter  244
surface sizing  275
survival rate  165, 168, 171, 173, 174
sustainable sago palm cultivation  191
sweet potato  253, 342
swelling power  253, 255
symplastic ion transport  140
syneresis  263
syrup  273

tapioca  344
tapioca pearl  260
temperate oceanic climate  43
*Tepung kue(h)*  265
testa  90
thatching  12
thermal characteristics  258
Thermo-plasticization  311
thick root  61
thickening  111, 113
thin root  61
thylakoid  45
total sugar content  206
trade  328
transmittance  254
transparency  253
transparent  253
transpiration  127
transplanting  158–160, 163, 164
transport  162
transporters  143

*Trigona* bees  123
tropical biomass  295
tropical monsoon climate  43
tropical peat soil  46
tropical rainforest climate  43
trunk  2, 12, 75, 164, 165, 292
trunk apex  112, 302
trunk formation  80, 110
trunk formation stage  95, 105
trunk growth rate  227
trunk height  96
trunk length  96, 118
trunked population per area  227
trunked sago palm  224
tuber and root crop  231
typhoons  46

United Nations  341
urethane foam  312, 313
utilization  259

Vanuatu  304, 307, 325
vascular bundles  64, 79
*Vespa tropica*  124
vessel  254
viscoelastic  253
viscosity  253

Wallace's Line  239
*warabimochi*  253, 268
warp sizes  274
warp sizing  274
water stress  138
water table  56
Weber's Line  239
weeding  157, 158, 184
weevil  314
wet sago  251
wheat  253
white dextrin  276
whiteness  258
wild variety  216
WRB  41

X-ray diffraction 253
xylem 64

yarn sizing 274
yellow dextrin 276